The Do It Yourself Homestead

BUILD YOUR SELF-SUFFICIENT LIFESTYLE
ONE LEVEL AT A TIME

Tessa Zundel

Puddleduck Press
Mountain Grove, MO

Copyright © 2016 by Tessa Zundel.

All rights reserved. No part of this publication may be reproduced, distributed or transmitted in any form or by any means, including photocopying, recording, or other electronic or mechanical methods, without the prior written permission of the publisher, except in the case of brief quotations embodied in critical reviews and certain other noncommercial uses permitted by copyright law. For permission requests, write to the publisher, addressed "Attention: Permissions Coordinator," at the address below.

Tessa Zundel/Puddleduck Press
111 Union St. Box #121
Mountain Grove, MO 65711
www.homesteadlady.com

Although every effort has been made to ensure that the information in this book was correct at press time, the author/publisher do not assume and hereby disclaim any liability to any party for any loss, damage, or disruption caused by errors or omissions, whether such errors or omissions result from negligence, accident, or any other cause. This book is not intended as a substitute for the medical advice of physicians. The reader should regularly consult a physician in matters relating to his/her health and particularly with respect to any symptoms that may require diagnosis or medical attention.

Book Layout ©2017 BookDesignTemplates.com
Carrot cover photo © 2016 Christine Dalziel, used with permission. All other photos © 2016 Tessa Zundel.
Cover and interior illustrations copyright © 2016 Jennifer Cuzzola/Scratchy Pixel.
Cover design © 2016 Rachel Arsenault

Ordering Information:
Quantity sales. Special discounts are available on quantity purchases by corporations, associations, and others. For details, contact the "Special Sales Department" at the address above.

The Do It Yourself Homestead/ Tessa Zundel. —1st ed.
ISBN 978-0-9976715-0-6

Contents

The Motivation of the Modern Homesteader 8

Homestead Personality Test 13

How to Use This Book 23

In The Homestead Kitchen 33

    For the Homestarter: Use All the Parts 36

    For the Homesteadish: Ferments for Starters 61

    For the Homesteadaholic: Preserving What You Grow 73

    For the Homesteaded: Make Your Own Stuffs 90

In the Homestead Garden 103

    For the Homestarter: Soaking, Sprouting and Microgreens 106

    For the Homesteadish: Food in Pots 114

    For the Homesteadaholic: Plant a Wellness Herb Garden 136

    For the Homesteaded: Growing Smarter With Permaculture 150

Green the Homestead 165

    For the Homestarter: Who Put the "Up" in Recycle? 167

    For the Homesteadish: Downsize and De-Clutter 177

    For the Homesteadaholic: Foraging 185

    For the Homesteaded: Energy on the Homestead 194

Livestock Wherever You Are 211

    For the Homestarter: Vermicomposting 214

    For the Homesteadish: Make Your Own Chickens 221

    For the Homesteadaholic: The Dairy Animal Question 231

    For the Homesteaded: Points to Ponder on Meat 245

Homestead Finances 259

    For the Homestarter: Homestead Debt 261

    For the Homesteadish: Just Save It 269

    For the Homesteadaholic: Wisely Spend 280

    For the Homesteaded: The Homestead Side Hustle 286

Family Times on the Homestead 303

    For the Homestarter: Dinner Together 306

    For the Homesteadish: Traditions 315

    For the Homesteadaholic: Chore Systems 325

    For the Homesteaded: Family Work 333

The Homestead Community 351

    For the Homestarter: Homestead Book Club 354

    For the Homesteadish: Taking and Teaching Classes 361

    For the Homesteadaholic: Seed Swap 368

    For the Homesteaded: Mentoring 376

The Prepared Homestead 385

    For the Homestarter: A Healthy Three Days of Preps 389

    For the Homesteadish: Water—Storage and Conservation 397

    For the Homesteadaholic: Health-Wise Food Storage 405

    For the Homesteaded: Off-Grid Cooking 417

You Have Arrived 427

Resources	430
Afterword	443
Index	445

In loving memory of my grandmothers, both of whom passed on this year. One who taught me to plant; the other who taught me not to fret too much over the weeds.

And

*To Father
Who inspired me with the idea in the first place.
Soli Deo Gloria*

"All truths are easy to understand once they are discovered; the point is to discover them."

-Galileo Galilei

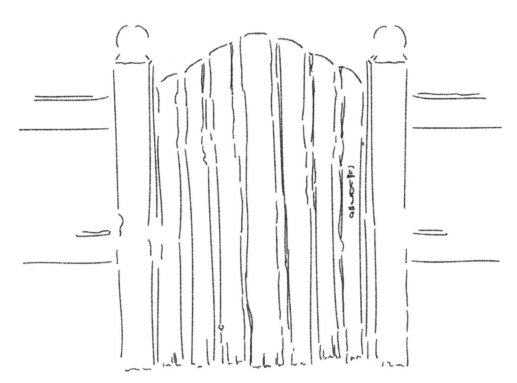

Introduction

I had finally made it to the public pool that summer with my kids—a rare treat in the life of a busy gardening and livestock-tending family. The mom sitting next to me mentioned she had a sister who'd decided to homestead, but that she hadn't really done any work yet. Another mom asked, puzzled, "What's homesteading?"

"Well," said the first, "you know, those people who have chickens in their backyard and grow vegetables instead of grass. At least, I think that's what my sister means. I don't know. She can't seem to get started."

Have you ever heard the term "homesteader" and wondered what it meant? In our uncertain times, many of us are trying to grow some of our own food, conserve energy, reduce waste, live frugally and learn important, hands-on skills; we're striving for a more self-sufficient lifestyle, but does that make us homesteaders? If it does, where is the guidebook that teaches us where to begin and how to progress?

The Motivation of the Modern Homesteader

I once had a reader of my blog ask if I could define modern homesteading for him. He referenced the U.S. Homestead Act of 1862, signed by President Lincoln, that allowed claim-seekers to (re)settle the American west by granting them large tracts of land to farm for little to no monetary cost. Hearty individuals and families came in droves to stake their claims and bring their patch of earth into production through food crops, animals, and hand-crafted products. Some failed miserably, and others succeeded, but the goal was always to create an oasis of self-sufficiency. They desired to provide for the very real and practical needs of survival, as well as the natural human yearning for peace, prosperity, and beauty.

The modern version of homesteading isn't too much different in its aim, although the tools and tricks of the trade have changed a bit over time. Modern homesteaders are working towards self-sufficiency in ways that are meaningful to them and that are feasible given the land they have. Some modern homesteaders live in apartments and townhouses where the amount of dirt they own for food production is negligible and relegated to clay pots and grow boxes in their windowsills. Still others of us are homesteading on tens or even hundreds of acres. Some homestead in large cities and others in towns so small you can hardly call them a wide bend in the road.

The reasons for homesteading are as unique as the geography and personal profiles of the homesteaders themselves. Though the original Homestead Act is no longer the law of the land, perhaps we are again inspired by the government to become more self-sufficient. We read the updates from Ready.gov and seek out ways to be more prepared in an emergency. Maybe we're concerned over our personal or national finances and find that a more conscientious mode of living would make our economic plans better secured. There are still others for whom the homesteading lifestyle is part of their family, ethnic or religious culture—so much so that to live any other way seems odd or impractical. Perhaps others are canning their harvest and learning to sew to prepare for the Rapture or a zombie apocalypse. Any one of these motivations, if meaningful to you, is valid and a sufficient place to begin.

Am I Homesteader?

The truth is, as much as we admire those who work in agriculture, and as much as we love to eat food, most of us really aren't interested in becoming full-time farmers. In fact, the agricultural industry is struggling to find new blood for the coming years. For proof, just look at the findings of an article in The US News and World Report from 2014 entitled, "US Farmers Are Old and Getting Much Older" that tells us that the average age of farmers in the US is around fifty-six. That average means that there just aren't a lot of young people getting into farming to replace those who are aging out of the business.

Although that same article notes that, as Bob Young, chief economist for the American Farm Bureau Federation observes, the age demographic of people attending farm meetings has dropped. Still, there are still a lot of barriers to young people becoming successful farmers in the U.S. (Not that I believe all is lost when it comes to the future of farming and the coming generations, but that's a long conversation for another time.) Farming is simply a lot of work. The work isn't necessarily intuitive either, since most modern Americans weren't raised on farms, or even in agricultural areas. Though most of our great-great-grandparents would have identified themselves as farmers, you and I are sometimes even unable to source our food, never really knowing how far it's traveled, or from where it originated. As unhappy as that might make us, we're just not sure what to do about it. So, we frequent our local farmers markets and wring our hands in trepidation at the tentative structure of our nation's food supply.

As a link between our farming heritage and our current, non-agrarian culture stands the modern homesteading movement. While purchasing from your local farmers market is always a good idea (I'm a huge fan of mine!), there's a lot more that we can be doing than just waiting for the next disaster to strike, or the next job loss to occur. We can learn to provide for ourselves not just food, but potentially all items of necessity, as much as we are able. We can grow crops and raise animals, reduce how much we consume overall and learn to re-use and repurpose what's left behind. We can create systems, in the home and on our land, that are self-sustaining so that, regardless of what happens around us, we can enjoy a certain measure of protection and normalcy. We can learn to be grateful for what we already have, increase our usefulness by becoming lifetime learners and show an example of provident living to the next generation. I'm convinced that much of the saving grace for the children of tomorrow will be found in how well we, their parents, have learned to be useful and productive today.

The main difference between a homestead and a farm is that, at its core, a modern homestead is developed with the idea of self-sufficiency in mind. Homesteaders are generous people who are involved in their communities, and many have developed action plans for assisting those less prepared for hard times. Indeed, my family's homestead is built on the powerful hope that we might be of help to anyone in need. However, the homestead is primarily created to provide for the family that lives on it, so that they might be self-sufficient in our modern culture. The goal is to live abundantly, frugally and practically so that we are neither a burden nor a drain on our fellow man, nor the earth on which we live. Unlike a farm, the homestead doesn't necessarily exist to create product to sell or trade, although homesteads often do produce such items. Unlike a police man, a teacher or a farmer, a homesteader is not a public servant. He is merely a good, solid citizen who is doing his part to be a responsible steward of the earth. By taking care of his own needs and the needs of his family, he strives to produce at least as much as he consumes, if not more.

That's not to say that a homesteader is some sort of veggie-growing, chicken-raising monk who cloisters himself off from the outside world, although some might choose to do that. Most homesteaders I know place a high value on community networking and associating with like-minded people, as well as maintaining very modern, parallel lives. Most of us have jobs, kids, hobbies, service work and soccer practice, just like everyone else. As my friend, Chris of Joybilee Farm (www.joybileefarm.com), says:

> "Homesteaders need friends and we need a community, a place to sip iced tea in the heat of the day, and get advice, cry over a loss, or share a victory. Self-sufficiency is not what you think it is when it comes to homesteading. It's more about placing yourself in the right community rather than removing yourself from the world."

Keys to Success

Whatever your personal motivation or past experience, wherever you are on this journey of homesteading, I hope that this book will prove to be a helpful and encouraging tool for you as you make your progress down the road to a self-sufficient lifestyle. You hold in your hands a homesteading guidebook and you should think of yourself as a true student with goals and outlines and accountability. There are so many valuable books available on the topic of homesteading these days; I own many and will continue to purchase, use and love them. When I was starting out, though, one of the problems I had as I devoured these books was that I got overwhelmed. In order not to leave out anything important, general homesteading titles often put in a little bit of everything. I was so inspired by what I read that I made great plans to do it all! The notion of doing it all makes me laugh now because I've learned to take a more measured approach. However back then, I wish I'd had a book that broke apart some basic homesteading principles and gave me tangible projects and suggestions for further study. I never did find a book quite like what I envisioned, so I wrote this one to fill that need.

That's not to say that I've got this homesteading thing locked up! I, too, will continue to need many mentors as I leap from learning curve to learning curve, riding along with my family in what often feels like an amusement park bumper car. We've had our definite failures and there are things we just plain would have done differently. But there is no doubt in my mind that each experience has been enriching in its own way. Although I wouldn't wish to spare you opportunities to grow through homesteading trial and error, I hope to point you in the direction of some quality methods, as well as tips and projects to produce success and confidence. I'm also hoping to be able to point out things you already may be doing and to give you permission to congratulate yourself. Lastly, I hope that the information, projects and plans included in this book will help you make and keep specific goals that will help you build your own Do It Yourself Homestead. Truly, this is a lifestyle in which you can engage no matter where you live, how many assets you possess, or how much experience you have.

Before we dive into the goals and plans, however, let's first cover a few of the mistakes you might want to avoid. The first one can be crucial: comparison. Please do not fall into the trap of comparing your efforts to others to an excessive degree. Some amount of comparison can be beneficial; examining what other people are doing and learning from them is important for the growth of your homestead. The best way to move forward and keep working smarter is to learn, learn, learn and experiment continually.

Having said that, however, imagining that all or even most other homesteaders in the world have it all figured out is a silly waste of your energy. I work hard. I work hard all the time. I'm the last to go to bed and when I wake up, I work, work, work all day long. And I never, ever, ever get everything done. My garden is full of weeds (we call them fodder for the goats), my sink is always full of dishes (which means I cook from scratch, right?) and I'm always walking around with straw stuck in my hair (my husband thinks it's sexy—I'm hoping). The point is: we're all just doing the best we can and falling into the trap of unhealthy comparison can be paralyzing to your progress.

That includes comparison to yourself. What do I mean? Raise your hand if you have a picture of what your homestead will one day look like. I think all homesteaders have that picture in our minds, both inspiring and haunting us. Sometimes we can allow the frustration at the reality of what our homesteads look like to cause undue stress. You are probably doing way more than you realize. *Anything* you do to increase your self-provision will be worthwhile, no matter what it looks like or how big it is. Do you also appreciate that every little task you accomplish to grow your self-sufficiency benefits your community? Well, it does, and we're going to explore that further a bit later. But for now, just know that you're doing alright.

FOR THE SKEPTICS

These are valid questions, after all.

Why grow tomatoes if you can just buy them at the store?

If you can afford to run the dryer, why wouldn't you?

Why preserve food from your yard when you can buy Organic, canned products if you want to?

How do you have time for all that?

Does this homesteading stuff really make that much of a difference?

I've heard some modern homesteaders say that their families and friends not only question what they're doing but are sometimes actively against it. I think of my grandparents' generation that came up through the Depression only to fight their way through a world war. Those were days of nothing to spare and forced frugality. Then the bounty of the ensuing decades came, and, for a time, there was prosperity and abundance, at least for some. Now, we are experiencing another swinging of the pendulum and some may ask with harshness why we're abandoning the lifestyle of apparent ease that was fought for so fiercely by those who came before us. The only answer to give them is that we do it because we must.

Either we're compelled by common sense in response to the uneasiness and unsteadiness of our economic times, or we feel a deep pull to return to a simpler lifestyle, closer to the earth. Still others of us feel a duty to protect the environment or enrich the legacy we leave to the following generations. Whatever our motivation for engaging in the modern homestead movement, the answer to the query 'why' is always the same—because we must. To do anything less would be proving untrue to ourselves and our times.

As others honestly question, laugh off, or even condemn what you're doing, keep in mind that change is hard for many people and few like to truly consider how unprepared they might be to live without their comforts. You, with the modern homestead heart, are in the very crux of a great societal shift. Things get cramped in the crux…and pinchy. Don't be surprised if you must work a bit to convince your friends and neighbors that you're not crazy, you're just a pioneer.

If it's your significant other or your children who are balking at the lifestyle changes, try to remember that you really don't want to go anywhere without them. Take it slowly, share what you're learning with them and try to keep it fun. Yes, work can be fun! Try to inspire them without requiring that they share your passion and purpose right away. Eventually, by the grace of God, the benefits of what you're doing will become apparent. Healthy meals do taste better after palates have adapted. It really is satisfying to grow your own food. There really are tangible benefits to learning to do it yourself.

A CAUTION

At no time do I ever want to give the impression that achieving your homesteading goals, whatever they are, will not require work. Creating a homestead lifestyle is about creating a home—a place of peace and refuge and safety. Home is a very important word to me, both personally and culturally. My entire adult life, and much of the preparation of my youth, have been devoted to the idea of creating a home for myself and my family. Types of families are all over the map these days and come in different shapes and sizes, but at the heart of each family is the home.

Home, to be maintained as a safe and wholesome place, requires the best you have to give and more. Creating a homestead will mean a lot of hours of manual labor at whatever level you're able to accomplish that. It will mean studying and improving your methods, perusing books, magazines, blogs, and YouTube media, and endlessly talking to local friends and teachers. You will have to integrate new ways of doing things into your lifestyle that will feel odd or even annoying at first. Please understand that as you work to change your lifestyle, your lifestyle will change; routines and norms will alter as you grow and improve. Take it at your level of growth, be true to yourself, but be prepared to work.

Homestead Personality Test

Are you familiar with the Myers-Briggs Type Indicator® or any of the other myriad of personality profiling tests (like ColorCode.com) online and in professional psychiatric practices? If so, then some of this information will be a repeat for you. Many of you already have a pretty good handle on what your personality strengths and pitfalls are and know to look for them. For others, these ideas will be completely new. Either way, I encourage you to keep an open mind while we discuss how your personality can affect and effect your homestead lifestyle.

I actually don't really care for the word *personality* when describing who we are in our core. After reading the fabulous book *It's Just My Nature*, by Carol Tuttle, I started using the phrase *energy profile*. Ms. Tuttle made a point that resonated with me about the word *personality*. I'm paraphrasing, but she basically pointed out that personality can be altered and effected by so many things like:

- environment
- upbringing
- family
- job
- social pressures
- major life changes or events
- trauma

- tangibles like economic status and safety

And so much more! Because our personalities can be altered to fit our circumstance, it can be dangerous to our well-being to stake too much on what kind of personality traits we exhibit. However, who we are inside, this light (or energy) we carry around as an intrinsic part of our soul is stable. There are core characteristics that are, simply put, who we really are eternally speaking. We can deny these energetic characteristics or forcibly live outside of them; nevertheless, they are who we are.

Still, *personality* is the word most of us are familiar with, so it's the one we'll use for our purposes here. To keep things simple, we will cover the four basic personality (or energy) types today. I didn't come up with this idea – in fact, neither did Ms. Tuttle, nor did the duo of Myers and Briggs. The "Four Temperaments," as they are referred to classically, can be traced back to Hippocrates. However, many philosophers, physicians and others have commented on these personality types throughout the centuries.

The following four personalities are listed according to similarity. That is, Sanguine and Choleric are the most similar to each other of the four. Also, Melancholic and Phelgmatic are the most similar to each other of the four. By those ancient thinkers these types were assigned the names:

- Sanguine
- Choleric
- Melancholic
- Phlegmatic

The color code assigns them colors (keeping the same order):

- Yellow
- Red
- White
- Blue

Ms. Tuttle assigns them energy profile numbers (keeping the same order):

- One
- Three
- Four
- Two

PERSONALITY TYPE DESCRIPTIONS

In brief, the personality types can be described as follows:

Sanguine – Yellow – Type 1

The Sanguine types can be described as active, happy, social, energetic and enthusiastic. Typically, Sanguine types prefer to be part of a crowd and are called extroverted. Being charming, outgoing and charismatic comes naturally to Sanguines. They are also more apt to take risks and try new things, rather than sit still and stick with what they know. They also have an ability to disconnect from and stop activities that they don't feel have value. Sanguines enjoy keeping things light and moving upward.

Choleric – Red – Type 3

Choleric types are also considered extroverted and very energetic. They are typically independent, entrepreneurially-minded, decisive and very goal and task-oriented. Cholerics do enjoy being part of a group but tend to favor deep connections with a handful of people; *friend* is a very special word to the fiercely loyal Cholerics. They are natural born leaders and can delegate well, being logical and quick-thinking. In spite of this, or perhaps because of it, Choleric types are extremely passionate about things and people in which they take an interest.

Melancholic – White – Type 4

Melancholics are deep-thinkers being extremely analytical, detail-oriented and ponderous. The typically have a studious nature and enjoy absorbing information. They are also self-reliant and individual, appearing reserved and quiet in a crowd. Melancholics are usually considered introverts and are deliberate and quiet in their behaviors. They are very perfecting, expecting quality and improvement in themselves, their surroundings and the people they know. This typically manifests itself in behaviors like tidiness and the offering of critique.

Phlegmatic – Blue – Type 2

Phlegmatic types are peaceful, calm, relaxed and easy-going. They are measured in their movements, graceful even. They are sympathetic and care a good deal about other people, though they are much less demonstrative than a Choleric or Sanguine type. Phlegmatics are considered introverts. They are also very good at making plans and visualizing solutions. Phlegmatics are great compromisers, usually able to find the solution that helps everyone.

Blessings of Your Homestead Personality

I have found that understanding my energy profile (personality) has positively affected nearly every aspect of my adult life. This knowledge has helped to guide me in my homemaking, homeschooling and homesteading efforts. In homemaking I am better able to accurately judge what I will make time for, how I will order my day and plan to accomplish the basic tasks of cooking, cleaning and living. In my homeschooling I am so much better able to plan and predict the performances and interests of myself and my children. How did I ever home educate before I understood these basic principles of my and my children's personalities?! On the homestead I can set reasonable, achievable goals because I can anticipate my own strengths and those of my family.

Some types have a very hard time accepting the concept of being defined by a type or title (a common issue for Whites/Type 4s and Reds/Type 3s). Others simply can't decide which type they are (a common problem of Blues/Type 2s and Yellows/Type 1s). No worries. Just ponder these ideas and see if you can find anything of merit here to help you homestead smarter, not harder. It's not a race, either. Or, if it is, the only person you're working with is yourself. I believe that someday – in a long, long distant future of eternal space and time – I WILL achieve complete balance between my types. All four types will be in perfect harmony and I will be perfect. (I prefer the Greek translation of that Biblical word which is *whole*.) But that day isn't today. So, I do the best I can on my homestead and begin from where I am.

Read the following and see if you can find yourself in these homesteading personality types. See the inherent strengths? See a few of the natural challenges that occur with each type? Both have value for you as work your home and land.

THE NON-MELLOW YELLOWS

You Yellow people are so full of energy! In fact, people often accuse you of pretending to be so happy and peppy but you're just full of natural joy. Your whole purpose in the universe is to bring people up; to bring light and activity and movement to any room or group of people. You're excited by information and projects. You are full of creativity and ideas. In fact, you're usually the one saying, "I have a great idea and it's going to be amazing!" Smothering your natural light and slowing down will NOT make the people around you happier, so never do it!

One of the things to look out for as a Yellow personality is that, in order to keep the flow going and people happy, you can accidentally become a chameleon. It is easy for you to change your own desires and plans to facilitate others. This can be called adaptability on one hand and becoming a doormat on the other. I should mention one more strength that can become a weakness on the homestead, if you're not careful. You have a fantastic ability to drop a project when it's just not worth your time or isn't working out. Sometimes, though, this can become a tendency to walk away from a plan, or even a person, when things get hard.

Be careful what you say yes to AND what you say no to – life isn't always fun. Things can get hard. But you can do hard things. In fact, you can laugh when hard things happen and that is such a gift to the universe and the homestead!

Homesteading Strengths and Pitfalls of Yellows:

- Creativity
- Energy
- Excitement
- Positive attitude
- Adaptability
- Ability to abandon a project if it's just now working out – this can be good, and not so good
- Difficulty concentrating on one project at a time
- Proclivity to let go of your own plans to facilitate another's – this can be good, and not so good

THE FIERY REDS

The Reds among us have a mantra – "I have a goal and you can get on board or get out of my way!" Reds know how to work, and they work hard. Being extremely task-oriented is both their strength and their caution in the universe. Reds need to watch out not to burn the candle at both ends while working the homestead. Reds are capable of making plans, usually a lot them, and seeing them through. They're interested in many subjects and usually have piles of books and materials tottering by their beds. Sometimes the plans they lay take a while to fulfill, but Reds will usually come back around to the original plan and get it done.

Reds, like all of us, have strengths that can also double as weaknesses. For example, Reds are passionate about so many things, but sometimes passion can override common sense, or even cause harm to relationships. Reds have an absolute sense of self-confidence, which can sometimes be understood as unhealthy pride. People will often be intimidated by your confidence and self-possession, so be compassionate and supportive of those who need a boost.

Be grateful for your nature - you are what you are, and what you are is awesome!

Homesteading Strengths and Pitfalls of Reds:

- Drive
- Excellent work ethic
- Ability to accomplish
- Curiosity and interest in a lot of topics
- Passion – this is a strength but watch out that you don't lash someone with it!
- An occasional disregard for advice or experience of others
- Can easily get overwhelmed with too many tasks and unhealthy expectations on the homestead

THE COOL BLUES

Ah, those soothing Blue types – what would the world be like if we didn't have their calming influence? I'll tell you what the homestead would be like: a big mess of half-baked projects and dozens of plans from those Yellows and Reds! Blues are soft, kind, empathetic and understanding. They seem to have an innate ability to understand people's (and even animals') hearts and minds. So much so that sometimes the feelings of a Blue can be easily hurt. They also tend to get overburdened with worry for others or circumstances. Sometimes Blues can seem shy, and perhaps some are, but unlike the first two types, Blues are simply able to hold still and hold their tongues. Blues are peaceful and quiet. This can sometimes lead Blues to allow others to "push them around" and have too much influence over them. Being accommodating and kind isn't the same thing as being a push-over.

Blues are very willing to set a reasonable pace for themselves and never take on more than they can handle. Blues are also capable of making detailed plans and thinking through a project before it's begun. In fact, Blues need to be careful not to get stuck in the planning. Don't just plan and plan and wait for everything to be perfect – a plan is made for action!

Say yes, get started, just do it – you're ready! The world needs your voice.

Homesteading Strengths and Pitfalls of Blues:

- Empathy and kindness
- Ability to understand the subtleties of projects and people
- Fantastic planning skills
- Naturally self-paced
- Need to watch out to not get stalled out with waiting and planning
- Watch out for worrying too much
- Don't let others push you around!

THE WISE WHITES

White personalities are often accused of being critical, but they're gift is a perfecting nature. Whites can analyze a plan and immediately see the holes and the problems that the rest of us can't see. You want a White personality quality checking your plans, I promise! Whites can be realistic about the vagaries of homesteading when it ain't all unicorns and rainbows. Whites takes the time to ponder, study and analyze – so much so that they really aren't fast movers. Don't expect a White personality to move like a Red or a Yellow one on the homestead because they have their own pace.

Whites speak directly and concisely and need to be careful not to be too short with people, or even impatient with other personality types. Whites love to organize data and set things in order. In fact, if you have tangled baling twine or Christmas lights, hand them to your White friends and they'll delight in getting those untangled for you. Whites need to be cautious not to put people and ideas in boxes that can't be changed and improved but keeping things in order is their forte. Listen to what the Whites have to say and let them help you improve!

Homesteading Strengths and Pitfalls:

- Analytical nature
- Capacity to plan
- Ability to take their time to really study out an idea
- Fantastic gift for seeing the flaws and weaknesses in any plan or idea
- Watch how you communicate with others; be helpful, not critical
- Be careful not to be too stringent and inflexible

RESULTS?

Remember that as you run through these ideas, you're looking for your dominant nature first. What is that personality that you typically lead out with and in which you feel most at home? What comes naturally to you? After you've done that, try figuring out your second, third and fourth types. You have all four in you just waiting to be explored and used to help you reach your full potential!

Please bear in mind that the types are neither good nor bad. Some attributes of each type are easy to identify as a strength or a weakness, but please don't feel these are BAD! Be grateful that God gave you the order types you have – it was meant to help you accomplish your work in the world.

You are exactly what you need to be.

WHAT'S THE POINT?

The point is to use this information to help your homesteading efforts. How will you make plans more efficiently now that you know your type? What projects will you begin, and which ones will you toss to keep true to your nature? Have you thought of your family – how do they test out? Will this knowledge change how you interact on the homestead?

Let this idea simmer a bit and then allow it to improve how you homestead moving forward. There are plenty of pitfalls and hurdles in homesteading; don't allow your biggest challenge to end up being you. Use your nature as wonderful tool, instead of fighting against it fruitlessly. Appreciate and honor your family members' personalities. What a wonderful collection of strengths and challenges you all have. Such a blessing.

THE HOMESTEAD ROADBLOCK

Time. Being a grown-up is one long experiment in how to manage time properly. We prioritize and adjust, and calendar and we still never seem to have enough time. I've lost whole years of my life trying to sort and pile and arrange my way through all that needed to be done, only to find that I still hadn't managed to accomplish even a fraction of it.

Take a deep breath. Before you jump into homesteading for the first time or take it to the next level, just exhale all those pent-up frustrations and anticipations that you can't possibly get it all done. You're right, you can't. Accept it now and move on with grace. *All* is a very relative word to begin with.

While we're on this topic, however, I will say that there's a great deal we CAN do, if only we will make the time for it. There are holes in everyone's schedules that are waiting to be filled with the quality and worthwhile work of homesteading. There are also activities in which we engage that may feel necessary to us now but, with time, will become burdens standing in the way of our homesteading progress. Don't be afraid to constantly re-assess your schedule and your personal time and be willing to dump activities and habits that are holding you back from using your time to the best of your homesteading ability.

To paraphrase former U.S. Secretary of Agriculture, Ezra T. Benson, when you make the self-sufficient lifestyle of homesteading a priority, everything else will either fall into place around it or drop out of your life. Embrace those adjustments and let them change you for the better.

Can I Homestead In an HOA?

Great question and the short answer is, yes, if you're willing to be creative! While I would never recommend you deliberately move into a neighborhood with the rules and requirements of a Homeowners Association (HOA), some homesteaders develop their homesteading dreams while already living in one and must adapt accordingly. The whole notion of an HOA vexes my core being, my very soul; to be frank, I'm not any fonder of CC&R's (Codes, Covenants and Restrictions). Which is probably why I removed myself from a city altogether and now live along a state highway in the middle-of-nowhere surrounded by pasture and forest. Where once I navigated by city landmarks, now I navigate by road kill. Still, though, not everyone can or wants to do that.

I can't pretend to be impartial on this topic, but I do understand that some homeowners are happy to have regulation in their neighborhoods to prevent specified unsightliness and curb contractually agreed upon improper behavior. If you have homesteading aspirations and move into an HOA, just do it with your eyes open and gather a lot of information beforehand. When I asked Amber, of The Coastal Homestead (www.coastalhomestead.com), what she liked best about living in an HOA she responded,

"Nothing at the moment, but when we first moved here twenty years ago, I liked the rules because the houses on both sides of us were rentals and we had A LOT of bad neighbors that had no respect for the homeowners. The rules helped us keep our sanity and gave us peace of mind."

As a homesteader here are some issues that may typically come up with your HOA, although specifics vary:

- Planting food crops in the front or side yards, or anywhere visible to passersby
- Hanging clothes on an outdoor line
- Livestock of any kind, including bees
- Building additions to your home or sheds on your property
- Selling homestead products from your home
- Having equipment like a compost bin or water barrel on site and/or where it's visible
- Similarly, off-grid energy equipment like solar panels

If you're looking for a way to homestead around the regulations in place, use Amber's experience as inspiration:

"There is a fine line between bending the rules and breaking them. I was on our HOA board for years and learned our By-laws and Restrictive Covenants backwards and forwards. We also contacted a real-estate attorney beforehand to clarify any questions.

"Our HOA animal restrictions state any animal (not just livestock) must abide by the rules [which are that they be] only on your property, not kept outside permanently and cannot be deemed a nuisance to the community (which needs a majority vote of the homeowners to be deemed a nuisance). Many HOA board members over the years have thought these rules only applied to livestock because they were specifically mentioned, until I pointed out the wording they were ignoring and insisted if they made one person [comply with those rules,] they had to go after everyone with a cat, as well."

So, Amber got her chickens and keeps them justified by complying with the technicalities of her HOA's rules and the assumption that no one is going to come after her unless they also want to take on all the feline owners. She further explains,

"First, our entire property is fenced in with a 6ft wood fence and lots of trees, so we have a lot of privacy. Our coop is connected to our house, which technically is an extension of our house, and we don't have a rooster, so they don't make a lot of noise.

Check with the local laws BEFORE you act, don't complain about it when you get fined or have to find your chickens a new home. Many cities/states will not be on your side in a court of law if you willingly break your HOA rules, then want to change them after the fact. Seek an attorney's advice beforehand to explain the restrictions and determine if there is any wiggle room to aid you in your homesteading efforts."

The point is, before you buy, be sure you know what the regulations are and if you're moving into an HOA, make sure that's what you want. If you already live in one and have developed your subversive homesteading desires after a home purchase, be prepared to think creatively. Read books like *Gardening Like A Ninja: A Guide to Sneaking Delicious Edibles Into Your Landscape*, by Angela England, that teach you to integrate your fruits and veggies into any garden style so that even your code enforcer will find them lovely. Then, maybe, you can work on getting everyone friendly with your chickens. Keep Angela's reassuring words in mind and apply them to the broader campaign of the urban homestead movement:

> "There is hope, of course. Many cities and municipalities allow gardens, chickens, rabbits, and other small livestock. In fact, as of December 2011, 93 of the largest 100 cities (by population) in the United States allowed keeping backyard chickens in some form or fashion. Often, cities will limit backyard flocks to hens only, or enforce a smaller size such as six hens per family within city limits. But even with limitations, the fact that so many of the nation's largest cities are friendly toward those seeking additional self-sufficiency is an encouraging trend."

- From *Backyard Farming on an Acre (More or Less)*

IS MY BUTT TOO BIG?

Every woman has asked that while looking at herself in a pair of new jeans. The fact is, size matters to us and the idea of being too big or too small or too _____ is something we think about even in homesteading.

Can you homestead in a space you consider too small? Of course! You can do anything to which you put your mind. As George Nash and Jane Waterman wrote in their book, *Homesteading in the 21st Century*,

"You don't have to live in the country to homestead—your mindscape is more important than your landscape."

Our first homestead was on a .14-acre lot with all of California's interesting urban zoning laws. Jessica homesteads (www.104homestead.com) on a ¼ acre and, in fact, the tag line to her blog is "Teaching you to homestead wherever you are." Permaculturist Amy Stross and her husband (www.tenthacrefarm.com) prolifically homestead on, you guessed it, a tenth of an acre and can inspire you to do the same. Ms. Bernie (www.apartmentprepper.com) homesteads in an apartment; PJ (www.survivalforblondes.com) does it in a condo. So, you see, you can homestead wherever you are because homesteading is really a lifestyle choice. As we are reminded:

"I always like to encourage people that it doesn't take land or a lot of money to get started homesteading. Oftentimes people get discouraged thinking that they'll never be able to live their dream due to their current circumstances. Believe me when I say you really can homestead wherever you are, right now! You don't have to wait. So much can be done indoors, without a square foot of land. You just have to get creative, be resourceful, and think outside of the box...I encourage you all to pick one self-sufficiency skill to learn and start there.

"Homesteading isn't an easy life, but it's one of the most rewarding things you'll ever do!"

-Kendra Lynne, Homesteader

Having said that, however, the ultimate goal of many (though not all) homesteaders is to achieve self-sufficiency. If you define that goal to include self-sufficiency in areas of energy, food production and waste, then, at some point, you're going to need at least a small bit of earth. Each patch of land is precious, and I feel it's a worthy goal for every homesteader truly seeking to be self-sustaining to procure some, even if it takes a lifetime to achieve it. Don't let the waiting for more land drag you down. Homestead where you are as if you'll be there forever. Just keep your land goals in the back of your mind as you progress.

For the purposes of this book, we will continue our discussion as if land acquisition (in whatever amount and type you define as being necessary for your family) is one of our many long-term goals. One of the common threads running through my interviews with the above-mentioned homesteaders and the numerous others I interviewed is their desire for more land than they have now—which, in some cases, is any land.

You look fantastic in those jeans; your butt is just the right size, but you're definitely going to want bigger land.

How to Use This Book

The Do It Yourself Homestead was written as an instrument of encouragement and education for the aspiring homesteader. I've been homesteading for over two decades and I still consider myself aspiring, so I divided the book into four levels of homesteady-ness (a completely made up word). These levels really aren't scientific at all but are, rather, based on a perceived level of difficulty that correlates to the subject matter. Each level has its own topic and instruction inside each chapter. The levels and information are applicable to all types of homesteader, but the hope was that grading the topics would prevent newbies from feeling overwhelmed, while keeping those with experience engaged and learning. The levels are designed to be a simple guideline as people read and use the book, but they're not gospel—you're free to read and engage in every level.

Bear in mind that this book is for the homestead *student*. That means that the reader is introduced to topics at their level and given enough information to get them started pondering, performing and perfecting. It is expected that the student will read dozens of books, scores of books, scads of books—NOT JUST THIS ONE. This book is not all-encompassing on purpose. It is merely an organized place to start. One of my favorite homesteading books is Carla Emery's fine publication, *The Encyclopedia of Country Living*; it is a fun resource and I love it. However, when I first happened upon that book I was a newbie homesteader and completely overwhelmed by the sheer volume of knowledge presented there, as much as I valued it. *The Do It Yourself Homestead* is not meant to overwhelm the homesteading student but to educate, encourage and inspire them, which includes pointing them in the direction of further knowledge through books, websites and other resources that have value.

Because this book is organized by chapters detailing independent topics, there's no need to read the book cover to cover if you don't want to because each chapter can be read on its own. Just pick the topic that speaks to you right now and go study that. At the end of every section is an action plan to be thought of as "homework" for the level of homestead student we have reading along. Remember this is a guidebook, kind of like a textbook. It's probably the most laid-back textbook you'll ever read, but it is meant to teach, and does require reader/student engagement to produce the most benefit.

Just a quick note of practicality: this book is over 400 pages long. 400. That's a lot of pages, and they're pages you're going to want to mark up and haul around with you to the kitchen, the bathtub and the garden. While the printer has done their job well, you may consider taking the book to your local office store and having them spiral bind *The Do It Yourself Homestead* for you. Similarly, you can have them take it apart and hole punch it for you for placement into a three-ring binder. I only mention this because I am extremely hard on my reference books and use them up until they can practically be composted. If you're like me, it may be worth the extra step to pull the book apart for ease of use.

Levels of Homesteadyness

There are four levels of homesteading experience reflected in the book. Like I said, this separation is in no way scientific. Simply try to find the level that best describes where you are on your homesteading journey. Begin there and move up and out as you're ready. Or go back to square one, if that's what you feel more comfortable doing. It's your homestead, so do it your way.

LEVEL 1 - **HOMESTARTER:**

This level is for you, the curious and courageous novice, who can see the benefits of cooking from scratch and conserving electricity and who wants to better understand how to consume less and produce more. You may not have gardened much but you'd like to learn how to produce and use your own food in some measure. You may not have space or inclination for a milk cow, but you are curious if there are other animals you might be able to raise. So many ideas may be forming in your mind but you're not sure what to do next. This is only the beginning and the beginning is the perfect place to start. At this level, a compost bucket or recycling bin may appear in your kitchen.

LEVEL 2 - **HOMESTEADISH**:

On this inspiring and exciting level, the plans for the garden are coming together, even though you're not quite sure if it's all being done correctly. Perhaps you're toiling away at preserving this year's harvest, but your spouse isn't quite on board with the idea of a beehive. Maybe you're starting to find like-minded people with which to network and take classes. It's possible you've even started to downsize your "stuff" or better organize your family. At this level, there's probably a chicken book by your bed and seed catalogs squirreled away in your desk.

LEVEL 3 - **HOMESTEADAHOLIC**:

You've caught the vision and are finding your way in the homestead lifestyle and what a fantastic and fulfilling place it is to be! Slowly the rhythms of growing food, tending some livestock, creating meals from scratch or simply playing around with herbs are starting to become part of your normal. You realize there's so much to do and it's all connected—sometimes you feel exhausted just reading through those homestead books piling up all around you, but you're so excited to know more! Quite often, at this level, you've located a local mentor who may be talking you into goats, or creating a seed bank, or making cheese.

LEVEL 4 - **HOMESTEADED**:

You've crossed over, my friend. Your vision, lifestyle and personal sense of mission all connect in the vision of attaining self-sufficiency and bounty to share. This is not a place of perfection in the sense of you never making mistakes or ever being "finished." This is a place of wholeness; a perfectly complete commitment to your future as a homesteader. At this level, you're no longer trying to maintain a "normal" life running parallel to your homestead life, although you may still work full-time off site. You follow the cycles of the seasons. You work home, family, land and animals. You still may not own your own land, but you've found a way to make it work through co-ops, local farms and homestead networking. This is it. This is how you define yourself when people ask the innocent question, "So, what do you do?" Before your profession even pops into your head, you answer, "Why, I homestead!"

Chapter Headings and Content

Once you find the level you'd like to begin with you can peruse the chapters and their contents. There are eight chapters, each with their own topic:

1. The Homestead Kitchen
2. The Homestead Garden
3. Green the Homestead
4. Livestock Wherever You Are

5. Homestead Finances
6. Family Times
7. The Homestead Community
8. The Prepared Homestead

The chapters have a section for each level of homestead student. All this information can be found in the Table of Contents. Each level of information inside the chapters is listed in order of "difficulty." So, the first topic will be for the *Homestarter*, the second topic will be for the *Homesteadish*, the third topic will be *Homesteadaholic* and the fourth topic will be *Homesteaded*.

Also included in each chapter are calls to action like brainstorming assignments in your homestead journal, projects and tips. These items are meant to provide further inspiration and experience with each topic. You don't have to feel pressured to do them all at once or even do them at all! This is your book and your homestead education; do what you feel you can, when you can.

 ACTION ITEMS:

These are specific ideas to be explored in a homestead journal. We outline things the reader can do now with the information provided and make suggestions on what to record. Maintaining a homestead journal is really a must as we go throughout the book. Get into the habit of writing in your journal every time you read, take a class, get an animal, plant a crop or fail miserably at some homesteading venture. It's all useful information to which you will want to refer. Please believe me when I tell you that you will NOT remember it if you don't write it down.

This section may also contain suggested links to Internet articles for further study. Follow the instructions included for each link to access that information.

DIY:

This section contains some kind of project that can be done with your own hands and brain using the information provided in the chapter or to provide for a future project on the subject matter.

This section may also contain suggested links to Internet articles for further study. Follow the instructions included for each link to access that information.

BUILD COMMUNITY:

This section is for suggestions on how you can gather like-minded people in meaningful ways to further your own education and provide a way for others to learn on the topic discussed. We get what we give so we learn to give a lot!

This section may also contain suggested links to Internet articles for further study. Follow the instructions included for each link to access that information.

BOOKS AND WEBSITES for each topic:
There are so many quality books and sites out there! Please know that when I give a recommendation, I've read or reviewed the book, or visited the site personally and for the purposes of my own education. Some resources have been more valuable to me than others and will be for you too, but all have added something to my education. However, I have NOT read every book nor visited every site that's of value. That's your job over a lifetime of homestead education. I encourage you to read EVERY book on your favorite topic and visit EVERY site for which you can spare a minute. (Remember that some sites and blogs stop being maintained over time.)

Read as many authors as you can for variety and perspective. Everyone's personal experience on any given topic is just that, *personal*. Just read, read, read and if you don't like a resource I've listed, chuck it and move on. Homesteading is still a worthwhile venture even if Tessa (that's me) likes a book or website you think is dumb. Homesteaders are a varied bunch of people and you're going to find something that resonates with you somewhere. In this homesteading niche there are homeschoolers and Democrats and foodies and Christians and raw milk drinkers and Canadians, and on and on and on. If you're reading or listening along to something and you run up against an opinion or a life experience that runs contrary to what you believe, just take a deep breath and relax. Listening to other peoples' points of view with respect and patience is all part of learning as a mature individual. Give it a minute or two and the author will most likely get back to the topic that brought you to the resource in the first place. Don't toss something, and mark it as useless to you, just because you and the author have different perspectives and philosophies.

A long list of websites is provided at the end of the book in the resources section, but we have also rooted out useful links online for you to use as you study the topics in *The Do It Yourself Homestead*. Online learning has become common place in our culture—we're always looking for a link! Because websites rise and fall all the time, we house the links covering various topics peppered throughout the book on our website at:

www.homesteadlady.com/homestead-links-page.

This enables us to check links every month, making sure they're still active. We'll swap outdated or broken links for shiny new ones when necessary. For readers of our print book, this will require that you go online and visit our site to retrieve the links outlined for you on the page built just for you. Happy reading!

HOMESTEAD LADY SPEAKS:
Homestead Lady is the name of my blog and how my current readers know me. This is one last place for me to share something on the topic that I think is of value. These musings are meant to be encouraging or uplifting in some way.

> Don't forget to check out the handy **RESOURCES** section at the back of the book for more homestead pleasure reading and perusing.

WOULD YOU LIKE FRIES WITH THAT?

This book was written to be a usable reference, not a coffee table sort of book. Make sure you read it with a pen in hand and mark it up as you see fit. I never feel like I've used a book properly unless I have notes and scribbles all over the text. To help cement what you're learning we've created other publications, including home education materials for those of you with homeschooled homestead kids; please see the website shop for details (www.homestead-lady.com/shop).

If online learning is something you're comfortable with, you may be interested in the courses built around the content of *The Do It Yourself Homestead* and other books. Be sure to check out our shop for those details, as they're available. I also invite you to participate further through continuing education like future books and webinars, as well as joining the conversation on our social media accounts and even sharing your insights for upcoming books. See the back of this book for ways you can continue to participate as you build your own Do It Yourself Homestead!

> **A QUICK NOTE**
>
> For the purposes of clarity, when I refer to a USDA certified Organic item or practice, I have capitalized the "O" so that you know I'm indicating that certification. I was using the term organic with its original meaning, as coming from the earth, long before the Feds decided to use the word for their own purposes. There was a time when Organic food was just called "food," no clarification needed. Ah, how I pine for those days.
>
> As Mary Jane Butters, Organic farmer and entrepreneur is quoted as saying in an article by Holly Funk at Organic.org,
>
> "I think we need to take back our language. I want to call my organic carrots 'carrots' and let [other farmers] call theirs a chemical carrot. And they can list all of the ingredients that they used instead of me having to be certified. The burden is on us to prove something. Let them prove that they used only 30 chemicals instead of 50 to produce an apple."
>
> Amen, sister.

However, I bow to the current vernacular for our discussions here, so watch for that capital "O."

The Homestead Journal

I've kept a journal on and off for as long as I can remember. It was part, not only of my family culture, but also of how I learned to study the sacred texts of my religion. For those who are spiritual, your holy works are those books given by your God to help you really know something. Something that can only be known by studying those sacred texts. They're meaningful and instructive because you spend so much time reading them that you eventually begin to put yourself into them—likening the stories to your own life. For those who aren't particularly religious, but who find great value in books you consider to be significant, you can experience similar personal growth each time you read; these books become classics and you become a better person.

But there is another kind of sacred text at your disposal. A good scholar will always take notes. Your personal journal becomes a record of empirical data, but it also becomes a kind of living scripture to you; something you can look back through and remember the times when you were wise. And foolish. As we move through life's vicissitudes, we record those profound moments of epiphany and clarity so that we can look back on them in darker, murkier times and remember things we once knew with such conviction. As pilot and religious leader Dieter Uchtdorf wrote,

> "Often the deep valleys of our present will be understood only by looking back on them from the mountains of our future experience."

Your journal is a text from which you can really learn something. A homestead journal is much the same but instead of recording ALL things you're learning and doing, you record only those things that pertain to your homesteading ventures. You'll be surprised at how many subjects you cover in your homestead journal—cooking, family, gardening, animals, energy, waste, foraging, herbs, and chores. (As it happens, all the subjects we'll be covering together in this book and even more.) Never make the mistake of thinking that something is too trivial on the homestead not to note in your homestead journal. We often come across some variety of vegetable we want to try or some recipe we know our family will love and we're convinced that we'll remember it. The truth is, we never do—or worse somehow, we only remember bits and pieces that aggravatingly never materialize into complete memories. So, we learn to pay attention and write it all down.

Another habit to be avoided with our homestead record keeping efforts is to neglect to create entries about the projects that fail. As someone who learns best when she messes something up, please believe me when I tell you that these failures are necessary to record; they're vital to the eventual success of your homestead. Former British Prime Minister Winston Churchill is credited with saying,

"Success is not final, failure is not fatal: it is the courage to continue that counts."

He also wisely noted that,

"Success is stumbling from failure to failure with no loss of enthusiasm."

Both thoughts are so important to internalize for the modern homesteader because they keep your forward motion moving vertical, instead of maintaining in the horizontal.

Even your successes aren't the final word on any subject because you continue to grow and challenge yourself, always moving higher and higher up your list of goals. In your failures, with no loss of enthusiasm, you note the process, analyze what you can do better and simply try again, always building on past experience. It will surprise you how far you've come, as you begin to record your homestead life, when you go back to read previous entries. While I discourage you from unhealthy comparisons, contrasting your beginner homestead-self to your more seasoned homestead-self will bring you joy and justified appreciation for all you've learned.

For those who are homesteading on their own, with limited or non-existent support from the people in their lives or with other special challenges, your homestead journal becomes a friend to you. It's so very hard to do this noble and worthwhile homesteading work when you feel overwhelmed or like you're in it alone. My heart goes out to you and I want you to know that I believe in you. No effort is wasted even if you, your journal and God are the only ones to note it. Through health challenges, setbacks, divorce, illness, death and depression, your homestead journal will be a source of strength and will provide you with the motivation to keep progressing. So many times, I've been ready to throw up my hands and walk away for good but, like a faithful companion, my homestead journal talks me down from the ledge and helps me get back to work.

SO, TO BEGIN TODAY

Procure some kind of journal—a plain lined notebook will do. Or, if you're creative and will appreciate the use of it more, a decorated homestead journal with quality scrap and clip-art. You can also use an online template to digitally customize your own kind of record.

If you prefer something digital, use an online resource like Evernote to record your homestead information. I enjoy the organic connection of writing with my own hand, so I prefer paper and pen. Besides, I haul my journal around all over the yard and kitchen, both notoriously messy places where I am not to be trusted with electronics. So, I keep a tangible journal and then later I often type up the gist of my notes into digital files, blog posts or books. Whatever works for you will be perfect.

If you are drawn to something formatted but you don't want to have to mess with setting it up, we have available for sale *The Do It Yourself Homestead Journal* at Homestead Lady's shop page (www.homesteadlady.com/shop). The journal correlates directly to the book, following each subject covered with a reiteration of every Action Item, as well as the DIY and Community challenges issued from each topic. The file is digital, so you can print what you need, when you need it. I suggest printing it all and putting it into a three-ring binder, filling in each section as you work with whatever topic you've chosen—you do not have to do the work in order.

Whatever journal format you choose, don't let the decision-making process delay you from beginning your homestead ventures for more than five minutes. It's not that big a deal what you write in as long as you write. As my friend Kathie reminds us in her book, the *Fiercely DIY Guide to Seasonal Living*,

"The journal is about you—you are not about the journal."

One of my favorite things to use my homestead journal for is making lists of things about which I have questions. My memory is so poor that I even forget what I don't know! I need to write those things down in my journal and am always so grateful when I do, because finding them later jogs my memory. Like any other important book, my homestead journal reminds me of the person I'm planning to be, as well as the goals and lifestyle I truly feel is right for me. Without that record, not just information, but something important in my soul would end up not being revealed to me. Homesteading is a way of life and my homestead journal is part of what keeps the life meaningful, personal and constantly improving.

With that in mind, let us begin!

CHAPTER 1

In The Homestead Kitchen

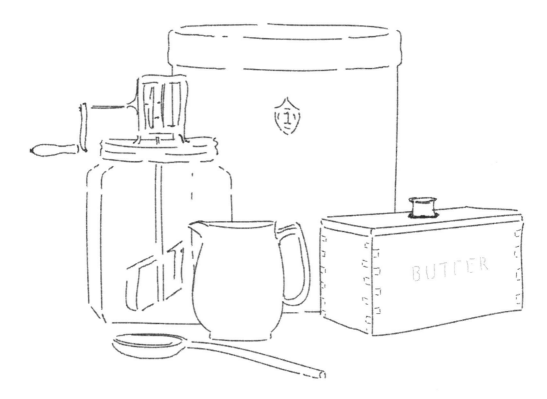

Great-Aunt Bert. My paternal grandfather's aunt was always referred to as "Great-Aunt Bert," and it didn't matter which relation was talking about her. Bertha is one of those spunky ladies that hang from my family tree like dignified, hilarious and powerful chimpanzees. Stories of Aunt Bert included tales of eluding Chinese check points on a Vespa; long, silver-grey hair that she swept up in ornate combs; rose hip gathering on the hills of San Francisco, and many more. One thing I remember in particular about Aunt Bert was that, as the story goes, every year she'd send tins of entirely inedible ginger snaps for Christmas along with newsy letters detailing the lives of people that nobody in the family seemed to know. She would write on and on as if everyone knew whom she was talking about, but the letters never made much sense and the cookies were as hard as hockey pucks.

What *was* understood was that Aunt Bert loved and cared about each recipient. My great-great aunt's cookies were entirely handmade in her own kitchen. Each edible gift was individually packaged in tins and lovingly wrapped for shipping. Every nonsensical letter was written and addressed by hand. Whenever I make a ginger cookie, snap or soft, I think of her and how she administered to her family. And I chuckle, feeling her love come down generations, even to me.

A kitchen is NOT just a room where we prepare food for our family to eat. A kitchen is a workshop where we dish up healing and love, plain and simple. As food critic and author, Craig Claiborne said,

> "Cooking is at once child's play and adult joy.
> And cooking done with care is an act of love."

But something has happened in our modern culture that has disconnected us from the recipes of our ancestors and their methods of food preparation, rock-hard gingersnaps aside. With that loss, in large measure, has gone our affectionate connection to the rhythms of cooking and eating, as well as our understanding that our food truly can be our best medicine. In this chapter, we're going to celebrate the homestead kitchen as a place of health, wholesomeness, frugality and, yes, love. As Sally Fallon observes in her book *Nourishing Traditions*,

> "Technology can be a kind father but only in partnership with his mothering, feminine partner - the nourishing traditions of our ancestors. These traditions require us to apply more wisdom to the way we produce and process our food and, yes, more time in the kitchen, but they give highly satisfying results—delicious meals, increased vitality, robust children and freedom from the chains of acute and chronic illness."

Here's what we'll be discussing in this chapter on what to do in the modern homesteader's kitchen:

HOMESTARTER level: Use all the parts of the foods that make their way into your kitchen. From strawberry tops to chicken bones, we've highlighted over twenty-five foods that can be used and repurposed in wholesome ways. This is a food-jam session!

HOMESTEADISH level: Fermented foods? That phrase may sound new and a bit odd to you but you're already eating fermented foods and loving them, although you may not know it. In this section we share three easy food ferments and what you can do with them.

HOMESTEADAHOLIC level: Preserving the harvest in its entirety is a weighty, but attainable goal for any homesteader. We review three basic ways to get as much food as possible onto your pantry shelves for the coming year, as well as guidance on setting some basic preservation goals for the home and garden.

HOMESTEADED level: Not so long ago nearly all the ingredients used to create meals for your family originated on your land and in your home. If you wanted flour, you ground grain. If you needed butter, you churned cream. Although we're accustomed to purchasing these items in a store now, it's often not too difficult to make them ourselves. By producing these staple items in our own kitchens, we control how they're made and with what quality they're created. We'll talk about making a few ingredients to bake up everyone's favorite treat, the humble cookie. Don't worry, no inedible gingersnaps will be served.

As with every chapter in this book, be sure to follow along at the end of each section as you reach the Action Items, DIY Projects, ways to Build Community, resource recommendations and parting advice.

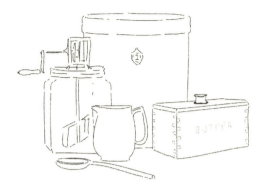

"Afterwards, they always had tea in the kitchen, much the nicest room in the house."

— Flora Thompson, from *Lark Rise to Candleford*

For the Homestarter: Use All the Parts

One of the skills that's been lost from modern homemaking is the ability our grandmothers and great-grandmothers had to use up "all" of something. Take the simple strawberry. I grow strawberries, pick strawberries and prepare strawberries for canning and dehydrating. But I always used the berry, not the tops – you know, that green part attached to a skimpy bit of leftover strawberry. It's a top, I thought; for what could you possibly use it?

Here are some simple things to do with strawberry tops:

- Keep them attached and toss the whole strawberry into your blender to make a morning smoothie, or even a batch of jam, with increased vitamins and nutrients—they are green, after all.
- Save the tops and put them in your livestock treat bucket—your chickens, rabbits, goats and pigs will all devour them.
- If you don't have animals, toss them into your worm or regular compost bin.
- Dehydrate the tops with bit of strawberry still attached and used for hot or iced tea to calm the tummy and boost the overall health of the body with their anti-oxidants.
- Soak the tops in some witch hazel for about a week, strain and dab some on your skin to clean and tone—be sure to do a spot check for an allergic reaction.

And that's just strawberry tops, folks.

The Picayune Creole Cook Book of 1922 tells us:

> "Do not waste anything in the kitchen. Our grandmothers scrupulously saved every piece of bone or fat, and these were utilized in making soft soap. The careful housekeeper will manage to keep out of debt and set a good table, with much variety, on a small allowance, by faithfully saving and utilizing the left-overs. Never throw away any beet, turnip or radish tops. They may all be cooked in the same manner as spinach au jus, or they may be boiled with salt meat, and make very good and healthful dishes."

Do you "scrupulously save every piece of bone or fat?" I know I'm not as conscientious as I might be. However, I have made progress! For example, take the extremely useful roasted chicken and how it can be used up.

TO MAKE BONE BROTH

1. Roast a high-quality chicken for dinner and, after the meal, scrape off every leftover bit of meat on the bones to make chicken salad for tomorrow night's dinner.

2. In the meantime, place the stripped bones in your slow cooker with one quartered onion and some garlic cloves. Cover everything with water and cook it on low for 12-24 hours.

3. Pour out the liquid, which is now a rich and healthy bone broth to freeze for later or use during the week.

4. Cover the bones with water again, adding 1 Tablespoon of quality vinegar and cook for another 12-24 hours.

5. Pour out that batch and save.

6. You may be able to draw out one more batch of broth with those bones but be sure to add the vinegar again, which pulls the rest of the minerals out of the bone material.

Homemade bone broth can be saved for soups, casseroles, stir fries, hummus and more. I usually freeze mine in 10-cup and 1-cup amounts so that I have some for soup (10 cups) and some for sauces (1 cup). It's a simple thing, roast chicken – but it's as good a place as any to start.

If you have picky eaters who baulk at leftovers, the best time to change their ideas about food is right now, this minute. Re-educating the youngest children will usually be the easiest— teenagers are much harder to work with than kids under five, though personalities vary. Just keep presenting new foods, stick to any food rules you've discussed as a family and try to decide on acceptable compromises. Especially if you've never done much in the way of re-purposing leftovers, give your family time to adjust and be patient. Never be afraid to say that the recipe is an experiment and get their honest opinions on new ideas. Try to weigh the useful, critical remarks against the rather useless, "I just don't like to eat new things" remarks. You could always use this quote from comedian Buddy Hackett, if they're whining:

"As a child, my family's menu consisted of two choices: take it or leave it."

An Alphabet of Kitchen Scraps

As Kris Bordessa (www.attainable-sustainable.net) reminds us,

"The biggest first step a person can take toward a more self-reliant kitchen, in my opinion, is learning to cook at home. Those with no knowledge whatsoever can learn to replace take-out food with meals cooked at home. Those who have some cooking skills but fall back on prepared meals due to a busy schedule can implement tricks like once-a-week cooking to prep for a busy week, freezer meals or learn to use a slow cooker. Meals made at home from whole ingredients can be much more cost effective than eating out."

At the beginner level of "Homestarter" the key is not to get overwhelmed with complex changes. We keep it simple to enjoy success and lay some good foundations. Here are some simple ideas for using leftover bits, as well as large quantities of ingredients. There's a list of ideas for every letter of the alphabet.

If you read an idea and are curious how to try it for yourself, check out the suggested reading found at the end of this section. Always determine if your local library has a book before you buy it – yes, the library has cookbooks! You may already practice some of these and might even have your own ideas to share with your friends and social media contacts. In our modern age, the Internet has become a very useful educational tool, especially for those learning to cook well. Several of the topics below have further reading provided in the form of links to specific articles online; as we explained earlier, these links are housed on our site at:

www.homesteadlady.com/homestead-links-page.

APPLES

I encourage you to try growing an apple tree or two when you're ready. Growing, or even picking your apples every year, will mean that you know exactly how the apple has been grown and whether they've had any pesticides or herbicides applied to them. You may find that your views on Organic vs. conventional growing evolve during your homesteading career, so it's good to start thinking about these topics even if you don't feel like you have all the answers today. Now then, what to do with extra apples...

- Make apple vinegar with saved cores and skins. Though not technically apple *cider* vinegar, this will process will produce a pleasant, light vinegar especially suited to salad dressings and marinades. If you ferment it for around a year, it will more closely resemble the apple cider vinegar you buy at the store.

TO MAKE APPLE CIDER VINEGAR

1. Collect apple bits from about ten apples (or you can use five whole apples), putting them in a large glass container and covering them with unchlorinated water.

2. Add a cup of sugar and stir until it's dissolved.

3. Remove any floating bits of apple or seed and cover the container with cheesecloth or muslin, securing with a rubber band. Allow the apples to sit on your counter for about a week; you'll see bubbles forming as the sugar turns to alcohol.

4. Once you can smell the alcohol, strain out the apples and pour the liquid into glass jars (I use my canning jars). Put fresh cheesecloth over the top and ferment for about a month more until it starts to smell and taste like vinegar.

5. A little sediment will appear at the bottom and an odd-looking, mushy thing will form at the top; this is called the "mother." All of that is perfectly normal and means your vinegar is doing well—Congratulations! Use simple pH strips to measure the acidity – it should end up between one and two. Always check for mold on the container in which you're fermenting and clean it out immediately, if you see any.

6. If the vinegar gets too strong for you, just dilute it with unchlorinated water and use in salad dressings, sauces, cleansing tonics and any other thing for which you might want delicious, healthy, homemade apple vinegar. For a stronger vinegar, ferment up to a year. Use the vinegar within a year.

- Make apple sauce in your slow cooker if you would like to do just a small batch and don't have a lot of time.

TO MAKE SLOW-COOKER APPLESAUCE

1. Core, peel and cut up as many apples as you have, filling your slow cooker.

2. Cook on low until apples become soft and mushy.

3. Use a hand mixer or a potato masher to pulverize remaining apple chunks and make a sauce

4. Add a little unchlorinated water if your sauce is too dry (I've never had to do that, though).

5. Cool your sauce a bit, then add raw honey and cinnamon to taste.

6. The best thing about using a slow cooker for making applesauce is that you don't have to stir it all day!

- Use young, green apples in place of commercial pectin in jams and jellies. Crabapples or quince work well for this, too.

> **TO MAKE APPLE PECTIN**
>
> 1. Wash and slice (but don't peel or core) about ten standard-sized apples (or about 3 pounds of crabapples).
>
> 2. Add at least four cups of water and boil the apples until they reduce in size to about half what they were. Strain your apple mixture through some cheesecloth or a fine mesh sieve to get out all the seeds and stems, etc.
>
> 3. Bring the mixture back to a boil and boil for another 15-20 minutes.
>
> 4. If you're going to use this homemade pectin for making jam soon, you can simply put it into any container with a lid and place it in your refrigerator. Or you can preserve it by processing it in a water bath canner.
>
> 5. To learn how to gauge how much homemade pectin to use, please visit our links page under The Homestead Kitchen section.

- Leave the skin on, slice and core with a rotating slicer, and dehydrate to make apple chips; you can even do this in your solar oven.
- Break up the cores between your livestock (including guinea pigs and hamsters if that's the only livestock you have) or put into the compost pile.
- Use as decoration during the holidays; slice and string to hang outdoors as a Christmas gift for the wild birds.

BANANAS

Botanically speaking, bananas are a berry and should be treated as gently as you would a blueberry. Once they're past their prime for eating out of hand, there's still a lot you can do with them.

- Save uneaten stubs and halves by freezing, then toss in smoothies and milkshakes to sweeten instead of sugar.
- Make banana bread with the fruit and feed the Organic peels to the chickens.
- Compost peels for a potassium boost in the soil.
- Bananas can be added to fruit kebabs or any raw dessert. Or, scrape out the middle of a banana still in its peel, fill the hole with chocolate chips and place the banana "boat" on top of the warm coals of your campfire for an alternative to S'mores.
- Ever made a banana face mask? Your pores will thank you, if you try it.
- Need trail food? Try dehydrating slices for banana chips.

- Fermented bananas can serve as a natural lure for bad bugs in the garden, as well as for beneficial pollinators.

TO MAKE A BUG LURE

1. Cut up an old banana and mix in 1 teaspoon of sugar

2. Mush it all together and place the mush in a shallow dish in the garden.

3. I usually add a little water to the banana dish for the butterflies and bees to drink. If you do this, be sure to place a small sponge into the dish on which they can stand without drowning.

4. You can manually collect and dispose of unwanted bugs each morning until the banana is gone. Leave the butterflies, of course!

CHEESE

You really don't need me to tell you how to eat cheese – it's cheese. Therefore, it is delectable. However, here are some cheesy things to ponder.
- Freeze hard cheeses you can't use before they go moldy.
- Use soft cheeses in sauces and for fondue night.
- Melt cheese wax and cover mild cheddars as part of your food storage program. Cheddar will continue to sharpen as it sits so start with mild cheese unless you really, really like sharp cheese.
- FYI, learning to make your own cheeses can be a fun pursuit—the soft cheeses like feta are the simplest to make. Bonus: if you prefer *raw* milk cheeses, feta is a traditionally raw milk variety.
- If you go crazy with cheese-making and make big batches of something like farmer's cheese, stuff it into any pita or pancake and drizzle with raw honey. You can also mix it up with fresh herbs and sea salt to spread on sourdough bread with dinner.

DIPS

Don't discount the dips, I say! Providing a tasty dip with a tray of carrots and celery is an easy way to get kids to eat their veggies.
- Dairy based dips can be repurposed into salad dressings and sandwich spreads.
- Salsas are perfect to add to taco meat and soups.
- Hummus makes a tasty filler in pita and is divine tossed into an omelet.
- Wash commercial dip containers and use in place of plastic storage dishes.
- If you make your own hummus dip, be sure to use the water from pre-soaking the garbanzo beans in your compost or garden.

- If you don't eat them all in time, feed aged hummus and other dips to the chickens.

EGG

My favorite food. Please note, that for all these ideas it's important to use eggs from humanely raised, clean-fed chickens and to wash any shells you use thoroughly.

- Extra eggs can be hard boiled on a stove top, baked in the conventional and/or the solar oven. This doesn't extend their life, but it does make them a versatile addition to sandwich fillings and salads.
- Make omelets, pudding, flan and custard.
- Bake eggs in muffins tins, freeze and use for breakfast sandwiches
- Similarly, scramble surplus eggs and freeze
- Save the shells, then wash, bake and crush them to put into your compost or feed back to your chickens as a calcium supplement.
- Wash, dry, crush eggs shells and add to homemade toilet cleaner where they'll act as an abrasive.
- Wash, dry and soak egg shells in water to create a calcium infused drink; similarly add lemon juice to make calcium citrate. Strain out shells before drinking.
- Wash, crush and add to water kefir to give the grains vital minerals.
- Wash and bag tightly to add to the laundry to whiten clothes.
- Blow out the eggs and use the shells to decorate for Easter.

> **BLOWING EGGS**
>
> 1. Using a *room temperature* egg, poke a small hole in each end with a slim nail.
> 2. Carefully blow out the egg from inside into a dish. Save the egg to use in recipes.
> 3. Gently wash out the shell by allowing soapy water inside through one of the holes. Cover the holes and softly shake the soapy water inside. Blow out the water. Repeat until water comes out clean.
> 4. Once dry the eggs can be dyed or painted.

FATS

Please note, I recommend you save fat only from humanely raise, clean-fed animals.

- Save all drippings from roasted meats for gravies and sauces.
- Refrigerate roasted meats to allow fats to gel, scrape gel into a container and save in the refrigerator to add to casseroles and pet food.
- Render leftover fat from butchering by cutting it into cubes, placing it in a slow cooker with a ¼ cup of water until it melts completely and, finally, straining it through cheesecloth. Store

- Save all bacon fat for frying and sautéing; in the American south, bacon fat's flavor is considered so indispensable it's simply called "seasonin."
- Make your own coconut butter which is especially tasty in desserts.

TO MAKE COCONUT BUTTER

1. Place five cups of coconut flakes in your blender.
2. Add 1 Tablespoon of melted coconut oil to add some cohesion to the flakes and make blending the down easier on your machine.
3. Blend for half a minute, stop, scrape down the sides, and begin again. Your coconut butter will probably still be a bit gritty, unless you have a high-powered blender, but your delicious coconut butter can be used in curries, stir fries, on pancakes and even in desserts.

GARLIC

Garlic is what Hippocrates might have been thinking of when he said, "Let food be thy medicine and medicine be thy food." Indispensable in the herb garden, medicine chest and kitchen – eat as much garlic as you can. (Unless, poor soul, you are one of the few who are allergic.)

- When you press garlic, scrape the leftover skin and add it to whatever you're cooking, or compost.
- Unwrap a clove of garlic and add it to poultry waterers to protect the water from bacteria and compost dry outer covering.
- Soak dry cloves in olive oil to use in cooking and for wellness.
- Sauté garlic scapes if you grow hard neck garlic and add to stir fries or infuse in oil to make salad dressing.
- Roast and freeze cloves for later use.
- Add to home-canned meals like soups and sauces.
- Dice and dehydrate for minced, dried garlic; slice, dehydrate and blend for garlic powder.
- If you grow garlic, hard neck garlics tend to last longer in storage but with any garlic, braid and hang to dry in a cool dark place.
- Compost any part you don't use.

HAMBURGER

Any ground meat, not just beef, or sausage is helpful to have in surplus.

- Save leftover hamburger to add to soups, eggs, taco filling, Stroganoff, chili, casseroles, spaghetti sauce and sloppy Joe mix.
- Use raw hamburger to feed your cats and dogs - remember to use only humanely raised, quality-fed meat.

- Use all fat from the hamburger in the dish you're preparing or reserve to add to homemade French fries or soup bases.
- Stuff some mushrooms with blue cheese, hamburger and rosemary for a delectable snack.
- Can ground meats in bulk in a pressure canner for go-to meal bases that will save you tons of time.

ICE CREAM

Please note, all dairy should come from humanely raised, clean-fed animals.

- Should you happen to have leftover ice cream, use it for milkshakes and ice cream cakes.
- If it's a plain flavor, add tasty mix-ins, a little more cream and refreeze to make a new ice cream—homemade ice cream is the best.
- Melt down and add to bread pudding in place of the called for milk and cream.
- Clean the carton, punch holes in the bottom and repurpose into a seed starting pot.

JAM

As a health suggestion, if you're not using home-canned jams, try to purchase jams that are free of high fructose variations like high-fructose corn syrup, natural corn syrup, isolated fructose, maize syrup, glucose/fructose syrup and tapioca syrup (not from corn, but also fructose).

- Collect all those jars with only remnants of jam left in them and make a batch of variety jam filled cookies—Hamentaschen, anyone?
- Something I learned in Russia, add one or two tablespoons of jam to hot water to make fruit tea.
- Toss any amount into a small saucepan on low, or double boiler on medium, heat to warm up for use on pancakes and waffles.
- While we're on the subject, add leftover jams to pancake batter for a bit of a sweet, fruity taste.
- Wash the jars and use as drinking glasses.

KETCHUP

I offer the same caution as for jam – if you're not using homemade, watch out for high fructose sugars and any fillers whose names you can't pronounce or recognize. Basically, this is a good rule to follow with any commercially produced food. Fillers and high fructose sugars are cheaper than their healthier equivalents, but "cheap" isn't the only metric that's valuable when it comes to our health.

- Add leftover bits of ketchup to chili, Stroganoff, stew, sweet and sour or spaghetti sauce.

- Because of the acidity, you can also use ketchup and a little salt to shine up the bottom of your copper bottom pans.
- Make a tomato-based salad dressing like Catalina.

TO MAKE CATALINA DRESSING

Combine:

½ cup of sugar (or ¼ honey)

½ cup of raw apple cider vinegar

½ cup light oil (like avocado)

¼ cup of ketchup

½ teaspoon of onion powder

1 teaspoon of paprika

½ teaspoon of basil

½ teaspoon of celery or parsley flakes

Blend well in an electric blender and add salt to taste.

LEMON

I grew up with two Meyer lemon trees just outside my bedroom window, so I must admit I have a special fondness for the wonders of citrus.

- Use half a lemon to detox your cutting boards and counters by running the lemon over the surface vigorously. (Do NOT use if you have natural stone countertops.)
- Add lemon juice (without pulp) to your humidifier to cleanse it when it starts to smell musty.
- Simply add a wedge to your daily water glass for healthy and tasty hydration – no sweeteners needed!
- Squeeze one tablespoon of lemon juice into your water spray bottle to clean out the nasty gunk and odors from your fridge. Bleh.
- Make an old-fashioned pomander with cloves and a whole lemon to deter moths in your closets and to decorate during the holidays.
- Use the peel of responsibly grown lemons to make lemon zest; dehydrate it to preserve and use later if you have more than you need.
- Save the seed from non-hybrid varieties of lemon, like the popular Meyer, and grow yourself a lemon tree. Visit our links page under The Homestead Kitchen section to learn how.

- Save and dry peeled lemon skins to make a lovely citrus tea throughout the year.
- You can also preserve lemon zest (or thicker grated lemon peel) in a high-quality sugar to enjoy lemon-sugar on cupcakes, cookies and in your morning herbal tea.

TO MAKE LEMON SUGAR

1. Blend on low the zest (or pith-less, chopped peel) of at least three lemons mixed with one cup of granulated sugar.

2. Remove from the blender and mix in one more cup of sugar.

3. Lay the mixture out on a cookie sheet to dry at room temperature for an hour or two. You may also dry this in your dehydrator on its lowest setting (citrus oils will volatize in high temperatures so keep it as low as you can).

4. Place in a lidded jar once dry.

- Preserve quartered lemons in sea salt for use in spicy stews, hummus and grain salads.

TO PRESERVE LEMONS

1. Quarter the lemons but keep them attached at the base so that they open kind of like a flower.

2. Rub down the inside of each lemon with sea salt (do not use table salt). Use as many lemons as it takes to pack a canning jar completely full—try to eliminate all air pockets. As you push down the lemons, juice will squeeze out and into the jar. This is good.

3. Cover and let your lemons sit overnight, shaking them down in the morning. If you don't have enough juice by then to cover your lemons completely, add some fresh lemon juice to make sure they're submerged.

4. Let these lemons ferment for about a month, continually checking to make sure that they're covered in the salty juice. To make things simple, store in the refrigerator after three weeks.

MAPLE SYRUP

I recommend you use the darkest grade, pure maple syrup you can find for these suggestions.

- Sweeten smoothies, frosting and ice cream—you really don't need much, so be sure to taste test.
- My favorite homemade milkshake contains only raw milk, ice, natural peanut butter, sustainably-sourced cocoa powder and maple syrup.

- Dry syrup on a flat surface on a low-heat setting in your dehydrator until crisp to make maple candy sprinkles.
- Add some pure maple syrup to sweet and sour recipes, BBQ sauce recipes, homemade baked beans and Crockpot pork recipes.
- Make maple sugared nuts or use in pecan pie to replace corn syrup.
- How about homemade maple marshmallows? Please see our links page for a recipe. You can get an idea of the marshmallow making process in this chapter by flipping ahead to the *Make Your Own Stuffs/Homesteaded section*.
- Bake it onto your favorite grass-fed bacon or glaze seasonal veggies like carrots and Brussels sprouts.
- If you enjoy warm drinks but want to avoid chocolate, simply add some maple syrup to your favorite warmed milk.
- If you buy dark syrup in the gallon jugs, save the glass containers for fermenting your own apple cider vinegar.
- If your kids have a hard time maneuvering the big jug over their pancakes, buy or repurpose a pint size version of a bottle, with a pour-friendly lid, and fill it from your big jug.

NOODLES

A note on gluten free noodles: once they're cold and a day old, GF noodles will have a different texture than wheat noodles. I always make sure to add my leftover GF noodles to dishes that will be warm or hot to ensure that they're as tasty as they can be. Just an FYI, I usually use GF noodles that are made from brown rice. I have yet to make my own oat flour noodles—I really need to get on that!

- Leftover pasta can be added to just about anything you might be preparing for dinner—toss it in soups, casseroles and even omelets and cold salads.
- A few noodles added to tuna salad give it great texture.
- Get the kids into the kitchen and show them how to make a simple white sauce flavored with curry or cheese and pour it over leftover noodles.
- The chickens will love them, especially spaghetti noodles—I think it's because they look like worms.

ONIONS

Part food, part medicine, onions should be eaten as often as possible. Here are some things to do if you end up with a lot of onions at once or have several that have started to turn but still have good parts you can slice off.

- Make your own dehydrated onion bits or slices in the oven or dehydrator for use in meat dishes, soups, casseroles—anything in which you'd use a fresh onion.

- If someone is suffering from chest congestion, make a hot onion poultice to relieve it; use caution when handling hot onions.
- Caramelize a big batch of onions, eat them and be so grateful you're alive to do just that; you can freeze any leftover caramelized onions.
- Yellow onion skins make a fantastic yellow to orange dye depending on what mordant (the thing that makes the color stick) you use. To make a simple dye, cram a medium-sized sauce pot full (and I mean full) of onion skins and cover them with water. Simmer on low for 30-60 minutes and then strain out the skins. To learn more about the process, check out our links page under The Homestead Kitchen chapter or read Chris Maclaughlin's fine book *A Garden To Dye For*.
- If you're using green onions, you can replant the root ends to grow more green onions sprigs.
- Whatever odd bits you don't use, put in the compost.

POTATO

I struggle to grow good potatoes, so I buy most of mine in bulk. Potatoes are starchy so be sure not to overuse them in a balanced diet.

- I often bake extra potatoes, when we have our potato bar dinner night, so that I can cut them up and fry them in the morning with some eggs for an excellent skillet breakfast.
- Leftover baked potatoes can also be re-heated in a pot of water and then, after the water is drained, be mashed like you would boiled potatoes.
- Leftover mashed potatoes make excellent potato pancakes for an impromptu lunch, as well as a divine, gluten-free topping for shepherd's pie.
- Here's a little trick I learned from my Tasha Tudor Cookbook: you can thicken soup and plump out cake batter with mashed potatoes.
- If you have a bunch of raw potatoes that you need to deal with, remember that they do keep for quite a while in a cool, dark place and are a great addition to, what I call, "living food storage." Please see the *Health Wise Food Storage* section in The Prepared Homestead chapter for more ideas.
- Another great thing to do with your raw potatoes is to grate them (a food processor or attachment to your mixer comes in handy for this) and dehydrate them for storage. My family loves hash browns but using raw potatoes makes them soggy, no matter how long I cook them; using dehydrated potatoes, they end up perfect.
- Similarly, potato flakes, dices and slices can all be preserved for your food storage.

QUINOA

I have a quinoa recipe book that I use and love and, from salads to dinner to desserts, I've found this to be one versatile little grain.

- Quinoa makes a great, protein-rich substitute for rice, if you're looking for one.
- Quinoa works its way nicely into any soup, stew or casserole.
- You can also add it to omelets, curries and quesadillas.
- Combined with rolled oats, it makes a great addition to homemade granola recipes providing a crunchy pop when you chew.
- The individual grains are so small that they're easy to swap for traditional gluten-containing ingredients like bulgur and couscous, if you're looking to go gluten free.
- If you have a quantity of the dry quinoa you'd like to use, you can grind the grain into flour with a blender and bake with it for an added protein burst.

PLEASE NOTE:

Any time you use quinoa, you MUST pre-soak it. Some recipes will say to simply rinse it very well, but I suggest you soak for two to four hours and then rinse it until you no longer see bubbles appearing in the water. Quinoa has naturally occurring saponins that will cause the taste to be bitter if they aren't first washed away. Soaking and rinsing will do the trick, though. If you soak your quinoa a bit longer (about six hours), it will easily sprout, thereby increasing its nutritional value even further.

To make flour from soaked and/or sprouted quinoa, simply rinse and spread out onto a cooking sheet or dehydrator rack with a lining to keep the small grain from falling through the cracks. Dehydrate in an oven on its lowest setting or your dehydrator for a few hours, stirring a couple times. Dehydrating won't take long because of how tiny the grains are. If you are moving quinoa and, at any point, some cat or child bumps you, causing the quinoa to spill and go everywhere (literally), sweep it up and give it to the chickens or put it into the compost bin as it's useful in both places. For more sprouting ideas please see the *Soaking, Sprouting and Microgreens* section of The Homestead Garden chapter.

RED SAUCE

Are we talking spicy enchilada sauce or sultry marinara sauce? Either is worth saving and using down to the last drop.
- Enchilada sauce can be used to flavor taco meat and any kind of taco soup.
- It can also be poured over chorizo and eggs or a boring baked potato.
- Use it to wake up guacamole and garden-fresh salsa.
- Marinate kabab meat in it for a spicy twist on a Mediterranean classic.
- Pour it over a roast and cook it on low in the crock pot all day for a fantastic dinner
- Use it in any casserole to spice things up. Have you figured out yet that I love casseroles for using up every bit of the quality ingredients I have lurking around my kitchen?

What about that Marinara sauce? Really, the same ideas apply. We make pizza all the time and I invariably end up with some red sauce leftover and I'm always so happy when I do. Because these two red sauces are so rich in herbs and spices, they make a wonderful compliment to what would otherwise be bland foods.

- Marinara is a great add-in for meat stews, as well as meatloaf.
- It can also go over eggs in the morning and be tossed over squash for lunch.
- Ever had spaghetti sauce on spaghetti squash? It's a great way to make a gluten free version of a family favorite.
- Braise meats with marinara.
- How about using marinara to make tapenade, ratatouille or red clam chowder?
- Either of these sauces can be mixed with sour cream to make a fantastic dip (start with 1/2 cup sauce to 1 cup of sour cream), and both can be mixed with leftover rice or quinoa for a zesty dinner side dish. Combine the two to make the perfect buffalo wing roasting sauce.

SOURDOUGH STARTER

Learning to use a sourdough starter is NOT too hard for you, I promise. It's keeping it alive that takes concentration – ha! Sometimes I completely neglect my starter and force it to the brink of death. Other times, I'm totally on my game and have starter coming out my ears. Feast or famine is my personality, I guess. What kind of sourdough starter parent are you?

- I often end up with extra starter and the most common thing to do with it is to make pancakes or waffles.
- You can also whip up some bagels, pretzels and English muffins using your favorite recipe. We need to stop thinking of using sourdough starter only for bread. Bread is still intimidating to some of us and there are plenty of other tasty treats to bake up.
- Starter can be added to any regular batter you have going from cookies to breakfast cake to banana bread. You can let your batter sit on your counter to ferment with the starter in it, or you can mix the leaven right in and bake immediately. If it sits, the starter will start to break down the grains in your batter (which makes them healthier), but it will change the texture of your finished product. The best thing to do is get good at using sourdough by reading Melissa Richardson's books, recommended at the end of this section. For a simple breakfast cookie recipe that calls for sourdough starter, flip ahead to the end of the next section, *Ferments For Starters*, under "Homestead Lady Speaks."
- Likewise, starter can be added to any yeast bread recipe you have in the works— compensate by adding a little less wet to your batter. You can make, what I call, random bread by using your extra starter, tossing in whatever flours (wheat with oat or rye) and fillers you have on hand (raisins, seeds, nuts). You can use any

sourdough bread recipe as a guide for amounts and make these random loaves sweet or savory or plain. If you're still a newbie sourdough baker, go ahead and add commercial yeast to your sourdough bread if it's called for in your recipe. It's ok to still be learning and experimenting, especially with breads.
- Extra starter can always be fed to livestock for a healthy, probiotic kick.

Sourdough is something that works its way into your lifestyle and, eventually, making a loaf of bread won't be as intimidating as it might be right now. I've found that using sourdough has somehow made baking bread stuffs easier—but that could totally be a personality thing or just a weird quirk of mine. Keeping a starter alive is work, but it has sweet rewards. For more information, please see the *Ferments for Starters* section of The Homestead Kitchen chapter.

TOAST

My kids are notorious for putting bread in the toaster and then walking off, forgetting about it until hours later.
- Stale bread and forgotten toast all go into my "bread pudding" bag in the freezer. I love, love, love bread pudding and am so happy when I have a big enough stash of these lonely bread pieces to make a batch.
- You can also make bread crumbs and croutons with leftover toast and stale bread. Toss some of that leftover marinara sauce from above over your homemade croutons, add some dry cheese and a green salad and you have a lovely dinner.
- If it's easier to just rip up your bread and dry it in large cubes, put it in the blender to turn it into crumbs when you're making meatloaf.
- You can also use your homemade bread crumbs to make a fry coating for onion rings, fish sticks and fried chicken—something akin to the old Shake and Bake®.

TO MAKE FRY COATING MIX:

1. Mix together in a large bowl 3 cups of fine bread crumbs, ¼ cup of avocado oil, 1 tablespoon of sea salt, 2 tablespoon dried onion flakes, 2 teaspoons paprika or other dried pepper, 2 teaspoon powdered garlic, ½ teaspoon each of basil, thyme and oregano.

2. Add some ground black pepper to taste.

3. Incorporate all the oil by mixing until it's no longer lumpy.

4. Use immediately or store in a lidded canning jar in the fridge for a month.

UNPASTEURIZED MILK

Otherwise known as raw milk. If you don't consume raw milk, then just skip this and go to the next item because this won't be a paragraph on convincing you why you should drink raw milk. The first thing to remember about raw milk is that it will culture, not spoil. So, you can use raw milk and raw milk products to infuse live, active cultures and probiotics into various food preparations.

- Pre-soak oatmeal (among other things) overnight with a little raw milk, raw milk kefir or raw milk yogurt and enough water to cover the rolled oats then simply cooking it up in the morning.
- Add a touch of raw milk to culture batters for breakfast baked goods and set them out overnight on your counter.
- If you're having trouble maintaining your milk supply while nursing for whatever reason, you can make a high quality raw milk formula using the Weston Price recipe to be found at our links page.
- I feed raw milk to my baby chicks for their first month; it can also be given to other livestock.
- Raw milk can also be used to make various simple dairy products like ice cream, mozzarella cheese and buttermilk. To make simple soured cream two ways, follow the instructions below.

SOURED CREAM – 2 WAYS

1. For a sweet, pourable cultured cream drain or ladle off one to two cups of cream into a clean canning jar.

2. Cover with a lid and allow this cream to culture (just sit there and do its thing) on your counter for 12 to 36 hours, until it thickens. You'll know the cream is ready when it is thicker.

3. If you let it culture past the thick cream stage, it will separate out from the whey and you'll clearly see that separation through your glass jar. This is more like a very spreadable cream cheese. Either product is very tasty! This method is preferred if you'd like to keep your cream raw, as opposed to pasteurized.

4. **NOT RAW**: If you want your sour cream thick and firm (more like what you find in the store), heat your cream to 175F/79C and keep it there for a few minutes. Cool it down to at least 120F/49C. Taking it to this high temperature means the finished product will no longer be raw, but at least you start with fresh cream and end up with sour cream free of fillers and other unsavory commercial ingredients.

5. Stir in a generous spoonful of previously made sour cream (this can be store bought or homemade), kefir or even kefir grains (strain these out after step 3 is complete). This will be your starter culture, which simply kick-starts your sour cream and gives it flavor.

6. Add a loose-fitting lid and let the cream sit on your counter for twenty-four to forty-eight hours. The longer it sits, the more cultured, and therefore more flavorful, it will be. The cream will thicken as it cultures. If you want to speed up the process, you can insulate the jar in several layers of towels in an ice chest or an insulated bag.

7. If you insulate the cream while it cultures, it should be done in about twelve hours. For the first twelve hours, this cream is more correctly called crème fraiche and is still slightly sweet. It becomes tart as it cultures longer, becoming more like the American version of sour cream.

VINEGAR

When I think of vinegar, I think of two basic kinds, although there are many different flavors of vinegar. Typically, I think of white vinegar, which is usually distilled from corn, or apple cider vinegar (or ACV) that I either make myself or buy with the mother still alive and intact to keep it as healthy as I can.

White vinegar isn't something we consume at our house, but it has A LOT of uses.
- Use it to cleanse just about any surface.
- Deodorize your garbage disposal or HE (High Efficiency) washer and use in place of fabric softener for your laundry.
- Make a fruit fly trap by filling a cup half way full of vinegar, placing a plastic wrap over the top secured with a rubber band, and poking small holes in the wrap.
- One-part vinegar and one-part water can clean glass surfaces, kill newly emerging weeds, clean a shower curtain, deter cats, and can be uses as a hair conditioner.
- Visit Vinegar Tips (www.vinegartips.com) for over one thousand ideas.

Live apple cider vinegar is something I use in recipes to add a little probiotic, especially to salad dressings and Asian dishes. I've learned to make my own apple vinegar as described under "APPLES" above, but you can also purchase Bragg's brand, which has the live mother in it.
- I use one tablespoon of ACV in my homemade pain-relieving concoction alongside one tablespoon of raw honey, a pinch of sea salt, a 1/4 teaspoon of baking soda and a squeeze of raw lemon juice.
- I put some in my livestock waterers to give the animals a boost in their gut. ACV can also balance the pH of a person's gut on the inside and tone our skin on the outside. Some people swear by its use as an aid to lose weight, but I'm guessing that comes from the pH balancing (just science according to me and nothing else).
- A teaspoon of ACV in a small glass of water helps my stomach feel steadier when I'm doing a yeast cleanse (which can cause nausea) or am car sick.

WHIPPING CREAM

There are two kinds of people in this world: sweetened whipped cream people and unsweetened whipped cream people. I fall into the latter camp but only because I use the rawest*, cleanest cream I can find to whip, which is already slightly sweet. If I'm preparing it for others, I usually sweeten my whipped cream with maple syrup or a little bit of raw sugar. If you buy whipped cream, try to find one that has the least amount of refined sugars and other gunk. If we don't have a dessert planned, sometimes I'll just grab some fruit and whip up some cream to plunk on top. Our panettone French toast on Christmas morning must have homemade whipped cream on it. If you have just cream, which hasn't been whipped yet, here are some ideas to use it up:

- Light cream, which has less butterfat content, is great to pour over berries, cereal, in your morning hot beverage or to mix with milk to make half and half.
- If your cream is technically whipping cream, with a higher butterfat content, besides whipping it, you can add it to soups, batters or mashed potatoes.
- If you have the additional ingredients, make ganache or bread pudding with your cream because both are divine.
- Any little bits of whipped cream left can go to livestock.

WHIPPED CREAM TIPS

* Raw cream can be a bit more difficult to whip than pasteurized cream but don't be discouraged. Because raw products are still teaming with beneficial life, they can have minds of their own sometimes.

To ensure success, pre-chill your whipping equipment (bowl, beaters or paddle) in the freezer.

Also, older cream seems to work better than new cream.

When skimming cream off milk, be extremely careful not to get milk mixed into your cream. You can use a cream separator to be sure of the purity of your cream, or you can skim off cream from the top of your milk container and wait twelve hours to see if you get any milk falling to the bottom of your container. If you have your own dairy animals, be sure to take note if you have whipped cream success or failures at consistent times of the year—what a dairy animal eats can greatly affect the resulting products.

XACUTI - (or any Indian curry dish)

Curry, like many dishes rich with spice, tastes better the next day because all those herbs have had a chance to mellow and mix together.

- I usually make pitas about once a week and leftover curry was simply made to stuff inside a pita or wrap inside a naan (the Indian version of pita bread).
- Add a bit of bone broth and/or coconut milk to your curry and you have a hearty stew that can be eaten for lunch. Leftovers turned into soup at lunchtime is a favorite at our house because it's a lot less boring than a sandwich.
- If you want to add extra international flair, add some seafood and make curry paella or toss your leftovers in soba noodles for some Japanese curry.
- If you have wonton dough on hand, fill each piece with leftover curry and fry it in coconut oil.

- Try a grilled cheese and xacuti sandwich.

YAM

Or sweet potato, which is it? Yam and sweet potatoes aren't even related botanically speaking but, in the States, we often use the terms interchangeably. Although I invite you to research the history of both noble root vegetables, which will take you from the American south to the shores of Africa and Asia, the bottom line is that yams and sweet potatoes are both delicious. Our family enjoys these baked in the oven or in the hot coals of an outdoor fire. On the off chance we have leftovers, or when I've deliberately made extra and hid them from the children, here's what we do.

- Use leftover yams or sweet potatoes to make a batch of breakfast pudding by blending 3 cooked, skinned tubers in a blender with about 1 cup of any kind of milk (dairy, nut or coconut) and enough honey and cinnamon to add taste. In mere moments you'll have a delicious pudding that even the pickiest eaters enjoy.
- Use up pre-cooked tubers by slicing them thickly and adding them to your kabab skewer. You'll want to use cooked yams or sweet potatoes for kababs because their high starch content requires a long cook time and you'll burn the other items on your skewer waiting for them to finish.
- You can also use leftover spuds to whip up a quick "sweet potato" casserole with a tasty, oatmeal crumble topping. Add butter and cream to mashed sweet potatoes as you would when making mashed white potatoes. Add the oatmeal crumble and bake until your topping is crispy, and the sweet potatoes are hot.
- How about yam bisque or sweet bread?
- Because of its soft texture, sweet potato can be used to moisten cookie and cake recipes much like you would use apple sauce. Fiddle around until you find a recipe that works well for you.
- One of our favorite ways to use up leftovers is to make sweet potato pancakes by mixing some mashed tater and some egg to simply fry a veggie pancake, or you can mix veggie puree into a grain-based pancake batter. This doesn't have to be complicated; mix egg and puree until you get something that will mostly hold together. Put maple syrup on it and you will have a fantastic breakfast.
- How about making a yam pie, much like you would a pumpkin pie?
- For another dessert, make a sweet potato pudding as described above and fill baked phyllo pastry cups with it, then top all with fresh whipped cream and pecans.
- Try making stuffed French toast with your leftover tuber mash.

ZUCCHINI

Zucchini (or zuke) is often the brunt of a gardener's well-meaning jokes. Have you ever heard that you should make sure to roll up your car windows at church during the summer, lest someone secretly off-load their extra zucchini on you? Well, you should because they will. If you're just beginning a garden, zucchini grows well in most areas and I suggest you plant one or two plants. No more than that, though, because you will certainly end up with more zucchini than you can eat or give away during the summer months. Fortunately, there are a lot of things you can do with zuke to use it up or preserve it.

- Zucchini freezes and dehydrates very well—simply grate to freeze, and grate or slice to dehydrate.
- When canning, don't plan to preserve zucchini whole, but you can make relish out of it and pickle spears (though those might get a tad mushy).
- Toss thinly sliced zuke in some avocado oil and grated garlic to dehydrate or bake for an awesome homemade chip.
- There's always stuffed zucchini for dinner followed by zucchini sweet bread for dessert.
- Zucchini can also be baked into any muffin, pancake, waffle or bread.
- If you're staying away from grain, you can make "noodles" from zucchini with a spiralizer tool, a mandolin slicer or a simple knife. To learn how to use a spiralizer, visit our links page.
- If you are using a spiralizer, be sure to save the oddly shaped bits that are left over from your zoodle making to add, thinly sliced, to your next morning frittata. Or stew. Or spaghetti sauce.
- Roll zucchini slices in breading and bake up like a fish stick or slice them into omelets and quiche.
- How about frying thick slices up like a burger and eating them for lunch?
- If you shave zucchini down thinly, you can easily eat it raw in any salad.
- It also makes a wonderful pureed soup and a topping for vegetarian pizza.
- How about trying your hand at zucchini fritters or zuke-based baba ghannouj?
- Add shredded zuke to any meatloaf or meatball recipe, as well as a stir fry, lasagna, casseroles or, of course, ratatouille.
- Your goats will enjoy the leftover plants once the zucchini stop producing in the cold weather of fall.
- Before frost comes to threaten late blooms, be sure to harvest them and fry them up after coating them in a simple batter—never had fried squash blossoms? Ooooh, you simply must try them!

ISN'T THIS A LOT MORE TROUBLE?

Well, it's a bit more work but I wouldn't call it trouble, despite our hectic schedules.

Honestly, I don't think most of us are evil, hate the planet, love to waste or are lazy or stupid. I think we're just a product of our modern culture which is a place wherein we can buy what we need in a store and can put what we don't think we need in a garbage can.

And we're busy. I think a case could be made for the argument that we're too busy, but I digress. We need to learn a better way of doing things not because we're wicked, but because we don't know what we don't know. I was and still am guilty of not using up every resource I've been blessed with in the kitchen and so I'm ready to explore this topic with you and figure out what we all might do better to "use it up, wear it out, make it do or do without," as our pioneer ancestors often quipped.

ACTION ITEMS: This week, observe five food items in your refrigerator or in your pantry and list in your homestead journal ten different ways to use each item. This is just an exercise to begin to train your brain to think of ingredients as multi-functional in a wide range of recipes.

Also, this week, write down every piece of food you throw away, compost or feed to the chickens, including amounts. Look at your list after a week and think of some ways you could have used that food instead of tossing it. If the food has gone bad, that's not a reason to skip over that item; ask yourself what you could have done with it earlier so that it went into your belly instead of molding and going into the compost. This is a hard one for me as I'm forever stashing stuff with the intention of using it in something later and totally forgetting about it. When I'm on my game, I keep a list of leftover items right on the fridge door - one list for each new week - so that I know what to use.

Sadly, I often forget to keep my list and my precious leftovers become chicken food. The chickens appreciate my forgetfulness, but they can eat bugs—I need to use my leftovers!

DIY: If an item from your trash list can be found on our kitchen tips list above, try one of those suggestions. For example, if you have a bit of spaghetti sauce left in a jar in the back of your fridge, try making meatloaf or Spanish rice with it. Write down what you're doing in your homestead journal, even if it doesn't turn out to be something you think is tasty. Try again. Create a section in your cookbook or start a new cookbook with your recycled food recipes.

If you have a child getting ready to go to college soon, or you know a newly married couple, your completed recipes would make a great gift after a few months of you practicing. Most young adults aren't trained to do this with their food, but it will save them money and time— two savings any busy twenty-something can appreciate. Type out your recipes nicely in final draft form on Word in one of their recipe templates so you can print them out beautifully before you gift them.

BUILD COMMUNITY: Should you decide to compile a simple cookbook or recipe card collection, do it with friends instead of on your own. Send out emails and messages to all your contacts asking for their best kitchen tips with an emphasis on recycling leftovers into new recipes. Ask your grandma, your pastor's wife, or the pastor himself if he's the one that does the cooking in the family. Make sure you get everyone's name with each recipe or tip so that you can credit back to them.

When you have a good-sized collection, gather as many of the contributors together as you can, asking each to bring a blank recipe book. Give each person a copy of the recipes you've compiled and assemble your cookbooks together, making them as fancy or as plain as you like. If there are stories or a history attached to any of the recipes, be sure to include those as they'll be a treasure to the reader. Some of us are naturally adept at this kind of food creativity and, by sharing what seems like a too-simple tip with others, we can truly be of help to someone who is trying but struggling to come up with new ways to use the same ol' stuff in the kitchen.

BOOKS TO READ: If you're not online much, never fear! You can learn how to do all the tips mentioned in this section by studying the books listed here. What's more, you'll learn to think like a foodie and learn to experiment successfully on your own. A great book to begin a real foods education is *Nourishing Traditions*, by Sally Fallon. This book will be of benefit to anyone looking to learn the how's and whys of using food to its maximum nutritional benefit. If there were a college level course in homemade, real foods this book would be required reading. Primarily a cookbook, it also contains information, including scientific data, on the benefits of a real foods diet and the detriments of a diet based on highly processed, refined foods. Bone broth, ferments and cultured dairy, organ meat and game recipes, how to properly prepare grains and more—it's all in this book.

Other titles for various needs and to explore more of the ideas listed in this section:

- Transitioning to homemade foods and making friends with your kitchen - *Homemade Pantry*, by Alana Chernila
- Whole or real foods cooking - *The Art of Simple Food*, by Alice Walker and *The Elliot Homestead: From Scratch*, by Shaye Elliot
- Whole or real foods cooking with family favorites - *The Feel Good Cookbook*, by Jonell Francis
- Recipes using produce from your garden - *Bountiful: Recipes Inspired by our Garden*, by Todd Porter
- Homestead food, recipes and production - *Independence Days*, by Sharon Astyk
- *Paleo diet for families - *Against All Grain*, by Danielle Walker

- *GAPS diet for families - *The Heal Your Gut Cookbook*, by Hilary Boynton; just a note: my family used this book for our GAPS cleanse and I guarantee it's as family-friendly as you can get while cleansing your gut
- *Raw foods diet - *Rawmazing Desserts*, by Susan Powers—apart from salad, I think everyone's raw food experiments should start with desserts
- To learn how to cook and bake with quinoa - *Quinoa 365: The Everyday Superfood*, by Green and Hemming.
- To learn how to use sourdough, or natural leaven - *The Art of Baking with Natural Yeast*, by Richardson and Wornock and *Beyond Basics with Natural Yeast*, by Richardson. Also, see the next section, *Ferments for Starters*, for more information.

*I don't necessarily recommend you live on these diets forever unless that's what makes your body happiest. For most of us, they can be helpful tools as we learn to prepare home-cooked foods that are healthy and delicious.

If you need a basic kitchen book - setting up your kitchen for the first-time kind of thing - try, *Saving Dinner Basics: How to Cook Even if You Don't Know How*, by Leanne Ely. Also, good to check out is *Conquering Your Kitchen*, by Annemarie Rossi, and *Zero Waste Cooking*, by April Lewis.

WEBSITES: For a foodie education and to learn basic principles that will enable you to use up all parts of your pantry, please visit:

- Weston Price (www.westonaprice.org)
- Nourished Kitchen (www.nourished kitchen.com)
- Whole New Mom (www.wholenewmom.com)
- Delicious Obsessions (deliciousobsessions.com)

HOMESTEAD LADY SPEAKS:

If making use of every scrap of food in your kitchen is not how you were raised, and you feel disinclined to start analyzing the volume of food that goes into your garbage can each day, please allow me to interject with a bit of encouragement! I know from personal experience that habits can change. A remarkable thing about human beings is that we can deliberately and consistently change our routines, habits and even the way we think. Don't try to overhaul every cooking method your mother taught you and don't dump all the contents of your fridge. There's a lot of merit in the skills you already have and the way you already do things. Just pick one thing this month to do better, that's all. Start by using more and throwing away less. No big deal.

As Angela England writes in her book, *Backyard Farming On An Acre (More Or Less)*,

"Surround yourself with delicious, high-quality food that you've invested your time and effort into and you will gradually change your eating habits by default!"

If this section felt too basic for you because you already do the use-up-leftovers, zero-waste cooking thing, great! On to the next level...

"When General Lee took possession of Chambersburg on his way to Gettysburg... Among the first things he demanded for his army was twenty-five barrels of Sauerkraut."

Editor, The Guardian (1869)

For the Homesteadish: Ferments for Starters

Have you read up on fermented foods yet? Naturally fermented foods are full of nutrients and beneficial bacteria—they're like superfood for your gut health. Ferment isn't exactly an appetizing word, but many of the foods you're already eating have been fermented to improve their flavor and digestibility. Sauerkraut, so prized by General Lee, is an excellent example of a tasty fermented food that can easily be made at home. Feta cheese and sourdough bread are also good examples. Not only can these foods be made at home to fill hungry bellies, but they can also be used to build up those bellies in healthy ways.

Many people suffer from damaged or "leaky" gut problems – just peruse the Mayo Clinic's discussion boards (www.connect.mayoclinic.org) on the topic to see how many! There are medications that those suffering from gastrointestinal issues can take, but a good place to begin soothing your upset tummies is with nourishing foods.

HEALTHY GUT LINGO

When talking to people who use wholesome foods as a kind of medicine, you may run into a few words and phrases that are unfamiliar to you. They may say something like,

> "My gut is so broken, I have yeast overgrowth everywhere! My joints ache, my stomach acid is going crazy and I have horrible headaches. I've got to do a yeast cleanse!"

If you didn't understand all of that, don't despair! I didn't know what most of that meant a few years ago either.

An herbalist friend of mine named Amy Jones began to educate me on exactly what a healthy digestive system should "look" and feel like. I soon realized that mine was badly broken and needed healing. The following is a much-abbreviated explanation of yeast overgrowth in the human body from Amy's website, Yaya Guru. I encourage you to visit her site, as well as others on this topic, and do as much reading as you can on if you're experiencing ailments that just don't seem to ever go away!

> **YEAST OVERGROWTH**
>
> It's the most underdiagnosed problem in the world yet is at the root of most ailments.
>
> Internal yeast, or Candida Albicans, is part of our body's "zoo." We need a certain level of it in order to be healthy. Yeast is responsible for eating our bodies away when we die. But when we have an overgrowth in our bodies, it can literally eat us alive.
>
> In the yeast state (a healthy state) candida is a non-invasive, sugar fermenting organism. But when it becomes fungal or mycelial – known as Candidiasis – it can quickly overgrow and produces rhizoids that have long, stringy, finger like root structures that burrow into the intestinal wall, leaving microscopic holes and allowing toxins, undigested food particles and bacteria and yeast to enter the bloodstream. The broken-down barrier also results in compromising digestion, vitamin and mineral assimilation, hormone balance, your immune system, and oxygen saturation. The ethanol that is given off by Candida parasites can be a major cause of fatigue and stamina reduction and likewise destroys enzymes that are needed for providing cells with energy. Ethanol also releases free radicals that hasten the aging process.
>
> *What Can Trigger Yeast to Overgrow and Good Bacteria to Die?*
>
> Refined flours, sugar and commercial yeast
>
> Processed food
>
> Alcohol, recreational drugs, prescriptions
>
> Corticosteroids and NSAIDs
>
> Stress
>
> Exposure via the womb and breastmilk if the mother has an overgrowth
>
> Heavy metals
>
> Artificial coloring, flavorings, sweeteners, and synthetic odors (i.e. perfumes, scented candles)
>
> Preservatives and synthetic ingredients in personal care products

For more information on typical symptoms of yeast overgrowth AND for information on how to restore healthy levels of Candida Albicans to your gut, be sure to visit Amy's site yayaguru.com. For her very informative article, "Taming the Yeastie Beasties" visit our links page at www.homesteadlady.com/links-page. Always consult your medical professional when appropriate.

Dairy: A Place to Start

My introduction to fermented, or cultured, foods came with dairy. I had dairy goats and I had all this milk that I needed to learn what to do with, and I needed to learn fast. I got a whole dairy education condensed into a few months that went by in a flurry of cheeses, dressings and kefir. Kefir is a collection of yeasts and bacteria that resemble a small mass of cottage cheese. When you feed them with milk, not only do they thrive, but they ferment the milk that covers them, filling it with probiotics.

I typically culture my kefir overnight, then strain out the grains and use the cultured milk in my morning smoothie. Every time I use my kefir, I feed it with new milk and let it sit for at least six hours, or overnight. I also use kefir in salad dressings, as a pre-soak for grains and I'll even make a soft, spreadable cheese with it. It has a strong flavor, but you get used to it, until eventually you crave it.

Yogurt is an easy dairy ferment to do yourself because it's simply heated milk mixed with a yogurt culture that sits for a few hours in a consistently warm place like an insulated cooler. Once you start making your own yogurt you'll slap your forehead and say, "I can't believe I've been buying this for so long when I could have been making it!" In the recommended reading, at the end of this section, there are several quality cookbooks that will help you explore the world of dairy ferments and teach you how to make items like yogurt. Internet searches, especially those on YouTube can also be very helpful. (What did we ever do before YouTube?)

> **TO MAKE YOGURT**
>
> Here's a quick recipe for yogurt:
>
> Place half a gallon of milk in a medium pot on medium heat. You may stir in 1 tablespoon of Organic beef gelatin to help with a firmer set and add nutritional value.
>
> Slowly bring the temperature to 180° F/82°C, stirring frequently. The longer you keep it at that temperature, the firmer your finished yogurt will be. A standard recommended time is twenty minutes; I usually get bored and wander off after about ten minutes.
>
> Cool the yogurt to 120°F/49°C.
>
> Place into two glass quart jars (or a half gallon jar) and stir in 1 tablespoon of yogurt, kefir or cultured cream.

> Incubate your yogurt for 18-24 hours in an ice chest, Wonder Box, box dehydrator, sink full of warm water, Yogotherm or any other place you can maintain the yogurt at temperature. Wrap your jars in a blanket if you need extra warmth.
>
> The yogurt will be done when it has set up and has achieved the tang you find tasty. For sharper yogurt, culture a little longer. Homemade yogurt might end up a little softer than commercial yogurt, but you're also not eating a bunch of fillers and ingredients you can't pronounce. And, the best part, you made it in your kitchen!

Sometimes, even if the cultured dairy is raw, your gut flora may be so out of balance that you are unable to eat dairy without dietary upset, fermented or not. Some people are lactose-intolerant but it's also possible they simply need to pacify their leaky guts before they can consume cultured dairy. Part of that soothing can come when people, struggling with an angry stomach, learn to consume grains and flours in a healthier way by pre-soaking and naturally leavening them.

Soaking and Sourdough

Not long after my kefir education, I was introduced to the very simple fermentation process that begins when you pre-soak beans, legumes, seeds and grains. Imagine that surrounding each seed is a kind of jacket made up of acids and enzymes that are designed to protect the seed from the degrading elements of nature, preventing it from sprouting and growing before the appropriate time of year. When you eat the seed, that jacket also prevents it from being fully digested by your body, typically causing stomach upset and an inability on the part of your body to absorb the nutrients from the seed. Pre-soaking grains, beans and even flours, removes that jacket. Sprouting is similarly beneficial, but we'll talk more about that in the *Soaking, Sprouting and Microgreens* section of The Homestead Garden chapter.

Natural leaven, or sourdough, is a great pre-soaking agent for flours because it's an effective way of breaking down that jacket of enzymes and acids with the bonus of being a rising agent in your flours. Some people get nervous about the amount of work involved with maintaining and using a sourdough culture but, for me, it really is so much simpler than anything else I've done to make baked goods healthier, even though I mess up a lot. I purchased my original sourdough starter from Cultures for Health and followed their instructions for activating it. Then, I realized that as helpful as their website and others' blog posts were, I was going to need a good natural leaven cookbook. Using natural leaven is a process and you can't learn it all from a book, BUT you can use a book to help guide you.

I've read several sourdough books now and my two favorites are *The Art of Baking with Natural Yeast* and *Beyond Basics with Natural Yeast*, both by Melissa Richardson (though her first was co-authored by Caleb Wornock). These cookbooks are for regular people (in other words, not full-time bakers and chefs) who want to learn to use sourdough as easily as possible. These books feature no commercial yeast at all, which will be of interest to those wishing to eliminate commercial yeast for gut health. Be savvy about added commercial yeast when checking sourdough cookbooks because so many include it in the recipes. Let Melissa show you the benefits of natural leaven and how sourdough functions, so you can work it into your daily life by reading her books.

I have no pretensions to be an expert in this topic, but I can happily share a few things I've learned about getting started with sourdough.

- Accept it now that you're going to mess up. A lot. Just get over it and move on because life is too short to worry about killing your starter or baking loaves that are so heavy they could double as weapons. You can do everything right, following all the steps in perfection, and still have the starter or the sourdough baked goods end up ailing or flat. Leaven is a collection of living organisms and living things have "moods," minds of their own and personal preferences. I often wish my sourdough starter could speak English, so it could tell me exactly what it needs!
- Sourdough starters are tough, really, and you'll be amazed at how many times you think yours is dead only to have it come back to life. Melissa can help you with resurrecting your starter, but she can also teach you good habits to keep your starter healthy and thriving. When you need a break, you can put it to sleep in the refrigerator, so you can go on a weekend trip or have a baby without worrying about your starter.
- If you're planning to make a loaf of bread, you need your starter to be riddled with bubbles and doubled in size. If you can't get your starter to double in size satisfactorily for bread making, try adding a little less water when you feed it. If you're starter is too wet, it can be too heavy to rise well.
- For sandwich loaves, try the cast iron bread pans because they're taller and narrower than regular bread pans, which is exactly what you need for a nicely shaped sourdough loaf. Also, use an Organic, unenriched white flour (or sift out as much of the bran as you can from home-ground flour) for sandwich breads to keep them lighter.
- If you start getting indigestion from eating your sourdough products, try doing a good feeding session for your starter and commit to feeding it more regularly moving forward. *Beyond Basics with Natural Yeast* has a trouble-shooting section that taught me more about this problem.
- Even though Melissa taught me to keep my starter in the fridge, I discovered that I just couldn't make it work for me in there. Did you know that in the world of fiber spinning there are two kinds of people? Fine spinners (who naturally spin out more narrow yarns) and bulky spinners (who naturally spin out heavier yarns); a good spinner can do both, of course, but one just comes more naturally. There are two kinds of sourdough people in this world, too: fridge starter people and counter

starter people. Turns out, I'm a fine spinner and counter starter lady. I'm also a night person. To each his own, just do what works for you after trying out all the good advice you read from multiple sources.

- If you must go off grains for a while because you suspect they're making you feel sick even after all this soaking and fermenting, go off them until your gut is healed. I did that for about a year and do so periodically if I feel like I've had too much. Any grain I prepare myself, though, I either soak and/or leaven before I consume it to make it easier on my gut. I'm always careful with how much grain I eat regardless of how it's prepared because my gut was so broken it still requires some special care. However, my children, who have been raised with healthier eating habits, usually do just fine if I leaven baked goods.

- Not really on the topic of sourdough, but if you go off grains and then leaven them later, and you're still having problems, I can recommend Hilary Boynton's *Heal your Gut Cookbook*. My family were super troopers this past winter when we all followed that book and went through her guided version of the GAPS introductory diet, which cuts out all grains for a time. We all did so well, and I was so proud of us. We'll do it again this year and maybe once a year until we're all sure that our tummies are strong and thriving. All that to say, everyone is different, and you know your body. So, do what it tells you to do even if you want to do something else. I knew sourdough was going to be an answer for our family but a whole lot of gut healing stood between me and it for a while. Take one principle at a time and healing will come. Our bodies want to be healthy.

- If the idea of sourdough bread is too intimidating, know that there are so many other things to make with sourdough. Even if you never get to making bread, there are pancakes, waffles, flat breads and tortillas, cakes, cookies, muffins, scones, biscuits, bread sticks and more. I also use sourdough to culture oatmeal as I soak it overnight and steady doses of sourdough culture are fed to my livestock for gut and rumen conditioning. For my family, I usually prepare some sort of leavened grain 2-4 times a week and bread is my least favorite thing to make, so I don't do it very often. I'm as precise with sourdough as I am with everything else; which is to say, not very. If I paid closer attention to detail and were more disciplined with my sourdough, I'm sure my loaves would end up light and perfect every time, like Melissa's. Since I'm not that way, I find a happy medium by baking pitas. Baking bread is part science, part psychology.

- If you're a prayerful person, pray over your starter. It is alive and could use a boost of blessings just like you and me. I'm convinced that God is tickled when we ask him for help with the "little" things.

To purchase a quality sourdough starter, visit Cultures for Health at www.culturesforhealth.com. They can also teach you exactly what to do with it once the starter arrives at your door. They'll hold your hand while you and your sourdough starter make friends.

If you're new to sourdough, I don't recommend you try to save a few pennies by capturing your own wild starter. (Instructions for this are all over the internet.) I don't recommend *anyone* try to capture a wild starter unless they have a local mentor to help them. The simple reason can be found in its designation – WILD. Even if you manage to lure in some wild starter, it may behave differently from what all your books and resources are telling you. Sourdough is a collection of living organisms and the wild ones often only want to break curfew, wear their pants slung low and party all night. Figuratively speaking, of course.

For an easy sourdough recipe to test your skills, using a civilized starter, see the end of this section under the Homestead Lady Speaks segment.

Kraut - I'm Not Eating That

The last cultured foods to cross my path have been fermented veggies and fruits. I'd read about them in the past, of course, but it wasn't until we did that GAPS (Gut and Psychology Syndrome) introductory diet as outlined in Hilary Boynton's book, *The Heal Your Gut Cookbook*, that I decided to try them. Her book instructs you to consume the juice from homemade sauerkraut near the beginning of the diet, and then you can start eating the kraut itself. The process of this introductory diet assists your body in cleansing and soothing your gut systems with nourishing foods and each step serves a purpose. I found her book instructive, indispensable and inspiring, but as I was first reading through it, I read about kraut and balked.

The only sauerkraut I'd ever had was from a can and there was simply no way on earth I was going to eat that again. I had faith, though, that homemade would be better. In fact, I went over to a friend's house whose family had done the GAPS introductory diet to ask for her advice. She gave glowing, yummy reports of homemade kraut and other fermented veggies, though I think my nose probably wrinkled up in misapprehension. However, I knew her to be an honest woman who would never lie to an unsuspecting kraut-doubter, so I took her word for it.

A few weeks prior to starting the diet I made up a bunch of kraut and we sat around taste testing it with some veggie chips. Some of us liked it alright and others loved it. My six-year-old tried it and didn't like it, but the toddler ate it well. I was amazed at my own reaction—I was in the "loved it" camp! I found myself going back to it over and over and, in no time, it was gone. It takes at least three days to ferment kraut on your counter; that was a long three days waiting for the batch to be ready and I quickly learned not to let us run out. I haven't done much more than kraut and some ginger carrots but I'm willing to try other things now.

> **TO MAKE BASIC KRAUT**
>
> Finely chop one washed, green cabbage. In a half gallon canning jar or fermenting crock, put down a hearty layer of chopped cabbage and a generous sprinkling of sea salt. Continue adding layers of cabbage and sprinklings of salt until you run out of cabbage. Allow the cabbage to sit for ten minutes so the salt can pull some moisture out of the cabbage and tenderize it. Pound the cabbage thoroughly with a kraut pounder or any flat, wooden instrument that will fit into the mouth of your jar or crock. Pound each layer and keep pounding until the resulting liquid has covered the cabbage.
>
> You may need to add a bit of water to finish covering the cabbage; water content of cabbage varies from season to growing practice to variety of cabbage. However, make sure that you've really pounded that cabbage before you settle for adding water. Your cabbage should be a darker green color than when it went into your crock and will have reduced in size by at least half.
>
> Using a glass fermenting weight, a piece of window screen cut to fit or any other thing that will keep the cabbage at the top submerged under the water, loosely cap your crock and place it at room temperature for at least three days, but seven is better and a month is best. Anything not submerged under the safety of the Lactobacillus-rich, fermenting saline solution will be subject to mold, so make sure you've got it all under a weight of some kind.
>
> Eat a little kraut every day with beans, eggs, salad, steak, chips or just about anything but dessert. Each properly fermented food we consume gives our stomachs a boost of Lactobacillus health.

Forgive me while I wax spiritual for a minute but there's a very famous Proverb that most Christians can quote to you even if church attendance isn't their thing.

> "Trust in the Lord with all thine heart and lean not unto thine own understanding. In all thy ways acknowledge Him and He shall direct thy paths for good. Be not wise in thine own eyes…It shall be health to thy navel, and marrow to thy bones." (Proverbs 3: 5-8).

There are so many blessings to be had by trusting the Lord with all your heart, but the only one specifically cited in this scripture is "health in the navel and marrow in the bones." For me, the journey to health was created spiritually before it was created in my body. I had prayed to know why I was the sickest healthy person I knew. I was earnestly seeking to find a better way; a way that was right for me and my family. I believe I was led to various principles of health and well-being step by step as I was prepared for them. Because the journey is personal, it won't look the same for everyone, so I share this little bit of information in case it might be useful to you where you are on your journey right now.

I will say, that most of these principles were not what I'd been taught in school and I needed to re-learn nearly everything I thought was correct about healthy living. Hence, the injunction to "lean not unto my own understanding." As I've let myself be prayerfully led to new ideas, I've seen how principles of wholesome nutrition and real foods are interconnected. You don't have to be worried that you're going to miss something vital, because it's all tied together, and you'll come to each principle as you're ready. Don't forget to bring your family with you as they are prepared—you don't want to go anywhere without them! The journey has required lot of work and faith, but I can report that, in very real ways, this journey has been "health to [my] navel and marrow to [my] bones." It will be for you, too.

> "If more of us valued food and cheer and song above hoarded gold,
> it would be a merrier world."
> -J.R.R. Tolkien

ACTION ITEMS: Pick one of these three ferments - kefir, sourdough, or kraut. Start doing it. Get your homestead journal out and write down the pros and cons of your choice and the materials you're going to need to get started. I'm not going to suggest a fermented food to begin with because you know your gut and your family better than I do. If you're a praying person, make it a matter of prayer or meditation—ask which ferment your body needs right now.

Be prepared to make a few bad batches along with the good. If you can't tell the difference, ask a ferment mentor or search the Internet for fermentation troubleshooting. Remember that these are live cultures (as in, they're living) and they're going to do what they're going to do—you just need to play nicely and make friends. Think of yourself as a host for the cultures and strive to provide what a host always provides for his guests—food, drink, a comfortable temperature, and a place to rest.

I constantly have something fermenting on my counters; it's a buffet of bubbles and bacteria. Sometimes my husband will come into the kitchen, pick up a bottle and look at it sideways. "What's this again?" he'll ask. He's such a good sport.

DIY: Host a kraut-making party! Send out two basic kraut recipes to your closest, real foods friends and open your home for an evening of kraut making. Have your friends bring the cabbage, containers and any other ingredients, while you provide the sea salt and some refreshments. Everyone can chop and pound and chat while you make a month's supply of kraut to keep in a cool, dark room. You'll be the crazy kraut lady on your block, but your gut will be healthy and strong.

The other thing you might want to do is, ironically, not a DIY suggestion, unless you're a wood worker. Someday you're going to make kraut (maybe you picked that as your first thing to try) and when you do, you're going to want a kraut pounder. When you're making kraut, an essential step is to pound the cabbage repeatedly until the salt you're using draws enough water out of the cabbage that it covers the top by about an inch. To do this, you need something strong, as wide as will fit into the mouth of your kraut container and made from material that won't be worn down by exposure to salt. I use the plastic tamper that came with my high-powered blender, but I don't like using plastic with salty foods and it's not going to hold up forever. On my to-do list, and you may want to add it to yours, is to buy a kraut pounder from some homesteading friends of mine at Homestead Chronicles – you can find them on our links page under The Homestead Kitchen chapter: www.homesteadlady.com/homestead-links-page.

DIY this one if you are a wood turner but, otherwise, do yourself a favor and buy this wooden tool so that you can make your own kraut effortlessly.

BUILD COMMUNITY: Offer to host a class on kefir at your house if your kefir-savvy friend will come teach your homestead group about it. She doesn't have to know everything there is to know about kefir to share what she's already doing and how it makes her gut feel. Make sure she shares a favorite recipe with you beforehand, so you can make it as a treat for your group to sample after class. May I suggest milk kefir ice cream? Or water kefir "soda?" It doesn't have to be just kefir, either. Maybe you have a friend who rocks sourdough bread, or the local extension has a lady who teaches homemade kraut classes and she'll come to your home for free. Follow your nose to the fermented experts and invite them to come teach! It's a great excuse to clean your house and eat good food with friends.

BOOKS TO READ: *Real Food Fermentation*, by Alex Lewin is a great read on the topic of fermented foods without the vulgarity found in *Wild Fermentation*, by Sandor Katz (a popular ferment title).

Either of Melissa Richardson's books on natural yeast will be helpful to you, but my current favorite is *Beyond Basics with Natural Yeast: Recipes for Whole Grain Health* because of the great troubleshooting information and new recipes.

WEBSITES: Cultures for Health (www.culturesforhealth.com) has everything you'll ever need to know about fermented foods - I especially love their dairy information. Be sure to check out their free e-books, blog and YouTube tutorials, as well as their newsletter. You can also visit Ferment Tools (www.fermentools.com) online for equipment, and their blog for recipes, troubleshooting, tips and more (www.fermentools.com/blog).

Holistic Squid (www.holisticsquid.com) and Wellness Mama (www.wellnessmama.com) both have quality information on real foods eating, including ferments and nourishing dairy. Butter For All (www.buterforall.com) is also a good source.

HOMESTEAD LADY SPEAKS:
To make sourdough as non-intimidating as possible, may I introduce you to an easy way to use it? You're not limited to sourdough bread loaves when using natural leaven. How about

some sugar free cookies for breakfast? I'm not usually one for fat free anything as I'm a firm believer in healthy fats and healthy proteins, like free range eggs. However, I was out of butter (how does that even happen?), and eggs, even though I have fat laying hens on my homestead. No big deal; we made these Sugar Free Cookies for breakfast without either of those things.

If you soak/sourdough your grains overnight, these take another five minutes in the morning to mix and 15 minutes to bake. Not bad for a wholesome, homemade breakfast.

SUGAR FREE BREAKFAST COOKIE

Delicious, chewy, wholesome breakfast cookies that are egg, fat, and sugar free.

INGREDIENTS:

1 Cup high fiber hot cereal mix or oatmeal

1 Cup whole grain flour

1 Cup sourdough starter

3/4 Cup applesauce

1/2 Cup raw honey or grade B maple syrup

1/2 tsp. sea salt

1/2 Cup shredded coconut

1/2 tsp. each of cinnamon, nutmeg and cloves

1 tsp. baking soda

1/2 Cup dried berry, optional

1/2 - 1 Cup fair trade chocolate or carob chips, optional - to stay sugar free, omit

INSTRUCTIONS:

Mix cereal, flour, and apple sauce and sourdough starter well. Cover and let sit overnight or at least six hours.

Preheat oven to 400°F/204°C.

Mix in rest of the ingredients and spoon in rounded chunks onto greased baking sheet. Bake 13-15 minutes or until slightly brown on top.

NOTES:

Because these have no fat in them, I suggest you eat them with a large glass of raw milk or a raw milk kefir smoothie to facilitate the absorption of all the amazing nutrients in this grainy mix. You can also add an egg to the recipe and just plan on a slightly larger cookie. Or, serve them up next to some free range scrambled eggs.

"Preserving food is everyday work."

- Sharon Astyk, from *Independence Days:
A Guide to Sustainable Food Storage and Preservation*

For the Homesteadaholic: Preserving What You Grow

Ah, canning season! This is where the rubber meets the road around August, in my growing zone, when the harvest is coming in by the bushel. If you're smarter than me, you'll be very diligent about food preservation all year round so that you don't burn out as quickly in the late summer because you've left it all until then. To help you gauge what and how much of each fruit, veggie and meat you would like to preserve on your homestead this year, let's take a few minutes to talk about goals.

Setting Preservation Goals

First, let's start in the garden by answering these three questions:

1. How much of each product will you need preserve this year? (In other words, how many cans of green beans, carrots, apple slices, jam, ground beef, etc.)
2. How much of your own produce do you want to *grow* for preservation? (For example, are you going to grow all the green beans you preserve?)
3. When will you plant your crops to achieve your preservation goals? (This means you'll need to decide when you want to harvest the crop from your garden for preservation.)

Let's take each question and discuss it a little to help you decide what kind of year you'll be having in the garden and the kitchen.

HOW MUCH PRODUCT WILL YOU NEED TO GROW THIS YEAR?

It's important to get a rough calculation of how much of each canned, dehydrated or frozen item you typically go through in a year. Using tomatoes as an example, for the past few years I've dehydrated sliced tomatoes instead of canning diced tomatoes. I re-hydrate them as I need to make sauces and such throughout the year, using just the right amount of water for whatever I'm making. I do still can spaghetti and barbecue sauce because we like having them on hand, but I prefer to cook with my dehydrated tomatoes. Because of the acidity, we don't eat a huge amount of red sauces and stews. So, I can usually get by with 6-8-gallon sized bags of dehydrated tomato slices a year, which is typically about two to three bushels of standard sized tomatoes. I also need several bushels for spaghetti sauce, etc.

Some vegetables I don't have experience with canning yet, like carrots. I'd like to can them this year because I've tried freezing and dehydrating carrots in the past and am not particularly pleased with the results of either, though both are very usable. Since I don't have experience using canned carrots in the kitchen, I'll just guesstimate how many carrots I need to grow throughout the season for preservation and see how it goes. Plus, with something like carrots, you and I can also factor in the amount of time they'll do well fresh in the cold storage of a root cellar or basement room. Root cellaring isn't a preservation method for the long term but keeping raw foods fresh as long as possible is something our bodies will thank us for in the dark days of winter.

If you're new to canning, try to calculate the number of commercially canned fruits and veggies you go through each year. This will help you logically guesstimate how much you need to plan to plant in your own garden and buy from the farmer's market to put up yourself in the coming year. Just make your best guess!

HOW MUCH PRODUCT DO YOU WANT TO GROW YOURSELF?

This can vary from year to year because sometimes the tomatoes do well and some years, not so much. Some seasons see you accomplishing all your goals in an organized fashion; other seasons, life just happens. It's certainly a wise idea to plan to grow some and purchase some from local growers. That's what local growers are for and they're good at it! If this isn't the year you grow all your own produce for preservation, no worries. Just do what you can do.

For example, as I write this, it's nearly the end of May and we're still in the middle of moving so I'm most likely going to be too late (given the zone I'm moving into) to plant my own tomatoes this year. Since our family can't go a whole year without dehydrated tomatoes, homemade barbecue, spaghetti sauce and ketchup, I'll be purchasing enough tomatoes for my preservation needs from local growers.

I may have time for planting some things for later harvest, though, and so the kids and I have been looking at what we might plant in our fall garden. Yes, the fall garden! Use all the seasons available to you and don't get too upset if you miss planting something in the spring garden that you really want to preserve for the coming year. There are some vegetables you can even grow over the winter if your weather isn't too extreme.

With a little creativity, in most growing zones, you and your family can harvest food crops year-round. It may sound like a lot more work but, bear in mind, if you have harvests "stored" and growing in your garden year-round, you don't have to work to can or dehydrate them. For example, if you want kale in February, all you need to do is go out to the garden and harvest some. If you're growing it, that is. Winter gardening takes some planning, but it's an excellent way to extend the fresh harvest of foods for your family. It can also take some pressure of your shoulders since you don't have to preserve fresh, growing veggies!

(Please Note: if you'd rather eat your own head than garden year-round, forget I mentioned it and move on to question number three.)

After you think about how much you want to grow, look at the size of your garden and make sure your plans will fit into the space you have available. If you don't have enough in-ground garden space, you may want to cover your deck with a container garden; if so, see the *Food in Pots* section in The Homestead Garden chapter. You may also want to investigate if you have a community garden nearby, or if you can rent some garden space from a local farmer. Perhaps a neighbor would be willing to let you use their backyard if their garden days are done. Don't give up – you'll find the growing space you need! Gardeners are really determined people, after all.

WHEN WILL YOU PLANT TO REACH YOUR PRESERVATION GOALS?

If you're growing your own produce to preserve, you need to be sure to plan your garden plantings accordingly. There are basically two ways to plant a "canning garden":

1. One option is to learn, live and love succession planting so that your produce is ready for harvest in intervals, with something maturing pretty much every week of the season. Succession planting is the practice of planting separate rows of the same vegetable, a week or two apart. For example, to keep beets ripening in the garden over several weeks, I plant one row of beets in one bed, one week. The next week, I plant another row of beets in another bed. And so on and so on, until the weather gets too hot for beets. This is a good method to employ in gardening, even if you don't preserve your harvest, because it keeps you in beets consistently, instead of having only one big bunch of beets one week of the growing season. This means that you always have something ready to be harvested and preserved in smaller batches. This can be a life saver for the busy person who only has an hour here or there to devote to dehydrating green beans or canning beets.
2. On the other hand, some of us might prefer to get it all over with at once so we plant several large beds of carrots that will, for the most part, ripen at the same time and need to be dealt with in a big batch. The key point to consider here is

that, when the harvest is ready to be preserved, you really need to clear a large chunk of time to get it all done at once. FYI, after your preservation batch is harvested, you can still plant more carrots to eat seasonally. It's completely up to you, your personality and what your schedule looks like.

I favor "trickle" preservation by using succession planting and indeterminate tomatoes that ripen at different times that enable me to can and dehydrate here and there. However, I'll honestly confess that I'm inconsistent with preserving food throughout the year. Every harvest season from August to November I'm a crazy, food preserving person who wishes she'd stuck to her plan a little more strictly.

I'm no better with big batches of foods to preserve, either, in the interest of full disclosure. I've had bushels of carrots or corn or beets to do all at once and, if I'm not careful with how I plan the rest of the demands on my schedule, I usually end up giving some product away, so it won't spoil. I can only move so fast through my canning and dehydrating because of all the other responsibilities I have. Aren't I blessed with work?! And it is a blessing of the highest merit. I just need to be disciplined with myself and stick to my plan!

So, are you ready to form a rough plan now? Need a few more things to think about? Here you go...

A BUSHEL AND A PECK

Like I confessed, I never preserve as much as I mean to during the year before the real crush of the harvest is on me in the fall. The early spring and summer both have their bounty and, as much as we consume fresh, there are still items that go to neighbors instead of being preserved because I don't have the time to can, dehydrate or otherwise preserve them. Because of this, I've learned to use the slower, "off seasons" of the garden to preserve what I'm able at a more leisurely pace. Early spring, late fall and even a few things in early winter can be preserves all year long/

Here are a few ideas:

1. **Leafy greens** like loose leaf lettuce, spinach, kale, dandelion, mustard, mache and more can be dehydrated. Keep their leaves as whole as possible and blend them to a powder when need to add to smoothies and soups. Why buy "green" energy drink mixes when you can just dry them from out of your garden?
2. **Herbs** begin to ripen early in the spring and go all the way until the heavy snows and I never save enough of them. Typically, air drying will work for most herbs, but a dehydrator can also be used if you prefer. Look for early chives, dandelions and emerging mints in the spring. Gather all the lavender, rose hips and golden rod you can in the fall. Enjoy every herb of summer in between!

3. **Meats** can be bought in bulk year-round and preserved for the seasons to come. Jerky can be made in the dehydrator and is a delectable snack, especially for family hikes and excursions. Canning plain meats, like chicken breasts and sautéed hamburger, saves time and keeps the meat preserved. Many of us use our freezers to keep large amounts of meat preserved but, in winter storms, electricity can fail and the freezers stop working. Home-canned meats are a healthy, affordable way to have preserved meat on hand. You can also put up stews and casserole fillings, with meat included, to have whole meals on hand.

4. **Foraged foods** can be saved in a variety of ways. For example, foraged fruits can be combined to make unique batches of seasonal jams and jellies. Early plums and cherries make a great combination. You can even throw in a little rhubarb for good measure. In the fall, cranberries and pear go well in homemade compote. Ramps, mushrooms, as well as chicory and dandelion root can all be dehydrated from the wild. Even the flowers can be harvested for jams—ever tried dandelion and forsythia jelly?

Following is a popular foraged flower recipe from Homestead Lady (www.homesteadlady.com) to try.

FORSYTHIA DANDELION JELLY

A foraged, golden-colored, spring-tonic- jelly for your toast that includes real vanilla and raw honey. Gather the blossoms for this jelly in the morning before the sun is too high and be prepared to make this jelly by afternoon. Wilted blooms are harder to clean and process, FYI.

INGREDIENTS:

About 9 cups of blossoms, that become 3 cups of tea

3 cups quality honey

2 Tbsp. lemon juice

1 Vanilla bean, cut open and scraped

1 box regular pectin or appropriate amount of Pomona's pectin (read the instructions in the box)

INSTRUCTIONS:

Pick about 8-10 cups of fresh blooms. This isn't an exact science. For this batch, the kids and I picked one-part dandelions to two parts forsythia blooms. You can just make this with forsythia blooms, but I like the pollen and color boost the dandelions give the jelly.

I don't bother to take the bottoms off the forsythia as they just taste "plant-y" but I do remove the green bottoms of the dandelions since they're bitter. I cut off the green part with kitchen scissors and keep the petals, pollen and fluff.

Pour enough boiling water over the flowers to cover them—at least 4 cups. The more water you add, the weaker the tea will be and, therefore, the less golden flavor the jelly will have in the end. Fiddle with the recipe a few times until you find what you like.

Steep the water and flowers for 4 hours or overnight.

Strain through cheesecloth to remove the flower parts, leaving a lovely, golden tea.

Put tea, lemon juice and the contents of a box of pectin into a large saucepan. Add vanilla seeds—you can also add the vanilla bean for the boiling. If you do that, simply remove the bean before you put the jelly into jars. Or, to be all hip and stuff, leave the bean in one jar and gift it to someone extra special.

Bring to a boil.

Add honey.

Bring back to a boil and boil for about 2 minutes. Ladle into hot, prepared jelly jars, leaving 1/4-inch head-space and assemble the lids.

Process in a water bath canner for 10 minutes.

Root Cellaring, Canning, Dehydrating, Freezing

Once we've set some goals, it's important to decide which method of food preservation we want to use for which variety of produce. If you've never preserved food before, then this is your year! I can honestly say that food preservation is one of the things I do that brings the most peace of mind and contentment on the homestead. I thrill when I need a food item and go "grocery shopping" in my basement. Seeing all those jars, bags, and containers of home-grown food waiting patiently on the shelves so they can nourish my family is pure joy for me. Like any other worthwhile thing, it's a lot of work and commitment. But, you're a homesteader, so work and commitment are your bread and butter.

ROOT CELLAR IT

A gardener sees the value in eating fresh foods seasonally, as they ripen, because the flavor and freshness translates into amazing home-cooked meals as each month progresses. As Gerre Gettle reminds us in his book, *Heirloom Life Gardener*,

> "…Prior to the 1960's, most fresh fruit and vegetables were only available from April through November, roughly. When grapes were ripe, people ate grapes…Even the hardiest of vegetables weren't looking so hot in the old root cellar by the time March rolled around."

However, because we love fresh produce so much, we want to cheat the winter as long as possible by keeping those fall apples and early winter parsnips, crispy and raw. Setting aside a space on your homestead to store fresh fruits and vegetables is a wise endeavor. Even if you only factor in flavor, completely ignoring the superior nutritional content of raw food, fresh just always tastes better!

An added benefit is that by using a root cellar (also called "cold storage") environment, we can take a bit of the food preservation pressure of our shoulders every year. As Nancy Bubel talks about in her book *Root Cellaring*, it's not as if we've given up on canning and freezing at our house. Like Nancy, we'd miss those item – pickles, catsup, barbeque sauce – if we didn't have them. However, a lot of the pressure of food preservation that descends upon us around August can be taken off our shoulders with properly root-cellared fruits and vegetables.

You can create a root cellar environment in your home, garage, barn or even buried in your garden if you're not lucky enough to have an actual root cellar. The key factors to keep in mind as you're reading up on this topic are:

- What space you have available
- The average temperature in that space
- The ambient humidity levels in that space

Some produce simply isn't suited for fresh storage, especially over several months. Each fresh item, from tomatoes to turnips, has its own requirements for humidity level and storage container or medium. For example, cabbage has a particularly strong smell that can be passed on to other items, so it should be wrapped in newspaper. While carrots do best stored in damp sand. Some items like onions and garlic last better if they've been cured in the summer sun for around two weeks before being placed into storage. Then there's the ethylene gas produced in spades by the likes of apples and tomatoes that causes other items to spoil rapidly if placed near them. Some vegetables like parsnips need to be held in temperatures just above freezing while others like squash do better nearer regular room temperatures.

The bottom line is, don't assume that all the items you want to store will play nicely together all winter, without first identifying their needs and accommodating them. For most of us, unless we're blessed to have an actual root cellar, this will mean getting creative with our storage space and having multiple spaces in and around our home dedicated to storing the harvest. We'll be those hip in-laws with potatoes stored in the closet in the coolest corner bedroom. We homesteaders are cutting edge, I tell ya.

For further reading on root cellars and over-wintering fresh produce, please read *Root Cellaring: Natural Cold Storage of Fruits and Vegetables*, by Mike and Nancy Bubel, and several other books on this topic—why read one book when you could read five, right? I urge you to learn all you can about the requirements of preserving your harvest in its original form for as long into the next year as possible. If possible, go visit the root cellar set-up of every friend or even vague acquaintance that has one. Root cellaring is not a long-term storage option (as I mentioned before), but it is an extremely valuable one.

> **A COOL DARK PLACE**
>
> To keep your preserved foods in good condition for as long as possible, you're going to need a cool, dark place. Basements and cellars, provided they aren't too damp, are ideal places for stored food because their temperature is in partly regulated by the earth into which they've been dug. If you have a handy basement room that has a duct for heat in the winter time, simply close that or stop it up in some way. If the room has a window, cover it up with reflective material and several layers of heavy blankets to keep out both heat and light.
>
> If you're concerned about moisture in the room you have available, there are commercial, 50 to 70-pint dehumidifying units available for purchase. You can also buy simple desiccant dehumidifiers that will reduce the ambient humidity with a highly absorbent material like silica. You can make your own silica dehydrator if you prefer. Place silica crystals (which can be purchased in the dried flower section of your local craft store), into a muslin bag and out the bag in a tray in your room. Silica, though it holds moisture in its crystalline structure, doesn't become wet so you needn't worry about creating a pool of water in your tray. Silica can easily be recharged, once it's reached its considerable water holding capacity of 40% of its weight, by placing it on a baking sheet in an oven set to 250°F/121°C for several hours. The easiest thing to do is purchase silica crystals with humidity indicators that turn color once they've reached their capacity.
>
> Most modern homes that are regulated by central heat and air systems aren't really in need of humidity control. However, humidity may be of concern for those living in older homes or in particularly humid areas like the American southern states. The easiest way to monitor temperature is to purchase a simple thermometer for your storage room. Likewise, you can buy a hygrometer to measure moisture.

BUILDING A HOMESTEAD ROOT CELLAR

What we don't have we can build! If you have the space on your property and the legal right to do so in your county/city, I encourage you to at last investigate building your own root cellar. The place my husband and I began our actual planning of our root cellar was with the simple but thorough publication, *Building a Homestead Root Cellar*, by Teri Page and Brian Thomas (of www.homesteadhoney.com). This book has some basic information about what a root cellar is, as well as what kinds of things can be stored in one. This beginner information is great, but my favorite part is the details of how they built their root cellar.

They give measurements, materials lists, cost assessments, time frames and so many other helpful tables and tools in this little book. They also talk about things they would have done differently – I love learning from others' mistakes, so I make fewer of my own. My husband and I are currently planning for our root cellar and so the excitement for it is fresh in my mind. Being able to store the harvests I work so hard for is going to make a world of difference to how my family eats, both fresh and preserved foods!

I know it will be a learning experience, but I've been learning for years as I've stored my harvests in random places. Our typical method with establishing a new system of anything is to start out with a rough DIY, and then move slowly into a more sophisticated arrangement. It's important to keep in mind that whatever we do, however imperfectly, is better than doing nothing at all. So, don't stress; build the root cellar when you're ready.

> "Each fruit or vegetable will have its own particular temperature and humidity needs. Of course, in one root cellar, it is hard to please all the vegetables! And other than moving food from the back chamber to the front, you really don't have much control over the temperature of your root cellar through the year.
> Don't stress too much about it, and do your best, and each year you will find which crops do best for your cellar. Food may not store as long in non-optimal conditions, but it will still keep for months."
>
> - Page and Thomas, *Building a Homestead Root Cellar*

CAN IT AND DRY IT

If this will be your first year canning, I suggest you keep it simple, but push yourself to try making different home-canned products. I challenge you to make a goal to put up at least one batch of:

- jam
- pickles
- green beans or corn

If you're feeling adventurous you can also do spaghetti sauce. Be sure to follow a canning recipe for each item and follow the instructions precisely, especially to adjust for variations in altitude. After asking your family what they'd like to try and considering your own ideas, if you don't want to do those things I've suggested, then go off on your own adventure. To can the items, I suggested, you'll need a water bath canner (for the jam, pickles and sauce as long as you haven't added meat or veggies) and a pressure canner (for the green beans and corn). You can just use a pressure canner for anything you would preserve in a water bath canner, but not all of us are ready to invest that much in a tool we're not sure we'll use.

If you want to start with the less expensive water bath canner, go right ahead. You'll also need jars, lids, rings, a good canning book and some other hand tools for working with the hot jars. If you don't have the funds for all that, skip the pressure canner this year and purchase the cheaper water bath canner, limiting yourself to jams, pickles and red sauce.

I also encourage you to try dehydrating because it's a lot less messy than canning. Not to mention that dehydrated foods store smaller and a lot lighter than their canned counterparts.

A dehydrator makes the process easier, but you can also use your conventional oven on its lowest setting, or your solar oven with the lid vented. You can also construct a solar dehydrator if you're handy with a drill. You can learn how to build one by visiting our links page under The Homestead Kitchen chapter.

Dehydrating provides some versatility to foods, too. If I dehydrate a batch of carrots, I can easily rehydrate them to use in soups, casseroles and stir fries. I can also powder those carrots in my high-powered blender to create a powder to add to smoothies, sauces, pasta dough and even ice cream. Because of how light and durable they are, dehydrated foods are ideal for emergency kits, as well as snack and trail food. By the way, homemade jerky and fruit leather are divine, in case you were wondering, and both can be made in a dehydrator.

Freeze drying is another way to make raw foods storable for the long term. Freeze-dried foods are taken down to temperatures between -30 and -50 degrees Fahrenheit, and then gradually warmed so that the water in the food evaporates out. Freeze dried foods are superior in taste and longevity and freeze-drying units are now available for the home preserver. We recently dug into our food storage during an economically challenging time and I can testify that the flavor of freeze dried foods is wonderfully preserved. Their superior texture also makes them much easier to cook with, in my opinion. The unit price for a freeze dryer is far above what you'll pay for even a top-of-the-line dehydrator, but there's no comparison when it comes to flavor and shelf-life. For more information on freeze drying units for home use please visit Harvest Right (www.harvestright.com).

FREEZE WITH CAUTION

I don't recommend conventional freezing very much, or very often, simply because I've lived where power outages are common all year round due to weather. I learned as a young homemaker to keep my freezer stocked but to never put all my eggs in one basket, so to speak. I freeze fruit during the harvest for use in smoothies but not much else. I buy half a cow once a year, and that could go in the freezer, but I'd much rather can a good portion of my meat, instead of relying on the freezer to store all of it. If you are going to freeze a lot of produce, try to stick to varieties that can be canned or dehydrated easily, in case they end up getting defrosted against your will by the first ice storm of the season.

The only equipment you need for freezing food is a freezer and freezer bags and/or freezer paper. It can be handy to have several cookie sheets on hand, too, for placing fruit and veggie slices onto so you can freeze them individually and bag them together later. This makes it easier to take out only what you need for each smoothie or soup.

Preserving Homegrown Protein

There are several types of protein that can be preserved, so let's look at a few.

A MORSEL ON MEAT

Deciding how Preserve meat, as with any home-preserved item, is going to come down to what method you want to use and how you're going to use the preserved meat. I submit that a pressure canner will probably come in handy for you at some point, so you can preserve broths, stew meats and even whole meals that include meat.

A dehydrator will also work for various meats including beef, venison, chicken, turkey, ham and others. Learning to make jerky is a great meat preservation skill to master. As Shelle Wells reminds us in her *Prepper's Dehydrator Handbook*,

> "Once you've learned the art of making jerky, you have the ability to raise your food storage plan to a higher level. Not only are you producing fruits and vegetables, but now you're also making a protein that is all-natural and free of additives, food colorings, and preservatives."

The real key to dehydrating raw meat is to remove all the fat you possibly can, because it will turn rancid as it's exposed to oxygen over time. You may find that you have good luck dehydrating meats that have been previously canned, meaning they've been through the pressure canning process. Often, these meats will rehydrate better, having a much better texture and flavor. To learn more about the practicalities of dehydrating many types a food, a fun website to visit is Backpacking Chef (www.backpackingchef.com).

For information and ideas on homestead meat production please see the *Meat* section of the Livestock Wherever You Are chapter.

KEEPING DAIRY

If you're looking to preserve the dairy that's produced on your homestead that can get trickier. I'm deliberately not stepping into the cow-pie that is the argument over home canned milk and butter—it's a discussion riddled with land mines of food safety. I will say, though, that arguably the tastiest way to preserve dairy is by making cheese. Dairy is highly perishable but even learning to make your own yogurt, butter and soft cheeses, like feta, can extend your milk's shelf life by a few days to a few weeks, depending on the product. Hard cheeses will give your dairy a much longer shelf life—some cheeses can be aged for years!

Learning to make cheese is a process and is, perhaps, not one you're ready to tackle at this phase of your homesteading career. No big deal, take your time. I mention the possibility only because it is a really tasty way to preserve your dairy harvest against a day of want. If you have digestive problems with commercial cheese I encourage you to try making your own with a variety of different milks, including raw. If you're just not sure cheese is in your wheelhouse, try making the feta (a soft cheese) recipe below. You may end up falling in love with making your own soft cheeses. After that, moving into making hard cheeses (which store longer) won't feel so overwhelming.

All the books and websites seem to say that the quick version of mozzarella is the easiest soft cheese to make. However, most recipes require the use of a microwave, which I don't own or like to use. Plus, I just stink at making mozzarella. I'm hit or miss with the texture of it for some absent-minded reason of my own. My friend Quinn from Reformation Acres has a fantastic easy mozzarella recipe that doesn't require a microwave. She's graciously allowed me to reprint it for you here. All of Quinn's dairy recipes are simple and straightforward and I highly recommend you peruse her site for other cheese making ideas – like her to-die-for mascarpone recipe.

MOZZARELLA CHEESE (THE EASY WAY), BY QUINN VEON

This recipe is used with permission from Quinn Veon, formerly of ReformationAcres.com and co-creator of the homesteading app, Smartsteader. Quinn knows her dairy and I'm always pleased with the results when I follow her recipes. This simple mozzarella recipe does NOT require a microwave to make.

Ingredients:

1-gallon milk

1 ½ teaspoons citric acid

¼ teaspoon non-GMO rennet, dissolved in ¼ cup cold water

⅓ cup salt

Instructions:

Dissolve the citric acid in the milk and heat it to 88°F/31°C degrees.

Stir in the dissolved rennet water, slowly, until curds form.

Heat the curds to 110°F/43°C degrees, strain them from the whey.

Knead the curds like dough to remove excess whey.

Heat the whey in the pot to 150°F/66°C degrees. Stir in ⅓ cup salt.

Dip the curd in batches into the hot whey.

Warm and stretch them until they are super stretchy.

Heat it once more and work it into a ball.

Run under cold water to chill and then wrap in plastic wrap.

Refrigerate.

As much as I love a good, homemade mozzarella, I still say that feta is so easy to make that every cheese newbie should start with it. Plus, because feta is traditionally a brined cheese, it can last up to several months in the fridge. If you can't find goat milk to make the feta recipe below, try your hand a mozzarella with cow's milk and don't listen to me whine – mozzarella is a fabulous cheese. You can make feta with cow milk, but the flavor is different, and I can't ever get it to brine well.

The following is a super-simplified step by step process for making feta from goat's milk and I encourage you to check out a cheese making book from the library and/or go online to find a tutorial, especially if you're a visual learner. And I *highly* recommend Ashforth's book should you decide cheese making is for you. So, if you have goat milk, try making feta; if you have cow milk, try making mozzarella. Either way, you'll eat hearty!

HOMEMADE FETA CHEESE

I adapted this recipe from Ricki Caroll's *Home Cheese Making*

INGREDIENTS:

1-gallon whole, raw or pasteurized goat milk

1/4 tsp. lipase powder diluted in 1/4 cup water and allowed to sit for 20 minutes (optional—use if making this recipe with milk other than goat's milk)

2 oz. prepared Mesophilic starter or one packet

1/2 tsp. liquid rennet (or 1/4 rennet tablet) diluted in 1/4 cup cool, dechlorinated water

3 tbsp. sea salt

1/4-1/3 cup sea salt, for brine (optional)—brining gives it a stronger flavor, though, which is what makes feta distinct

1/2-gallon water, for brine (optional)

INSTRUCTIONS:

Combine the milk and the diluted lipase, if desired; heat the milk to 86°F/30°C.

Add the starter, stirring to combine; cover and allow to the milk to ripen for 1 hour.

Add the diluted rennet and gently stir with an up-and-down motion for several minutes; cover and allow to set at 86°F/30°C for 1 hour. You can set your pot into a hot water bath in your sink and monitor the temp a couple of times in the hour to make sure it stays close to 86. Don't freak if it falls below; just add some warm water to the bath.

Cut the curd into 1/2-1-inch cubes.

Allow to set, undisturbed, for 10 minutes.

Gently stir the curds for 20 minutes.

Pour the curds into a colander lined with cheesecloth.

Tie the corners of the cheesecloth into a knot and hang the bag over the sink to drain for 4 hours or more, depending on the temperature in your house and how sharp you like it. You can hang it anywhere clean in your house, but the kitchen is a logical choice. Remember to put a bowl underneath to catch the whey. Massage and/or knead the feta every now and then to encourage more whey to drain out.

Untie the bag and cut the curd into 1-inch slices, then cut the slices into 1-inch cubes.

Sprinkle the cubes with the salt to taste; place in a covered bowl and allow to age for 4-5 days in the refrigerator.

For that strong feta flavor and to preserve the feta up to three months in the refrigerator, make a brine solution by combining 1/3 cup of salt and the water. (FYI, feta made with some store-bought, pasteurized milks can disintegrate in brine. Feta is traditionally a raw milk cheese.)

In eleven easy steps you have cheese where there was no cheese before. You are that amazing.

IS THAT IT?

People have been preserving food throughout the centuries in various ways. Although modern science has done us a great service by indicating which methods are safest, we aren't limited to canning, freezing or dehydrating when it comes to putting up food. There are several ancient methods like salting, sugaring, brining, fermenting, curing and smoking that can be done safely and allow for a greater variety of flavor in our food stores. These methods come in particularly handy with meats and cheeses. The techniques range from quite simple to slightly complex, requiring special equipment and varying lengths of time. When you're ready, start exploring this level of food preservation. These traditionally prepared and preserved foods offer so much in the way of taste and food security.

For further reading, try *Preserving Food Without Freezing or Canning*, by Deborah Madison, Charcuterie, by Michael Rhulman, and *The Art of Natural Cheesemaking*, by David Asher to begin your education.

> "One cannot think well, love well, sleep well, if one has not dined well."
>
> — Virginia Woolf, from *A Room of One's Own*

ACTION ITEMS: In your homestead journal, write down your answers, or at least your brainstorming, of the questions posed in this section. Decide which method of preservation you'd like to try this year. Pick one to five pieces of equipment for your food preservation method of choice and start pricing them out. If you're going to need to save money to buy them with cash, start now. Also, if you know you'll need to purchase some or all the produce you've decided to preserve for your family this year, start making a list of local farmers or quality bulk grocers that you want to patronize.

To start guesstimating amounts and types of produce to be grown or purchased in bulk from local growers, as well as calculating the amounts needed for your family's year's supply of food, follow this link to download and print ten pages for your scribbling calculations on our Food Storage and Home Preservation sheets which you can find on our links page.

DIY: For your food producing garden, especially if it's large, learning to make your own seed tapes can come in handy. Seed-tapes are created for planting crops like carrots and onions that are planted in high volume but have very small seeds. When you plant crops with small seeds, typically you plant too many in once space because they get away from you. If that happens, you need to go back and pull, or thin, out extra seedlings which can be tedious for you and detrimental to the plants left behind. Making seed tape that gets planted directly into the ground enables you to properly space out each seed on the tape, ensuring that you won't need to go back and thin. To make your own seed tape, you'll need unbleached toilet paper (or other thin paper like newspaper or paper towels), seed and some flour or corn starch to create a paste for attaching the seed to your paper. We have some detailed, online instructions for you on our links page here: www.homesteadlady.com/homestead-links-page.

BUILD COMMUNITY: Do not, under any circumstances, try to go through the whole canning season alone! This is a great time to involve your homestead group, church group and neighborhood friends as you learn together new methods of food preservation. If you are lucky enough to have an experienced friend or neighbor near you, then they become the super star of your preserving season as you gather people in your home to listen to your mentor teach you about the wonderful world of jam, the exotic world of jerky and the mysterious world of home canned stews and sauces. Even if it's just your BFF and neither one of you knows what you're doing, you need to have a food buddy this preservation season!

Please also remember to plan and preserve a little extra for those who may be in need during an emergency. Having to turn away a hungry child or deserving neighbor will be more difficult than we can perhaps realize in our abundance. Never mind, for now, that your neighbor should be preparing against a day of want on their own.

As Sharon Aystk wrote in her book, *Independence Days*,

"Ultimately, our own security in both a pragmatic and a moral sense depends on not having our neighbors go hungry either."

BOOKS TO READ: For making food storage and preservation a part of your homestead life, I highly recommend the *Independence Days: A Guide to Sustainable Food Storage and Preservation*, by Sharon Aystk. Similarly, *Traditional Kitchen Wisdom*, edited by Andrea Chesman, will take you from gardening to preserving to even beekeeping to get your kitchen as sustainable as possible.

I have several canning books and I find value in each one of them but my new favorite for straight forward methods and fun recipes is *The Organic Canner*, by Daisy Eli. I'm especially looking forward to trying her Branston Pickle recipe this year. Another fun one is *Food in Jars*, by Marisa McLellen, which focuses on small batch canning of year-round foods.

For dehydrating, my favorite book is *The Ultimate Dehydrator Cookbook*, by Tammy Gangloff. Also handy, especially for quick reference is Shelle Well's, *Prepper's Dehydrator Handbook*. For root cellaring, *Root Cellaring: Natural Cold Storage of Fruits and Vegetables*, by Mike and Nancy Bubel. For general food preservation, where several methods are discussed (canning, freezing, salting, smoking, drying, and root cellaring), my favorite is *Putting Food By*, by Hertzberg and Greene. Be sure to purchase the newest edition to keep abreast of new food safety recommendations as they develop.

For cheese making, *Home Cheese Making*, by Ricki Carrol is a great one but a few of the recipes can be a bit vague for the newbie. Another good one to have on hand, especially for those interested in learning to make your own rennets and other cheese making supplies, as well as simple cheeses, is *The Art of Natural Cheesemaking*, by David Asher.

WEBSITES: Simply Canning (www.simplycanning.com) for canning information—seriously, you can't ask a canning question for which Sharon doesn't have an answer. Homespun Seasonal Living (www.homespunseasonalliving.com) for growing, harvesting and preserving foods, fruits and herbs. Backpacking Chef for tips and tricks on drying food (www.backpackingchef.com) and Rockin' W Homestead (www.rockinwhomestead.com).

DVD: "At Home Canning for Beginners and Beyond" by Kendra Lynne is great for beginners and seasoned canners. It would make a lovely gift for the new homemaker or homesteader.

HOMESTEAD LADY SPEAKS:
The process of putting food by will put us in some great company with fathers and mothers throughout the ages, who've spent their time and effort to provide for their families for the year. Our society is no longer agrarian based and with supermarkets open nearly every day of

the year, and some of them, every hour of the day, it might be a struggle to imagine a time when our food preservation efforts may save the lives of our family members in a time of want. Our family has had lean times of job loss or job reduction when we only ate three meals a day because we'd preserved our harvests in more abundant times.

The truth is, in our time, as in all of history, anything can happen. Any food we preserve is better than preserving nothing; every extra jar of jam is one more taste of happiness and home that your family gets to enjoy because you planned ahead. Even if you and I aren't worried about food scarcity, the simple fact is that home grown and home-preserved food just plain tastes better and is better for you. (Read *The Heirloom Life Gardener*, by Gerre Gettle for more musings on this topic.)

When I lived in Russia, the women there opened their kitchens and their hearts to me as they talked about their dachas, small patches of earth in the country on which their families had been raising food for generations. Each woman knew how to save seed, cook from scratch, use every scrap of food and put up every piece of produce to protect their families against the want of the winter time. For these women, food preservation is a part of their yearly cycles and without it, there is hunger; with it, there is abundance even when other things are scarce. And the flavor of those foods they shared with me? Nothing like it. There was a vegetable relish that I tasted that was some combination of eggplant, wonderful, onions and heaven—I have yet to recreate it. Not having the presence of mind at the time to ask for a recipe, I'm still pining for it after all these years. Now that's powerful food!

> "...The kitchen will not come into its own again until it ceases to be a status symbol and becomes again a workshop... you and I will know it chiefly by its fragrances and its clutter."
>
> - Phyllis McGinley, Author

For the Homesteaded: Make Your Own Stuffs

There are scads of pantry products you and I can learn to make ourselves. I'll never forget the first batches of crackers and chips I made at home. They're as easy as cookies but it never occurred to me before to do anything but buy them. Then there was the first batch of marshmallows I made from water, beef gelatin and honey—it was as if I was the first human to make fire! From bouillon to mustard to pickles, most of the items in our pantry could be made at home; consider that, not long ago, they were. We're going to focus our discussion on making the baking supplies typically needed for a batch of cookies in the hopes that it inspires you, as it has me, to think about what else you can make yourself.

Making my own baking supplies saves money, to be sure. But, it also helps me control the ingredients of my staple foods, which means I can keep them as healthy and nourishing as possible. As author Michael Pollan encourages us,

"Don't eat anything your great-grandmother wouldn't recognize as food."

Well, she'd certainly recognize the procedures (apart from our fancy, modern equipment) and products below—and she'd be proud of you for giving them a try. So, before you plan your next batch of cookies, have a go at some of these ideas...

Flours

A most basic ingredient for quality cookies is flour. Growing your own grain is perfectly possible but it requires a good deal of space to end up with a measurable harvest. For example, you can expect to harvest twenty pounds per hundred square feet of wheat, give or take. You can grow grain in square blocks instead of rows to save space. Grains are fancy grasses, so they typically grow well, if you have the correct season and temperature for them. But, they do require some special equipment considerations for harvesting and threshing (removing the outer material from the grain), even in the backyard homestead. Growing and harvesting grain you've grown yourself is a very rewarding experience and you will quickly learn why our ancestors valued each grain, never wasted a single one and would consider us absolutely insane for feeding it to our livestock! (After all that manual labor, why feed it to the animals?!) To learn more about growing your own grains, please read the easy to understand and very well-formatted book, *Homegrown Whole Grains*, by Sara Pitzer.

Don't want to grow your own grains? No worries! You can also purchase grains in bulk and keep them in your long-term food storage. The wheat I use most often for baking is white wheat. However, I keep a wide variety of grains including Kamut (an ancient variety of wheat), millet, oats, quinoa and several more. To make your own flour, you will need a grain grinder. I have both an electric and a manual one. There are many grain mills on the market and, I'm not gonna lie, they're a financial investment. I have a Victorio hand mill and a Blendtec electric. Both are of good quality, but I encourage you to do your own research before you buy.

Being able to control the quality of my flour (freshly ground flour is far more nutrient dense) and create flour from my wheat storage became important enough to my family's health that I saved a bit here and there until I could afford a quality mill. I would suggest you not bother to buy a poorly made grain mill, because it will just break. I use my electric mill but a manual one works beautifully, just more slowly. As Barbara Swell writes in her book *A Garden Supper Tonight*,

> "You can't even compare store-bought grains to those you grind yourself. Not only will your baked goods scream with flavor and health, you'll love your new shapely grain-grinding muscular arms."

If you choose to use a manual grinder, that is; I wimp out and use my electric one. I keep a manual grinder for the same reason I keep emergency fuel—just in case.

ENOUGH IS ENOUGH?

How much wheat would you need to grind weekly to keep up with the demands of your household baking? If you haven't really thought about it before, let's consider together. Here are some common breadstuffs you can be making at home every week:

> Bread loaves
> Pitas, naan or tortillas
> Cookies, cakes, bars
> Pancakes, waffles, sweet breads
> Breadsticks, chips, crackers
>
> That list doesn't include other uses for flour like feeding your sourdough starter or thickening gravies.
>
> Work backwards from that list – which of these items are you baking each week? How many loaves or batches of each item? Or, similarly, how many hungry people are you feeding each week? Take seasonal changes into account, too. For example, we bake a lot more in November and December because of holiday treats than we do at any other time of year. So, I make sure we grind extra in October and through the holidays to keep up.
>
> I like to get the grinding done in one usually hour-long session once a month, as opposed to hauling out the grinder and making a flour mess every week. Because I'm grinding so much wheat at once, it's important that I stop after each batch and let both the flour and the grinder cool off. I let the flour aerate for at least ten minutes before I close it up. I let the grinder cool as I carefully brush out all its crevices and parts with a medium-sized paint brush reserved for just this task. Keeping your grinder clean will extend its life and help it to function efficiently. Read the instructions for your specific grinder on how best to do this.

To make your home-ground flour a little more like the all-purpose flour you buy in a store, learn to sift out the bran. You can use a sifter especially made for this purpose, or you can use a fine-mesh sieve in your kitchen. To use a sifter:

1. Measure fresh ground flour needed for your recipe into a sifter or fine-mesh sieve.
2. Over a bowl, work the mechanism of the sifter or gently shake the sieve.
3. You'll begin to see that the flour (endosperm of the grain) is collecting in the bowl while the bran and germ of the grain are collecting in your sifter.
4. Repeat this process, reserving the bran in a container for use in homemade granola bars or any whole grain recipe, except bread. Extra bran will weigh down bread dough. Bran is particularly dense because most of the fat from the grain is stored in it.
5. Keep extra bran in the fridge and use within a week or two. Use the sifted flour immediately or within a week or two from the pantry. Store in an air-tight container.

I advise you only bother to sift out the bran for a specific recipe like favorite cookies, a special cake or light sourdough bread. Simply sift the amount of flour you need for that recipe. Usually two siftings will produce an airy, all-purpose flour.

I don't have a lot of extra time in the kitchen, so I usually only do this for special occasions. It's completely worth it to enable me to avoid chemical-laden commercial white flours, though!

> **GLUTEN FREE FLOURS**
>
> The most commons gluten free flours are:
>
> White or Brown Rice Flour – makes lovely, crispy waffles
> Millet, Amaranth or Quinoa Flour – like each other once ground
> Oat flour – be sure to purchase certified gluten free, or grind your own whole oats
> Chickpea flour – often used in Indian cuisine
> Masa Harina or corn flour – these are NOT the same thing, FYI
> Coconut flour – used sparingly because of its very high fiber content
> Sorghum, Buckwheat and Teff – rich flavored, almost nutty
> Almond flour – rich in protein and very commonly used in Paleo and other gluten free recipes
> Cassava flour – both a grain flour substitute and a starch commonly found in commercial gluten free flour mixes
>
> Any of these can be ground at home, though commercial products may produce a finer product. An electric grain mill is only built to handle *grain*, so be careful what you put into yours. Rice, oats and grains like Teff should be just fine because they're dry. You'll need a manual grain mill, food processor, high-powered blender or nut mill for something like almonds and coconut.
>
> To learn more about what an electric and manual grain mill can handle, visit willitgrind.com and watch as both an electric and manual mill handle various foodstuffs. This demonstration is provided by WonderMill®, but most mills function about the same.

Fats

Time was I used Crisco for fluffy baked goods; now I wonder how I survived those days. Quality fat is a life sustaining substance that contributes to our health and quality of life. Fat is sacred, according to The Old Testament (see Exodus or Leviticus, in *The Holy Bible*). Butter is typically used in cookies and is one of the easiest dairy products to make yourself. You need high quality cream for good butter. If you're stuck with store bought cream, it will all be ultra-pasteurized and, therefore, not really that healthy but will still make tasty butter. If you have access to milk straight from the animal, then you can create a wholesome butter to use in and on everything.

If you harvest your own pigs or have access to fresh lard, learn to render it for baking and cooking as it is simply without peer in the world of healthy baking fats. Leaf lard is the best, and highly coveted by pie crust makers around the world.

TO MAKE BUTTER:

Place 1 quart of cream into a half gallon mason jar with a lid, a food processor or a butter churn.

If using a mason jar, begin to agitate the cream by shaking the jar.

As you shake, the fat will begin to bond and the whey (mostly water) will be left behind.

The first thing you'll produce with all that shaking is a kind whipped cream that, as you continue to shake, will begin to form chunks.

As you shake further, the mixture will "break" as the yellow butterfat clumps together and the white whey sloshes around on its own.

When using a jar to make butter, you'll know this has happened when you feel the thunk of the butter against the lid of your jar. Also, if you're using cow cream, you'll be able to see the butter that's forming easily because it will be a lovely yellow color. Goat cream stays white and gives you white butter, FYI.

After your butterfat has bonded, pour off the whey into a separate container and add it to breads, soups and casseroles. It's also suitable to feed to chickens or pigs, or to add to the compost pile.

Put your butter clump into a bowl and run some water over it as you press it into the bottom of the bowl. You can also do this just by holding the butter in your hands under the water. This process removes extra whey that's been trapped inside the butter. As the water in your bowl becomes cloudy, dump it off. The whey won't hurt you if you eat it, but it can make the butter sour over time. Since I use raw cream, I don't worry too much about getting all the whey out because raw dairy doesn't spoil, it cultures. I prefer a bit of culture in my butter, so I'll mush it around for a few seconds to get out any pockets of soggy whey, and then I just let my butter alone. We go through it so fast that it rarely has a chance to culture at all, to be honest. If I'm using pasteurized cream for some reason, then I'm much more diligent about getting the whey out of my butter so it doesn't spoil.

After you're satisfied that your butter is as whey-free as you can make it, stir in a bit of sea salt to taste if you'd like.

You can simply serve the butter up right away or wrap it in parchment paper and refrigerate or freeze. I like to measure out the butter in half cup amounts just in case I need to bake with it. You can make it fancy by pressing it into a butter mold that will stamp a pretty pattern into your butter.

If you're using a food processor or electric butter churn, the process is the same, but the agitation is performed by the machine. Old fashioned, upright butter churns are hand powered but are designed for larger volumes of cream—as in, gallons. This process is the same whether you're making butter from one quart of cream or one gallon.

If you're using your stand mixer to make butter, be prepared to cover the bowl completely with a towel or plastic wrap because the cream and whey will slosh all over the place.

Please believe me, all over the place. I'm still waiting for the manufacturer to create a cover for the bowl for us weirdos who like to make our own whipped cream and butter.

Sugars

There are two kinds of sugar that are, I believe, most feasible for homesteaders to make themselves. The first is maple sugar, or more simply, maple syrup. Is there anyone that can resist the robust flavor of pure maple syrup? To harvest your own, you're going to need a few things, including the right trees. Sugar maples are the most common tree tapped for syrup, of course, but almost any maple will do. Other kinds of trees can also produce syrup, including several kinds of walnut and birch. Sugar sapped trees grow in very specific climates, typically ones with cold winter temperatures. If you live in a warmer climate, skip this section and move on to the honey discussion.

You'll also need certain equipment to tap the trees and process the sap into syrup, which is a fun but somewhat time-consuming process. A lot of the equipment you can make or improvise yourself, though, and tapping/processing is a great activity for a bunch of friends to do together; you can make a day of it and hold and old-fashioned sugaring off party (for more information, see *Mrs. Sharp's Traditions*, by Sarah Ban Breathnach, suggested reading in the *Traditions* section of the Family Times on the Homestead chapter). To learn more about the ins and outs of tapping and making maple syrup, please read the very worthy book, *Maple on Tap*, by Rich Finzer. Rich shares real life experience, including the flops, and a lot of practical advice on equipment and procedure.

If you don't live where you can tap trees or are simply not inclined to do so, have you thought of the humble honey bee? In my opinion, the best and most healthy sugar to make yourself is honey—well, the bees do all the work. I prefer to keep my honey raw so that all the beneficial nutrients that are present in natural honey remain intact, particularly those that are thermo-sensitive and die off in the extreme heat of pasteurization. However, even when I do heat honey, by baking with it, I feel it's a healthier and certainly more sustainable source of sweet than commercial sugar. Honey is healthier sugar that I can "grow" on my homestead.

To learn more about keeping bees on your homestead in a natural way, please read Ross Conrad's, *Natural Beekeeping*. From bee to honey, Mr. Conrad will get you where you need to go in a wholesome way.

Back to the topic of table sugar. I'm not going to tell you to combine molasses with table sugar to create brown sugar. But, I will say, please toss out all your table sugar just as soon as you're ready because it simply has no redeeming features. Once you wean yourself off white sugar, it will most likely start to taste metallic and quite nasty. If you want a "brown" sugar, try succanat, rapadura or coconut sugar, all of which taste like, but much better than, brown sugar. Plus, they're a slightly healthier, less processed option than white sugar. They are a lot less sweet than commercial brown sugar, but as you lessen your addiction to white sugar by not eating it much, you may discover that you prefer a little less sweet.

For more information on the various, alternative, healthier sugars available for purchase and to learn how they're processed, please visit our links page under the Homestead Kitchen chapter and look for the Wholesome link: www.homesteadlady.com/homestead-links-page. If you need to make your own (white) powdered sugar, I suggest you use Organic cane sugar—which is still a crystallized, processed sugar, but not quite as refined as table sugar. Eh, besides, this is a treat, right? If you use succanat or coconut sugar, you end up with a powdered sugar that tastes like maple and is brown. You may also discover you have a strong desire to make old-fashioned, whole grain donuts onto which you can slather homemade frosting. Either way, anyone with a blender can make their own powdered sugar.

> **TO MAKE POWDERED SUGAR:**
>
> Measure one cup of your favorite sugar into your blender body—do one cup at a time to avoid overheating your sugar and making it sticky.
>
> Quickly turn the blender to its highest setting and blend until fine powder forms.
>
> Stop after twenty seconds to air out the sugar (it will become tacky if over-heated) and to bang the sides to move more granules down to the blades.
>
> Blend until all granules are powdered. This will take several times. If you want an even finer powder, you can remove your sugar to a mortar and pestle and manually grind it a bit more.
>
> Be sure to let the sugar cool down all the way before you store it.
>
> By making your own powdered sugar you can control what kind of sugar is used—you do NOT have to settle for store bought. I can never quite get mine as finely ground as commercial powdered sugar but that is a small price to pay, in my opinion, for better taste and a cleaner product.

Leaven

We've already discussed sourdough and kefir in the *Ferment All the Things* section in this chapter. Both ferments are biological leavening agents that can be grown and maintained in your home. You may also try making your own baking powder with 1 teaspoon baking soda, 2 teaspoons cream of tartar and 1 teaspoon of non-GMO corn starch. I'm not going to mix carbonic and sodium hydroxide to make baking soda, but if you want to, I won't try to talk you out of it.

While not technically a leavening agent, salt stabilizes baked goods and you can learn to harvest your own sea salt with a little reading, lots of practice and access to the sea. Sea salt is preferable, in my opinion, to table salt because it is less processed, likely has more trace minerals and elements, tastes better and you end up using less of it because its flavor is stronger. I'm probably never going to learn to harvest salt, FYI, and so I store bags of it. But you're most likely cooler than I am and will figure it out.

Vanilla, Sprinkles and Fun Stuff

I don't live where I can grow vanilla beans, but we can buy Organic vanilla beans and make our own vanilla extract—easy peasy! My biggest problem with commercial vanilla is that it's tinctured in alcohol which damages the flavor, in my opinion. I tincture my vanilla in Organic vegetable glycerin and make half gallon batches, several gallons at a time.

> **TO MAKE VANILLA:**
>
> Score 4-6 vanilla beans and scrape their contents and the beans into a glass jar.
>
> Cover the beans with 1 part filtered water and 1-part glycerin.
>
> Steep in a cool, dark place, shaking occasionally, for at least 3 months—I go as long as I can because we like the flavor strong.
>
> Strain out the beans and store your vanilla in an amber bottle, if possible, and/or a cool, dark place.
>
> You can use your spent beans to make vanilla honey by placing them into a container of raw honey and infusing (letting it sit) for at least three months. Once the honey is flavored, remove the beans and compost them. Use your honey to flavor teas and treats.

You can also make your own candy sprinkles to control the ingredients and keep them free of dyes, carnauba wax (which is botanically based, but often bleached), and any other item you may not want to ingest.

> **TO MAKE CANDY SPRINKLES:**
>
> Mix up your favorite royal icing recipe (1-pound Organic powdered sugar, 3 egg whites, 1 teaspoon fresh lemon juice).
>
> Add any color and flavor that you want, using vegetable dyes and essential oils if you prefer healthier ingredients.
>
> Using a piping bag and the narrowest, plain tip you have, pipe the frosting out in long lines on a piece of wax paper and let it dry at least overnight.
>
> Once dry, cut up the long strips into small bits.
>
> Store in a dry container in the refrigerator for up to three months.

A cupcake just isn't a cupcake without a little decoration! There are several types of edible adornment that you can make at home, with the bonus that they will be free of potentially harmful commercial dyes. To make herbal sugars that have both color and taste for decorating your cupcakes, I suggest you start with rose petals and lavender.

> **TO MAKE HERBAL SUGAR:**
>
> Take 1 cup of organic cane sugar and combine in a blender with 1 cup of rose petals to make pink sugar crystals. I grow rugosa roses and their petals are highly fragrant with deep color. I recommend them for many reasons, not the least of which is that they make incredible snacks and sweets.
>
> If you'd like to make purple crystals, use one cup of sugar and about 1/2 cup of lavender.
>
> You can do this with calendula petals (orange), hibiscus flowers (red or other colors), mint (green) and many more herbs.
>
> Experiment with the amounts of herb until you get the color and taste you like.
>
> Lay out on a plate to dry for a few days.
>
> Store your herbal sugars for up to a year in a cool dark place, though the flavor will begin to fade over time.

Guess what else? You can make your own chocolate chips and caramel. For more details on those ideas or any of these ideas, please visit our links page at www.homestead-lady.com/homestead-links-page. There are so many people on the web sharing their make-your-own recipes and ideas. If you aren't online, find a great-grandma somewhere to keep as your baking mentor because it's possible she did a lot of this as she was growing up. You can also experiment on your own!

> "Beyond basic cooking skills, learning to replace standard pantry items with a homemade option is another way to reduce reliance on the supermarket. Items like mustard and mayonnaise and salad dressing are all very easy to make at home. Bread is another staple that is surprisingly easy to make."
>
> - Kris Bordessa, Attainable Sustainable (www.attainable-sustainable.net)

ACTION ITEMS: Get your family's favorite recipe—dessert, main dish, soup, whatever. Look at the ingredients and write them down in your homestead journal. Try to source each one back to its point of origin; ask yourself what plant or animal produces that ingredient and where you purchase it. Brainstorm ways you might make that ingredient on your homestead instead of having to outsource its production. Write down anything that pops into your head, even if you can only come up with a way to manufacture part of the ingredient yourself.

Pick one item from your list and make a goal to attempt its production this month. Write down what equipment you might need to assist you. Think about making a big batch and storing it after you get the hang of if the first few times. Try to plan out how much you'll need to make of this ingredient every month to satisfy your household demand. Can you do it? Do you want to? It would take me less time to produce the chocolate chips my family needs each month compared to the amount of butter we consume. However, butter is a simpler process. What's it like at your house?

DIY: Get an education on fats by visiting Weston Price (www.westonaprice.org) and reading at least three of their articles on healthy fats. Take copious notes and add your own thoughts and observations. Be sure to write down questions that you have. Bear in mind that, while they are a fantastic resource for all interested foodies, some of the Weston Price articles can get very technical because they are educating the nutritional and medical community, as well as regular people like you and me. You're not a dope if you don't understand every single word.

Cross reference what you read with other information that's out there as you find it. Write up what you've learned in the form of an exposition paper or editorial opinion piece. Keep it on file in case anyone ever gives you smack about eating pasture-raised bacon.

Now go do the same thing for sugar.

BUILD COMMUNITY: It's cookie time, y'all! It's time to gather girlfriends, school friends or book club members and have a healthy-ingredient, real-foods bake-off using recipes that feature as many handmade ingredients as possible. Really strive to earn bragging rights for your completely handmade cookies. Or pies, or whatever. People love to eat treats and hang out with friends, so a handmade bake-off is a great vehicle for making-your-own. Plus, it's super fun and rewarding. Be sure to let everyone vote and have a prize for the winner—maybe a copy of Alana Chernila's fine book, *Homemade Pantry* or Shaye Elliot's *The Elliot Homestead: From Scratch*.

While you're already gathered, be sure to bake an extra batch for your sick neighbor or an exhausted single parent. Cookies taste better when you're gifting them to someone else. I'm not very creative when it comes to care packages and sick baskets, but I discovered a little e-book called *Gift it From Scratch*, by Kathie Lapcevic. This little book is a cookbook containing 60 recipes designed specifically for gift giving, whether that's a hand-delivered basket or a mailed care package. The book is short but sweet as it offers real foods recipes, upcycled presentation ideas and simple wisdom. A great community building tool, especially if you happen to have extra cookies on hand.

BOOKS TO READ: *It Starts With Food*, by Dallas and Melissa Hartwig is a prose style book of nutritional information to get your brain rolling along the lines of whole foods and real ingredients. As with all these books, take away what you're ready for, don't be afraid to question and accept that sometimes, for your body, you'll disagree. Good to read *Nourishing Traditions* at the same time.

Some of these other cookbooks you'll recognize from the *Use All the Parts* section earlier in this chapter, but they bear repeating:

Homemade Pantry, by Alana Chernila, or her newest, Homemade Kitchen, can help you transition your kitchen to a homemade one slowly but surely.

The Art of Simple Food, by Alice Walker (and other titles by her), *The Elliot Homestead: From Scratch*, by Shaye Elliot and *The Nourished Kitchen*, by Jennifer McGruther are all wonderful, inspiring and educational whole foods cookbooks.

The Feel Good Cookbook, by Jonell Francis, and *Whole Foods For the Whole Family*, by Roberta Johnson are both written for real people with real families. I've tested recipes from all these books on my family and can report that they are both delicious and nourishing.

For growing your own grains, *Homegrown Whole Grains*, by Sara Pitzer. For wholesome beekeeping practices for the novice and seasoned beekeeper alike, please read *Natural Beekeeping*, by Ross Conrad. For a maple syrup harvesting education try *Maple on Tap* by Rich Finzer.

WEBSITES: For from-scratch cooking, practical advice, and simple food wisdom, try Gnowfglins (www.gnowfglins.com) and Fork in the Road (www.forkintheroad.co). If she's not taking a blogging break while writing another fabulous title on natural leaven, The Bread Geek (www.thebreadgeek) can help you out with sourdough and whole grain cooking information. Pantry Paratus (www.pantryparatus.com) can supply you with equipment and products but they also have a great from-scratch blog.

HOMESTEAD LADY SPEAKS:

Remember those marshmallows I mentioned? Want the recipe for one of my favorites? These simple, homemade marshmallows are double the delight because they're made with organic gelatin and honey. There are no fillers, dyes, corn syrup or artificial anything. Homemade marshmallows do require a good mixer, but even a hand-held electric one will do the job well. This is NOT a difficult kitchen DIY so don't be intimidated by the steps.

I never thought I liked marshmallows until I made some at home from honey. I didn't realize it was just the store-bought variety that I didn't like. Turns out homemade is better.

PUMPKIN SPICE MARSHMALLOWS

A taste of fall in gooey goodness. FYI, all homemade marshmallows basically follow the same method so feel free to nose around online for other plain or flavored recipes.

INGREDIENTS:

1 cup pumpkin puree

1/2 cup cold water

1/2 cup Organic beef gelatin

1/2 cup cold water

2 cups honey (I use raw honey, but it doesn't stay raw because of the high temps)

1/4 tsp. sea salt

1 tsp vanilla

1/2 tsp. nutmeg

1 1/2 tsp. cinnamon

1/4 tsp. cardamom

INSTRUCTIONS:

Prepare ONE 8 1/2 x 11 casserole dish one of two ways: you can grease the dish with high quality fat and then dust it (powdered sugar, coconut flour, cocoa are just a few ideas) OR you can simply place parchment sheet to fit (including up the sides) inside the dish. The parchment paper won't lie down until the marshmallow is weighing it down but once they're dry the paper will just peel off the marshmallows.

Put the gelatin and first 1/2 cup of water in the bottom of your mixing bowl—stir slightly to make sure all the gelatin is submerged in the water. This will cause the gelatin to "bloom" (or poof up a bit).

Once the gelatin has completely soaked into the water and bloomed, add the pumpkin and mix thoroughly in bowl.

Heat the honey, other 1/2 cup of water and salt in a medium saucepan on medium heat until boiling.

Gently boil honey while constantly stirring, until candy thermometer reaches 225°F/107°C. The honey should bubble, froth and turn a deeper caramel color.

Once the temperature is reached, immediately remove from heat and slowly stir honey mixture into the pumpkin mixture. Turn your mixer on low/medium (use the whisk attachment) and drizzle the honey down the inside of the bowl at a slow, steady rate.

Mix to incorporate, stopping to scrape down the sides of the bowl a few times.

Once honey and pumpkin are mixed, put the collar on your mixing bowl (if you have one) and/or place a towel over the top of the bowl to prevent splashes. Believe me, you do NOT want to scrape marshmallow goo off your fridge. Or off your toddler.

Turn the mixer to high and watch for it to magically transform from slop into marshmallow cream—this can take anywhere from 8-20 minutes.

Spoon the marshmallow cream into your prepared dish with a greased scraper. If you forget to grease it, your kids will happily suck off the sticky goo that's left. Waste not, want not.

For softer marshmallows, let them set up for 4-6 hours. For dryer marshmallows, let them set up for 8-24 hours.

Cut into squares using a greased pizza cutter for streamlined sizing OR greased cookie cutters.

Dust your marshmallows in powdered sugar, cinnamon, cocoa, coconut flour, or roll them in chocolate chips, chopped nuts, and cookie crumbs, or anything tasty that will hold still long enough. This step is optional but if you don't dust them, they will be stickier to handle. Not a problem if you're just going to devour them anyway.

NOTES:

The key to successful marshmallow making is to prepare ahead—get your pan, utensils and all your ingredients out and ready to go.

You can make marshmallows with a hand mixer, but you must stand there while it mixes and that can take a while. However, you're a tough guy who makes his own marshmallows, so you can do it.

CHAPTER 2

In the Homestead Garden

Delphiniums were what did it for me, I think. My paternal grandmother lived in a very old house with a very old garden in a very old section of Sacramento, California. When her mother built the house (with her own two hands) there was enough space in the yard for a garden. I don't know what that original garden looked like, but I could wax rhapsodic about the garden that formed under my grandmother's hands once she inherited the house. I don't remember any food being produced in that garden but, as it twisted and turned and surprised me every season, I came to call it the Secret Garden.

Plumbago tumbled over everything, Delphinium shot their showy heads out of mixed beds and an impressive Australian Tea tree flanked one of the entrances to the garden, its bright pink, paper-like blossoms creating a welcome mat. Every perennial and annual flower you can imagine poked up from one or another of the seemingly random beds that Grandma had carefully planned out and nurtured. One summer, she taught me how to lay a brick path and allowed me to help her (how clever of her). We carefully leveled the ground, laid down sand and placed each reclaimed brick. If I remember correctly she begged the free bricks from some obliging workmen at a demolition site. She hauled them home in her hatchback, her white head bobbing back and forth from car to garden, car to garden, as she unloaded them all.

Gardening is in my blood (Gardiner is my maiden name) and I've been tending small patches of earth in pots and backyards for as long as I can remember. But I think it was Grandma's Secret Garden that made me a gardener, down in my soul; much the way a creatively inclined child becomes an artist after seeing Michelangelo's David.

Some of my gardens have included raised beds almost as far as the eye could see. Other gardens I've tended in containers on pallets placed over stony, weed-infested ground that I simply didn't know how to combat at the time. I have always grown something. So, when your homesteader's heart cries out to garden, please know that I understand exactly how you feel.

Whether you're living in an apartment or on several hundred acres, a garden can be a reality for you. Indeed, a garden will be a reality for you simply because nature finds a way. For my homesteading friends who have struggled with a black thumb in the past, fear not. Even if gardening isn't your "thing," there are ways to produce fresh fruits and vegetables that suit your personality and nature. Gardening is for everyone—it is no respecter of persons. You *can* grow your own.

My biggest piece of advice is to begin right away and seize the day. My grandmother's garden is gone now—the victim of urban progress. Where there once was magic, now there is a parking lot. Don't wait for more land or better equipment or free time. Begin to garden where you are because, as the old saying goes, the best time to plant a tree is twenty years ago. The next best time is now.

Here's what we'll be discussing in this chapter on gardening for the modern homesteader:

HOMESTARTER level: Honestly, one of the easiest places to grow food for your family is in your kitchen—you're in there all the time, right? Learning to sprout seeds and/or to grow microgreens is something that every single person has space and equipment for. All you really need is a container, seeds and some water, and you're good to go. We'll talk here about the basics so that you can be enjoying fresh greens by next week.

HOMESTEADISH level: Even if you have the space for an in-ground garden, sometimes growing a garden in containers just seems less intimidating. If you're a small space homesteader, container gardening is a must. Although you can grow nearly any food-producing plant in a pot, we're going to focus here on a few types to begin with, some special considerations for container gardening and some potential problems of which to be aware. Use every garden nook and cranny by learning to grow food in pots!

HOMESTEADAHOLIC level: In pots or in ground, growing herbs is so rewarding. Plants that produce food and medicine?! Bring it on. Planning and planting your culinary and wellness herb garden is just about deciding what herbs you want to use. Then, you learn how they all grow so that you can take your plans and make them a reality. Herbs are newbie-gardener friendly, too, being relatively low maintenance. They are also a lovely addition to even the most highly manicured gardens. In this section, we'll be walking you through some initial herb garden planning, suggesting resources and giving tips along the way.

HOMESTEADED level: For some of us, who have gardened a certain way all our lives, when we think of gardening we imagine hours of back-breaking weeding and battling bugs for a few measly ears of corn. It does NOT have to be that way! Want to learn to grow smarter using the plants themselves to create self-sustaining cycles of nourishment and growth in the garden? Permaculture is all about creating those cycles on your land without dredging up dirt and yanking out weeds. There are ways to work with nature and still get the results you want on the land you already have, no matter where you are. In this section you can learn the basics of permaculture, the love language of the garden.

As with every chapter in this book, be sure to follow along at the end of each section as you reach the Action Items, DIY Projects, ways to Build Community, resource recommendations and parting advice.

*"Don't judge each day by the harvest you reap,
but by the seeds you plant."*

- Robert Lewis Stevenson, Author

For the Homestarter: Soaking, Sprouting and Microgreens

Gardening doesn't have to be intimidating, so if you've had a black thumb in the past, take a deep breath and trust me here. If you're just starting out with the idea of growing fresh foods for your family, we're going to begin our gardening ventures nice and slow with sprouting seeds and growing microgreens. There are a lot of different ways that we can enjoy fresh greens, even indoors and year-round, but these two methods are the simplest, in my opinion.

Soaking and Sprouting

I must be completely honest here. I miserably fail at sprouting. I'll elaborate as to why later, but I wanted you to know that up front. I really wish I could manage to do it successfully because it is one of the easiest and best ways to increase the nutritional value of nuts, seeds and grains. Plus, it's a wonderful way for people with limited growing space to still be able to grow fresh food. However, I'm such an airhead that sprouting is too much detail for me. Just because I'm a ninny doesn't mean you're one, though, and I can honestly recommend sprouting whole heartedly. With more practice, someday I'll be good at it, right?

SOAKING AS A PRE-REQUISITE

As you're reading along in your copy of *Nourishing Traditions* by Sally Fallon you'll come upon information explaining the benefits of soaked and sprouted grains, nuts and seeds as well as one on snack and finger foods that has tasty soaked-nut recipes. Nuts and seeds are a wonderful source of wholesome fats and vitamins, being not only delicious but nutritious when properly prepared. *Nourishing Traditions* teaches us that all kinds of seeds and nuts, including grains, contain several enzyme inhibitors that can tax our digestive systems if consumed in large amounts. Nuts and seeds become easier to digest and their nutrients more readily absorbed by our bodies if they've been soaked in water (from a few hours to overnight), rinsed, and then dried in a low oven or dehydrator. This process eliminates the bulk of those harmful enzymes. After the process is complete, nuts and seeds can be eaten raw or prepared in your favorite recipe.

If you think about it from the nut's point of view, those enzymes protect the nut in nature from sprouting prematurely. Not until the right amount of damp mixes with the right amount of warmth in spring is it safe for those enzymes to break down in nature and allow the seed to sprout and grow. Without bothering to grow sprouts or microgreens, you could stop right there and simply make sure you get into the habit of pre-soaking all nuts and seeds. You could go super-healthy and start soaking your flours, too; pick up your copy of *Nourishing Traditions* to learn more about that.

FROM SOAK TO SPROUT

If you're already soaking your grains, seeds and nuts and would like to increase the nutritional value of those soaked items, then learn how to sprout them. Here are some fun facts about the benefits of sprouting:

- One of the most important vitamins that's produced during sprouting is Vitamin C. Humans can't manufacture their own Vitamin C and so it's really important that we know where to find it in nature. Yay for sprouts that have it in spades!
- A whole host of other vitamins are created by sprouting, as well.
- Phytic acid, which inhibits the absorption of many important nutrients like calcium, is neutralized with soaking and sprouting.
- Half the battle with our food is to pack it full of nutrients. The other half of the battle is to eat food that allows our bodies to absorb those nutrients. Sprouting enables both actions: provision of nutrients and absorption of nutrients.

Sprouting merely takes soaking a few steps further. By soaking, you convince the seed it's time for germinating; sprouting gives it time to do just that. No matter what kind of grain or seed you use, the process of sprouting is the same for all; the only thing that varies is the length of time needed to sprout.

> **BASIC SPROUTING STEPS:**
>
> Cover a handful or two of seeds with water and soak them overnight.
>
> Rinse them and lay them out on some kind of tray with good air circulation and drainage and rinse them twice a day.
>
> The seeds or grains will swell and sprout until you see small, green leaves or grass.
>
> You can separate the sprouts from their seed hulls with a good rinse and some shaking, or you can just eat them, too.
>
> Store leftover sprouts in your fridge, for up to a week.

You can also perform that same sprouting process with a mason jar.

> **MASON JAR SPROUTING IN BRIEF:**
>
> After soaking the seeds and nuts overnight, rinse and drain as before.
>
> Place the rinse seeds and nuts in a mason jar, about half full.
>
> Rinse and drain twice a day and watch for sprouting.

There are commercial sprouting trays and cups you can buy, and I've tried a few of them with varying degrees of success. I do better with stackable trays that have drainage holes cut in the bottom.

There are a few things that should not be sprouted, so be sure to do your homework. There are A LOT of blog posts, YouTube videos and books on this topic that can teach you the nuances of success with sprouting. See the suggested resources at the end of this section for more information.

SPROUTING PERKS:

1. Sprouts are sproingy. Even though lettuce is crunchy, sometimes my youngest children just won't eat it. Sprouts have a bouncy, crunchy texture that can only be described as sproingy. The kids dig it.
2. Sprouting is relatively quick and painless compared to laying a fall garden outside so that I can harvest winter greens while the snows fall. In my climate, I must cover my fall/winter garden with horticultural fabric to protect it from the cold – no small task! Plus, and this is a big one, you don't have to weed your sprout tray. Not true in the garden!

3. You can't get more local than your kitchen counter; bypass the store entirely and grow your own sprouts.
4. Sprouts enable you to have a lot of variety in your greens without taking up too much space.

As cool as this method is, there are a few drawbacks. One such issue is that you must remember to rinse your sprouts *twice a day*, no forgetting. If you struggle with details, as I do, then this part of the process will really be a chore for you. This is, without exception, where I fail with sprouts. I inevitably neglect one of the rinses and the sprouts either dry out or start to mold. There's an automatic sprouting tray you can purchase that does all that for you, and it looks cool, but they only sell it to people with money. So, for now, either I remember to rinse or my sprouts rot.

Which brings us to the other typical issue with sprouts and that's bacteria. Sprouts are susceptible to salmonella and other forms of food poisoning. Some sprouting literature will suggest you rinse your sprouts with diluted amounts of bleach. Others might suggest you steam or cook them fully before you eat them. For most foodies, using bleach isn't going to be something we want to do. For raw foodies, the beauty of sprouts is in the raw nutrition they provide, so cooking isn't really appealing either. (Please note that soaking doesn't present the same amount of risk if you soak for the normal amounts of time; usually between 4 and 24 hours.)

Don't freak out, though! Well-maintained sprouts will be healthy and tasty; just keep rinsing and refrigerate the sprouts once they're fully sprouted. OR...if you want all the added nutrition of sprouting but are struggling with your sprouts, or just aren't into vomiting or bleaching your food, may I suggest...

Microgreens

Microgreens are seeds taken fully to the next phase of growth, the true leaf stage, and are grown in a tray of soil. Similar in appearance to sprouts, they are only slightly different overall.

PERKS OF MICROGREENS:

1. Microgreens offer a bit more nutrition than sprouts because they grow in soil and can absorb minerals from it.
2. Microgreens also have time to photosynthesize which produces even more nutrients.
3. Microgreens reduce your risk of food poisoning compared to sprouting. Why? Because microgreens are grown in a tray of soil, are kept only damp and use sunshine to grow.

4. If you've ever grown anything from seed, you can grow microgreens. If you've ever tucked a child into bed after a bath, you can grow microgreens. If you can tie your shoes, you can grow microgreens.

TO GROW MICROGREENS:

Fill a tray that has drainage holes with an inch or two of damp soil. You don't have to have drainage holes if you're worried about mess; just be very sure not to over water.

Pick a seed—say sunflower seeds.

Spread your sunflower seeds on top of the soil in your tray; you don't want them on top of each other, but you don't need to sit and space them individually like you would if you were planting them to use in your garden later. Sprinkle them on like salt and pepper.

Then cover them with the appropriate amount of dirt according to the planting depth measurement on the seed packet. For sunflowers, that's about 1 1/2 to 2 inches.

Keep the soil damp by misting it a few times a day (do NOT let the soil dry out) and provide either a sunny window or a grow light as the seedlings begin to emerge.

For sunflowers, cut the microgreens to the dirt line once one set of leaves has appeared on each sprout. Sometimes the second set of sunflower leaves can be bitter.

For most other microgreens, once the seedlings have two sets of leaves (on most seedlings that's one cotyledon leaf set and one true set) they have earned the title of microgreen and you can harvest them similarly with scissors and eat them up.

Start a new tray of microgreens.

To help prevent damping off and other fungal issues in seedlings, sprinkle a generous dusting of cinnamon over your seed tray just after planting. Be sure to keep the soil damp but not wet, and that will go a long way to maintaining healthy and steady growth.

Also, growing microgreens is a great way to use up older seeds you might have in storage; since you're sowing them so close together, who's to care if not every single one germinates?

Microgreens must be regrown from new seed every time, but they can be stored in the fridge for a few days once cut, just like sprouts. For whatever reason, microgreens are just easier for me. I guess I'm better about remembering a seed tray on my grow shelf (which is set up all year round) than I am about remembering trays to be rinsed sitting on my counter. There's so much brewing, fermenting and soaking on my counter that it's no real wonder to me that sprouts just get lost in the shuffle. I encourage you to begin soaking grains, as well as mung or lentil beans immediately. Then start sprouting or growing microgreens right after that, whichever seems easier to you.

The beauty of both these methods is that they can be done any time of year. If you live in a place with cold winters, you might really appreciate something fresh and green come January. Microgreens cost a pretty penny at my health food store and I just laugh when I walk by them, knowing how easy they are to do at home.

Sprouting is a great emergency preparedness skill to have. Chinese boatmen used to carry nuts and seeds with them for sprouting on long voyages so that they could be assured of essential vitamins and minerals as they traveled. Several hundred years later, even Americans were looking at sprouting during World War II to use rationed foods more effectively. In our times, it seems prudent to include sprouting grains along with water in our food storage program. For more information on healthy foods you can store against a time of need, please see the *Health Wise Food Storage* section in "The Prepared Homestead" chapter.

SPROUT YOUR OWN FLOUR

It's perfectly possible to make your own sprouted flour by soaking and sprouting your grain kernels before you grind them. Once they're sprouted even slightly you can dry the kernels in your dehydrator or oven set on its lowest heat. Once dry, you can grind the wheat in your grinder as you would any other grain. If you've ever paid the high price for sprouted flours in your local health food store, this simple, homemade process sounds pretty sweet.

I'll be honest again—I tried that. Once I had sprouted grains, I dried them, filling the nine racks of my dehydrator and spilling who knows how many precious kernels of wheat as I processed them. Then, I diligently sent those grains through my grinder. But because of their sprouted tails it took about three times as long as it would normally to grind my flour because I had to manually push them through. I just don't have that kind of time! Plus, I only got half as much flour for my efforts.

Bottom line, for me, grinding my own sprouted grain is not something I'm going to do until I have a lot fewer children at home and a lot fewer things to do every day. Until then, I'll bite the bullet and buy sprouted flour when I need it, gratefully paying someone else to go to the trouble of grinding sprouted wheat. Or, I just pre-soak or naturally leaven my baked recipes, negating the need for sprouted flour.

 ACTIONS ITEMS: Start soaking all grains, seeds and legumes now before you consume them. This is just an easy, no-brainer way to increase nutrition. Then, pick one of these methods (sprouting or growing microgreens) to try this month. Even though I stink at it, I suggest you start with sprouting because it doesn't require any special equipment outside of a mason jar and some screen. See below.

DIY: Before you buy any special sprouting equipment, make a simple sprouter with a canning jar, a rubber band and a piece of screen or fabric. Soak, drain and sprout all using the canning jar. This is a very simple method to use and it will help you learn if sprouting is for you. You can also figure out how to use sprouting trays if that turns out to be easier.

The first sprouting dish I used was a black, seed starting tray with a lid. Those trays have slits already in them to allow water to escape, so I put a solid one underneath. I experimented with larger seeds first like sunflowers and peas, so they didn't fall through the cracks. Remember, sprouts don't need soil. So, I simply turned out my seed onto the trays, covering them with the clear, plastic lid. When I decided I liked sprouting in trays better than jars, I bought a reasonably priced set with good ventilation and lids. I still use the black trays to sprout seeds for my livestock.

BUILD COMMUNITY: If you know of a family that's struggling financially offer to come over and teach them how to sprout to get the most bang for their strangled buck. Sprouting increases the nutritional content of every seed, bean and grain; nutrient dense food is more filling. For more information on that, please read *Nourishing Traditions*, by Sally Fallon. You may not be able to afford to pay your friend's mortgage or even buy them groceries, but you can share what you know to contribute to their well-being. Every little bit of fresh, green food helps when you're down to beans and rice. Believe me, I know.

BOOKS TO READ: *Indoor Kitchen Gardening*, by Elizabeth Millard covers sprouts, microgreens, wheat grass and even growing herbs all on your kitchen counter. *The Complete Guide to Growing and Using Sprouts*, by Richard Helweg is simple and straightforward, making it easy to understand the process of sprouting.

WEBSITES: Sprout People (www.sproutpeople.org) is a great site for all things sprouting and should have the answer to just about any sprouting question you have. They also have information on growing microgreens. They sell kits and seeds.

If you have chickens (or any livestock since the principles still apply), check out The Frugal Chicken's Feeding Your Hens Right e-course (www.feedingyourhensright.com) where she explains both fermenting and sprouting, and a lot of other important tips, for your feathered friends.

HOMESTEAD LADY SPEAKS:
Sprouting is a talent that will stand you in good stead once you obtain livestock, if that's one of your homestead goals. I remember the first time I toured a local grass-fed meat farm and saw their impressive sprouting set up. A good portion of the year, Shayne and Kristen Bowler, of Utah Natural Meat (www.utahnaturalmeat.com), pasture many of their animals in verdant mountain fields. Once they come down to the Bowler's urban farm, they are still fed fresh greens in the form of sprouted barley. The Bowler's system allows the monocot grain to fully sprout its single stem so that the sprouted mats produced look like rectangles of grass.

Each animal is fed according to its dietary needs, but there is no other grain supplementation going on with these healthy, thriving herds and flocks.

You and I probably can't afford and don't need the very large system that the Bowler's require to bring their customers the high-quality meats on which rely. However, we can put together a sprouting area on a much smaller scale to provide the benefits of soaked, fermented and/or sprouted grains to the livestock we tend. For more information on the Bowler's system and for ideas on what you can do at home, please visit our links page under The Homestead Garden: www.homesteadlady.com/homestead-links-page. If you're on Pinterest, you can also use our links page to find the board we created to gather various posts on fermenting, sprouting and fodder for livestock (as well as growing our own grain). From chickens to cows and everything in between, your animals will reward your efforts with improved health and full bellies. They are what they eat, after all.

"To plant a garden is to believe in tomorrow."

- Audrey Hepburn, Actress and Humanitarian

For the Homesteadish: Food in Pots

A homesteader's life revolves around food, am I right? We're usually to be found planting seed, growing veggies, pruning fruit trees, tending livestock that produces or becomes food, bringing in the harvest, preserving the harvest, sharing the harvest, cooking with the harvest and, in general, raising our own food. People with a homesteader's heart are often to be found reading nourishing cookbooks. We put more research into how to build up soil fertility than we did our thesis or our baby's name. How to turn a pig into bacon is the major theme running through the several books we have stashed around the house. We homesteaders take food very seriously.

Have you ever selected a plant for your garden solely because it's the major ingredient of a recipe you want to try, and your local farm just doesn't grow it? Do you ask for things like pressure cookers and heirlooms seeds for your birthday? Of course, you do, you're a trend-setting, hipster homesteader. Food is cool. Growing your own food, even cooler. As author Angela England once quipped,

"You might be a homesteader if you ate kale before kale was cool."

You don't need to have access to acres of land to get started on your vegetable garden. Some of the most amazing, food producing gardens I've seen are staged in pots, and some of the most talented gardeners I've met are container gardeners. If you're just getting started with vegetable gardening, growing in pots can be a simple way to experiment with what you'd like to grow and to learn what will grow with ease in your climate. If gardening doesn't happen to be your favorite part of homesteading, growing in containers can give you the flexibility you need to suit your garden to your needs.

The following suggestions are divided into two sections:

1. *Annuals* that need replanting every year
2. *Perennials* that continue to grow year after year.

There are some plants we commonly grow as annuals, even though they are technically perennials. Tomatoes are a good example of that. More on that later.

HANDY TIPS FOR GETTING STARTED

Now, I'm not going to say the vegetables suggested here, either annual or perennial, are fool-proof in a container garden. A fool can do a great deal of damage to any plant. However, after three decades of gardening in pots and in the ground, I can testify that these veggies can hold their own in most instances and usually perform in a way that pleases. Be sure to read up on the topic of container gardening using our suggested resources and your favorite books. You'll want to know as much as possible about these veggies so that you enjoy success. These plants can all be planted in the ground if you have the space. However, if you're interested in growing food producing plants in *containers* there are a few special things you need to know a good deal about, like fertilizing and watering (see the *Food for the Garden* sidebar further on).

While we're on the topic of learning useful things, your *growing zone* (for American gardeners) refers to the USDA's illustrated map of the different climates for each area of the country. Find that map online or in quality gardening books. Canada and Great Britain have their own, which can likewise be found online or in regional books. For planning and planting purposes, it is necessary to know your zone when reading gardening material or looking at seed catalogs. Plant only what is zoned to grow in your area, especially when you're just beginning, and you will save yourself a lot of gardening headaches. As you mature and become more practiced as a gardener, it will be even more helpful to know your *first and last frost dates*, your *lowest winter temperature* and your *highest summer temperature*. Don't worry much about that right now, just keep it in the back of your mind and know your zone.

Just a note, I ALWAYS suggest planting more than one of anything you're trying to grow because as you're learning sometimes plants just up and die on you. (To be honest, that happens even after you've been gardening a long time.) Also, if the family is involved in your gardening ventures, you'll see quickly that very young children can't differentiate between a weed and a cucumber seedling, and accidents happen. Children want to help you and please you but sometimes they get a little too exuberant. I once lost an entire bed of spring carrots to a group of excited kiddos who realized that if you pulled, out came a carrot! A carrot that was too young to harvest but a carrot nonetheless! Life happens and I'd rather my children were excited gardeners before being exact gardeners.

Annual Veggies for Pots and Containers

Here are just a few annuals to consider for your potted kitchen garden. Remember that this list is not exhaustive, so don't be afraid to experiment with any vegetable you'd like to grow in a container.

1) Tomatoes (*Solanum lycopersicum*) - Everyone's favorite veggie! Tomatoes are perennials in their native habitat which is why their root systems are so strong. This is also why, at the end of the season, you give yourself a hernia pulling them out of the dirt. A big root system also results in them being harder to kill than, say, a cucumber. Tomatoes are also incredibly satisfying to grow—there are hundreds of different shapes, sizes and colors to choose from.

Some of our favorite tomato varieties are:
1. Sun Gold cherry (hybrid) is simply one of the best things you'll ever put in your mouth.
2. Green Zebra (heirloom) or any of the striped tomatoes are a blast to grow because they're gorgeous.
3. Black Prince (heirloom) or any black tomato because the color is vibrant, and the taste is rich.
4. Mortgage Lifter (heirloom) because it has a great story behind it and because it produces tasty tomatoes very reliably.

I don't grow many hybrids because we save seed but there are lots of great varieties to try, and I'm sure you have your favorites. If you have a heated greenhouse or live where winter temperatures are warm, you can grow the same tomato plant indefinitely. However, most of us grow tomatoes as annuals due to their high sensitivity to frost.

SHOULD YOU START YOUR TOMATOES FROM SEED?

If you've never grown tomatoes before, do NOT start your tomato plants from seed this year. Seed starting is part art, part science. It is a skill you WILL master, but not yet. Instead, wait for your local nurseries to put out their stock of tomato seedlings in early summer and pick the variety that looks delicious to you. You can also order quality plants online.

FYI, a seedling is a baby plant. To be worthy of planting a seedling must have two to three sets of true leaves (if we're talking tomatoes), a strong stem that is relatively straight and a lively green color. Be sure to check the underside of the leaf, too; if it's purple or yellow, pick a different plant. Tomato plants begin to appear in most nurseries around late spring/early summer and should be planted when all danger of frost has passed.

If you are a seasoned gardener, but you have yet to try growing plants from seed, maybe this is your year to start a few seeds. Try some of those tomato varieties that you've seen in the catalog but can never find in your local nurseries. Growing your own vegetable plants from seeds gives you access to a huge variety that you've probably never been able to grow before. It will also save you money, if you're looking to plant a very large garden. Nursery plants, even young ones, can quickly add up in cost, but a packet of seeds is only a few dollars.

Pot Size: Determinate varieties (plus support like a cage or bamboo stick) will fit into an 18-inch pot; an indeterminate tomato (plus its support) will fit into a 24-inch pot.
Sun: 6-10 hours or full sun.
Special Notes: If you want to harvest tomatoes from the garden as soon as possible, an early variety that produces well for me is Early Girl. It's not the tastiest tomato around but at that time of year, I'm so starved for fresh tomatoes, that anything tastes fabulous. (I usually get tomatoes by mid-July in zone 5 with that one). Two favorite mid- to late-season ripeners are Cherokee Purple (heirloom) and Jet Star (hybrid). I've grown literally dozens of tomato varieties and have found something to love about each one, so don't be afraid to experiment.

A LITTLE LINGO:

Determinate varieties of tomatoes are varieties that are bred to grow to bush-sized. They stop growing when fruit sets on the terminal or top bud, ripen all their crop at or near the same time and then die. These are great to grow if you preserve your harvest, especially by canning, and would like to have reliable, large crops.

Indeterminate varieties of tomatoes are "vining" tomatoes that will produce fruit until killed by frost. They can reach up to 10 feet although 6 feet is more normal. They will bloom, set new fruit and ripen simultaneously throughout the growing season. These are great for consistent fresh eating all season.

2) Green Beans *(Phaseolus vulgaris)* – I pretty much love anything Elizabeth Enright wrote but I laughed out loud when we stumbled on the following quote during our family reading of *Thimble Summer*.

"'Green Beans never know when to stop!' said Garnet's mother in annoyance."

Green beans are reliable to a fault. They're trustworthy—you plant the seed, it grows, it produces fruit. A lot of it. And sometimes you wish they would just quit it! Green beans can be used in so many ways and they store beautifully.

There are lots of other things you can do with them if you don't prefer them raw or steamed. They're about the easiest thing on God's green earth to can. They also freeze and dehydrate with ease. We made up a batch of dilly beans when we did our pickles last year and they were, in a word, fabulous. My five-year-old gulps down those dilly beans so fast most of us risk losing a hand if we get in her way.

SHOULD YOU START YOUR BEAN PLANTS FROM SEED?

Yes! Beans (and any other legumes like peas) don't transplant well. Beans and peas have very delicate stems when they're at the seedling stage and, if you bend them wrong, you could end up killing your plant. For that reason, you really won't find them in the nurseries come spring. No worries—beans grow so well from seed that you can direct sow them into the pot.

Green beans can be planted when all danger of frost has passed in the early spring; they can also be planted late summer for a fall harvest before the frost comes back. When planting green beans, first decide what kind of space you have to grow them in because there are basically two methods of growing them. (These are estimations only; every plant does its own thing.)

1. You can grow a BUSH bean variety that gets to be about 2-foot-tall and 1 foot wide.
2. Or, you can grow a CLIMBING variety that will get upwards of five feet tall but only about 6 inches wide.

A LITTLE LINGO:

Direct sow refers to planting a seed directly into your prepared garden soil at the appropriate time. The other option is to plant the seed indoors in a grow tray to put into the garden as a seedling.

If you have a sunny deck that isn't very spacious, you could choose to grow green beans on a trellis or up a wall or fence. Climbing beans are beautiful, vertical producers and from just a handful of plants you could be enjoying fresh green beans all summer long. On the other hand, if you don't want to mess with a trellis, bush beans are the squatty and prolific sisters to those climbers.

I usually grow bush beans because I have small children and small children like to climb things. I've had a few trellises plummet to the ground while my back was turned. It doesn't take long for children to get to an age where they understand the rules of the garden and our family should be ready to use trellises again soon. Trellising is an extra step but it's an efficient use of space and quite often you can grow shade tolerant veggies like lettuce underneath your trellised beans.

Pot Size: 8"-12," as always, with good drainage; you can use small gravel in the bottom to improve drainage in any pot. Bush varieties will need more container space than pole.
Sun: 6-8 hour or full sun.

Special Notes: Green bean pods will start to bulge as they reach maturity. Snap or pinch them off the plant with care so as not to cause damage to the mother plant. Harvest them often to produce and mature as many beans as possible throughout the growing season.

3) Radishes *(Raphanus sativus)* - I don't like radishes and I can't make myself like them no matter how many varieties I try. So why do we grow them? Because they grow so easily and quickly, neighbors and livestock love them, and because they can be direct sown into the garden or pot. And because they're absolutely gorgeous. So many colors, kinds and shapes.

The French breakfast radishes have been the most palatable at our house. However, taste is such a relative thing that I suggest you try whatever variety sounds good to you and your family. Radishes are great crops for inter-sowing. Inter-sowing is the practice of planting quickly maturing plants alongside slower producing plants so that by the time the slower paced veggies are ready for more space, the quicker ones have already been harvested. When you're growing in pots, you want to be sure to use all the available space as best you can.

Because radishes are so small, they are particularly well suited to container growing if there's space to develop the bulbous radish beneath the surface of the soil. Radishes don't much care for high temperatures so plant them in the spring and fall, missing the intense heat of summer unless you're growing them in the shade of another plant. Make sure your containers don't over-heat or dry out because the radishes won't thank you for that.

If you have crops that you discover you really don't care for, feed them to the goats and chickens. Or give them as gifts to neighbors. Or try a new dish (like roasted radishes) to see if you like to eat them that way.

Pot Size: Really anything will work if it's deep enough to accommodate the length of the variety you're growing. Radishes can be inter- and under-planted with many veggies. You can plant them alongside slower growing veggies like beets and they'll be ready to harvest before the beets get big enough to be annoyed by the closeness of the radishes.

Sun: 6-8 hours but can take partial shade.

Special Notes: Radishes are cool season vegetables that produce smaller, sweeter globes in the spring and larger, stronger flavored globes in the fall. There are early and late season varieties so feel free to try several. Since radishes develop so quickly, it's important to harvest them right when they're ready to avoid them becoming pithy and losing flavor.

4) Pumpkins *(Curcurbita pepo)* - Oh, please grow a pumpkin or two! Pumpkins are simply magical vegetables to have in the garden. They do take up space, although growing the smaller varieties up a trellis is very doable, if space is an issue in your container garden. (I think space is an issue in every garden, quite frankly—how much of our garden brain is tied up trying to figure out what's best to do with the space?!) What you trade on space, you make up for in the usefulness of the plant.

Pumpkin vines are an impressive sight. The flowers are gorgeous (and edible!), and the fruit holds the promise of autumn recipes to come. I thrill at the sight of a pumpkin "patch" on a porch, whether it's the first flowers of spring, full of fat bees taking pollen baths, or the glow of orange in the yard as the vines die back and the pumpkins shine as if already carved into jack-o-lanterns.

I usually start my pumpkin plants indoors, but you can direct sow them into your containers, too. There are so many different varieties of pumpkin to grow—teeny, tiny ones and giant ones; culinary varieties and decorative varieties. And all different colors, too—orange, red, blue, white, green! My favorite pie pumpkins are usually any of the Cinderella's-coach shaped varieties like the Long Island Cheese pumpkin. Also tasty are the pink banana squashes but they can get large, so keep that in mind if your vertical growing space is tight.

Whatever you do, give each pumpkin plant a large enough pot. You may not decide to grow them every year but do it at least once. If you have the space, plant several so that your children's friends can come for a pumpkin harvest and carving party at your house.

Pot Size: At least ten gallons as their roots are quite vigorous. Be sure that the pot is broad and deep. The pot must also provide good drainage because pumpkins don't like wet feet.

Sun: 6-8 hours or full sun.

Special Notes: When you fertilize, be careful to not give pumpkins (and several other veggies like tomatoes) too much nitrogen or all they'll grow is leaves. Also, be sure you have enough pollinators about to fertilize your flowers. If you aren't sure, learning to hand pollinate pumpkins is not hard at all and it can be an interesting project to do with kids. It might be worth some extra credit at school!

5) Peas *(Pisum sativum)* - Traditionally, spring peas are thought to be the best, but I prefer planting peas in the fall garden. In my climate, summer can start with a vengeance and often comes while my pea crop is delicate. Summer heat and dry winds can wreak havoc on the pea harvest and I usually end up disappointed in the amount of peas I'm able to grow in spring. Don't get me wrong, I plant peas every spring but it's mostly because the kids just love them and can't garden without them.

Peas are super kid-friendly plants, right down to the seeds which are a medium size and can be seen and handled by children with relative ease. Hunting for pea pods becomes a morning ritual once the fruit is set. Many a time I've had to repeatedly call my kids in from the pea patch for the start of school because they're so engrossed in their pea treasure hunt.

Again, lots of varieties to choose from but the big choice is:
- Do you want climbing vines that need to be staked?
- Or bush varieties that hit 2-3 feet and then stop growing?

All peas can be eaten when they're quite young, just pods with itty bitty peas inside. Some varieties are meant to be harvested when the peas are a good size, shucked out of their shells and steamed to perfection with lots of butter. Peas, and even the tender growing tips of the curling pea vine, are pleasing to the young and old alike; babies on solids can gum them and cool teenagers can eat them with ease.

Pot Size: Almost anything will work as long as it has drainage holes and is at least 12" wide. I've planted peas in an old colander (a child's garden project) and they produced well with a trellis. The bigger the container, the more plants you can grow and the bigger the harvest.

Sun: 6-8 hours or full sun.

Special Notes: They do like even dampness and this may require daily, repeated watering during the potted growing season so be sure to stay on top of your fertilizer. Also, peas dislike heat so much that they'll stop producing once summer weather sets in, whereby you know it's time to pull them up, compost them and plant something new in their pot. Be sure to plant peas again around August so that you can enjoy them in the fall garden.

Another bonus of pea plants (or any legume) is that they can pull nitrogen out of the air and work with bacteria to fix it in nodules on their roots in the soil. Nitrogen is such an asset in the garden. Perennial legumes like clover and alfalfa fix much more nitrogen than snow peas, but all legumes are useful this way. To learn more, go to our links page under The Homestead Garden: www.homesteadlady.com/homestead-links-page.

6) Lettuce *(Lactuca Sativa)* - Super easy to grow and reseeds readily in many climates, loose-leaf lettuce is a winner in any container garden. I don't mess with heading lettuce like Iceberg in the garden because loose-leaf varieties are so much tastier and easier to grow. Reseeding means that you allow a certain number of your lettuce plants to go to seed and then you let the seed go where it will on purpose. You can also allow your lettuce to set seed and then clip it off to save and plant deliberately, if you don't want lettuce popping up everywhere. Don't worry about all that if you feel overwhelmed by the idea of saving seed this year. You can certainly direct sow lettuce or you can start it indoors.

Lettuce is beautiful, semi-frost hardy and tasty. I've gotten my kids to a point where I have them pull up a head, rinse it off, plunk it down on our table and rip off some leaves to augment their dinner. Some use dressing, some don't. We grow loose leaf varieties like Freckles, Buttercrunch and Oakleaf. Plan to continuous sow lettuce so that you always have new batches coming up behind as you harvest older plants. Eat all the lettuce you can before it flowers because once that happens it turns bitter. Adding a bit of shade will stave off the flowering process, but once you hit the warmer weather of summer it's time to let the lettuce give way for other crops. Never fear, lettuce does well in the fall garden, too.

Pot Size: Three to four loose leaf lettuces will fit nicely in a 14" pot but remember that lettuce can also be interplanted with other veggies in larger pots. If shallow-root lettuce is planted alone, find a container that is wider than it is tall.

Sun: 4-6 hours, but it can take partial shade.

Special Notes: Lettuce is over 90% water and so it's imperative that you keep the soil in your pot evenly moist, but not soggy; a clay pot is often very effective at regulating moisture. Lettuce is a great plant to grow indoors, on a sunny counter.

FOOD FOR THE GARDEN

When I was a beginner gardener, I must confess that I found fertilizers intimidating. Back then, all I knew to use were commercially produced fertilizers that each claimed some miracle result for a specific group of plants—roses, vegetables, perennials, etc. To feed the variety of plants I wanted to grow would have cost a fortune, so I bought what I could and then forced my plants to do without. I really didn't know what else to do for them. I wasn't made of money!

Happily, I continued to read and attend classes. Eventually, I learned all about the wonderful world of compost, manure and mulch. If you think about how plants thrive in their native habitats, it is never accomplished with commercial fertilizer. The habitat takes care of the food and water for each plant. When we garden, we are usually taking plants (like tomatoes) out of their native habitat and plunking them down into our gardens. We require that they produce food for us whether or not the environment is one in which they'd naturally grow. We become responsible for providing their food and water in place of their native ground. I discovered that it was far easier, cheaper, and more beneficial to my garden's natural systems to feed my plants with steady doses of composted manure and composted kitchen and garden waste. Green manures, like barley and oat grass, also proved themselves very useful. Green manures are crops grown specifically to a certain height and then tilled into the soil to feed it. To conserve and control water with any crop, a healthy layer of mulch (like shredded bark) does the trick wonderfully.

I am not going to outline the feeding requirements of each plant, as there are other fine publications that do that splendidly (see our resources at the end of this section). I will say, though, that natural fertilizers and mulching are so beneficial to container gardeners. Growing a garden in containers requires a greater attention to detail when it comes to feeding and watering than in-ground gardens. The reason for this is because a plant's roots are limited to foraging food in the pot in which they are growing. Vegetables and fruits are heavy feeders because it takes a lot of energy to produce vegetables and fruits!

If a container gardener must rely solely upon commercial fertilizers to feed their potted gardens, they may very soon run out of cash to invest in the project. Besides savings, a bonus of using these more natural methods is that you can produce them yourself. You can compost anywhere you live with worms (see the *Vermicomposting* section in the Livestock Wherever You Are chapter) and even produce your own mulch if you have access to the wood chips and leaves. If not, both compost and mulch may be available from your city's green waste program at discount prices.

The point is, don't wait to start your container garden because you can't quite keep track of all the touted functions of commercial fertilizers. Or because you're worried about being able to afford all the ones you think you'll need. Certainly, don't do what I did and starve your garden because you can't provide them all. Add quality compost during the growing season and a few inches of mulch to each pot. Then observe if your crops might need more.

Keep reading those gardening books and look for the term N-P-K to learn the function of each element in your plant's food. Particularly helpful for container gardeners is something called compost tea. To watch a super simple video on how to make compost tea, please visit our links page under The Homestead Garden and make some soon: www.homesteadlady.com/homestead-links-page. Don't get paralyzed by the science, you and your garden will reach an accord with each other over time.

Incidentally, N-P-K refers to Nitrogen, Phosphorous and Potassium, which are all elements that are important for soil conditioning, although they're not the only ones. You probably already knew that because you're a garden smarty-pants and you know more than you think you do. You've totally got this!

Perennial Veggies for Pots and Containers

The beauty of a perennial vegetable pot is that you don't have to plant it year after year. If the plant or its root system gets too large for the container, then you'll need to transplant it into a larger one or divide the roots, creating new plants. However, apart from careful feeding, these perennial veggies will just keep going and going.

If you live in an area with particularly cold winter temperatures (zone 5 and below), be prepared to pay special attention to the roots of your perennial veggies. To avoid freezing the roots, bulbs or tubers you can relocate your potted perennials to a basement or cold storage room as the frosts of winter come on. If you don't have the space for that, wrap your potted perennials in some kind of insulating material, grouping them together for warmth. Be sure to remove the material in spring as the weather warms to avoid cooking your roots. Perennial veggies take some maintenance in the container garden, but they'll reward you with rich flavors and continuous harvests.

1) Sorrel *(Rumex acetosa is common sorrel; Rumex scutatus is French sorrel)* - Related to the tasty, but very-difficult-to-grow-in-a-pot, rhubarb, sorrel is a tart, lemon-flavored herb whose leaves are easy to enjoy in soups, stews, salads and sauces. High in Vitamin C, the leaves are a great cool season crop but, like lettuce, the leaves turn bitter once the weather turns warm. A cut and come again plant that can be continually harvested until the season ends, sorrel, like most herbs, requires little extra care or fuss. Remove flowers unless you want to save the seed as they'll cause the flavor of the leaves to turn bitter. Plan to divide the plants every two to three years, which is standard for potted, perennial plants.

Incidentally, most herbs make fantastic potted plants so while you're mastering tomatoes, be sure to toss in a basil plant or two to make your gardening life complete.

Pot Size: Grows just fine in a 6" pot but can use larger.

Sun: 6-8 hours, but it doesn't care for intense heat.

Special Notes: If birds are bothering your seedlings, cover the pots with lengths of hardware cloth or bird netting until they mature. Once cut, sorrel leaves can wilt rapidly, like loose leaf lettuce, so be sure to use them quickly after harvest. Store any leftovers in a cool place like the fridge. You can dehydrate and powder sorrel leaves for some extra greens in winter smoothies.

2) Scarlet Runner Beans *(Phaseolus coccineus)* – Technically a tender perennial, there may be some zones where the runner bean dies back to the soil surface or dies back completely in winter to simply reseed itself come spring (if you let seed drop in the container). These beans are beautiful! The seeds are usually purple or pinkish with black spots and the flowers produced are typically bright red or similarly lovely colors. There are several varieties of runner beans and they all vary a bit in flower and seed color.

Like other beans, runner beans require even watering, pollinators in abundance, even amounts of composted fertilizer and a lot of picking to keep the plants producing. Beans are a good candidate for a self-watering pot because of their constant need for water and the fact that you have more to do in your day than stand around watering your garden. (See our DIY tip at the end of this section for more information on creating your own self-watering pots.) All that water makes them vigorous. As the runners reach the top of your trellis structure, pinch back the tips of the vines to encourage the plant to stop climbing and start sending out more side shoots and, consequently, making more beans.

Pot size: Try 18-20-gallon plastic tubs, with rope handles for easier dragging when the season is over. Be sure to drill drainage holes in the bottom of any container you use, if they're not already there. You will want to make sure that whatever trellis you use will also fit in the pot.

Sun: 6-8 hours or full sun.

Special Notes: Runner bean vines are strong and vigorous, covering a small structure in a season, so they're perfect for growing up playhouse walls, tepee trellises, or a privacy screen in a balcony garden. Plus, the beans are completely edible picked fresh and really yummy dried like any other soup bean. Refried runner beans, anyone?

3) Jerusalem Artichoke *(Helianthus tuberosus)* – Planted in the cool weather of winter or early spring, these are a tuberous rhizome crop that look like an iris tuber/ginger root/potato cross. The tops produce sunflower-like stalks whose flowers die back every winter, popping right back up in late spring. Be aware that Jerusalem artichoke tubers proliferate easily and abundantly. This can be a great thing for an edible garden because you can easily produce a large crop. The tubers that you harvest each fall can be knobby, which gets annoying when you're trying to clean them and prepare them for eating. Try the Fuseau variety for fewer knobs.

Jerusalem artichokes taste like a cross between a potato and an artichoke when cooked; which is probably how they got their name since they're in no way related to artichokes. They taste a lot like a water chestnut when raw. Without the starch content of a potato, these veggies are often used by diabetics to re-create their favorite potato dishes. You can peel them for boiling and mashing or leave their skins on for sautéing. Jerusalem artichokes can cause stomach upset in some people (they are not GAPS diet approved, as an FYI) so be sure to go easy with them in the first recipe you try, just to observe how your gut responds.

The flowering stalks, which can reach up to six feet and beyond, are quite useful in the container garden providing both forage for pollinators and a wonderful privacy screen for any patio, low-lying fort or fairy house. Also, once the stalks have died back (at which point you know the tubers are ready to harvest), they make excellent play swords for backyard imaginings and great kindling for the fire. Also, the biomass this plant produces makes it a useful green mulch source. Chop up stalks and leaves of Jerusalem artichokes and place them around your fruit trees as you would wood chips. They are a quality forage for livestock, too. However, you use them, you'll need to cut these strong stalks from the tubers before you can harvest.

You won't get quite as many tubers if you let your Jerusalem artichokes flower but, trust me, they'll still produce very, very well. There are so many that it's almost impossible to find and gather every single tuber in the fall. This means that you end up leaving behind seed tubers for next year's harvest. For these reasons, it can be a wise garden decision to confine their exuberance to pots, even if you have space for Jerusalem artichokes in your in-ground garden.

The best place to store tubers is in the dirt because they don't last nearly as long as a potato once harvested; two months in cold storage is all I've ever managed before the tubers begin to go soft. Bring your pot into the garage to prevent the soil from freezing and dig out tubers as you need them during the winter, keeping the soil damp but not wet. You can dig up all your tubers to inspect them for rot or bugs and simply put them back into the soil for storage.

Pot Size: Start with one five-gallon pot per planted tuber and it will produce many more during the season, like a potato. Unlike a potato, though, you do not need to hill the plant as it grows.

Sun: 6-8 hours or full sun.

Special Notes: Jerusalem artichoke tubers look a bit like a ginger root crossed with a potato. Like a potato, they have eyes (or buds of growth), but like the ginger root, there are quite often multiple branches of tuber all connected. Sometimes you'll get individual tubers that look like single, small potatoes. To prepare a tuber for planting, break apart the branching arms and each one becomes a seed tuber to plant into a single, large pot. You can also divide particularly large, single tubers by cutting them in half, leaving several eyes on each side—just like you would a potato.

4) Garlic *(Allium Sativum)* - We grow this vegetable as an annual because we dig it up every year, but the bulbs technically function as perennials. If you've ever grown garlic before, you know what a delight it can be; if you haven't, you're in for a real treat. Healthy garlic bulbs should be gently split into individual cloves and planted in the fall, flat bottom-side down, into any suitable container. During winter the clove will mellow in the cool soil and by the next spring send up a slick, green stalk that very much resembles an onion (they're from the same plant family).

As the shoots begin to die back in the fall, you know it's time to harvest the garlic that has now developed an entire bulb from that one clove you planted clear last year. I never quite get all my garlic bulblets out of the soil at harvest time and end up with random baby garlics the next spring, which I consider a great bonus. This process creates a perpetual garlic harvest. Each year you can dig up and then replant mature bulbs for a new year's harvest. Any baby garlic bulbs that are left behind in your container can serve to deter pests for other plants. It does take a full year to grow a crop of garlic, but it is indispensable in the kitchen for flavor and nutrition.

There are two kinds of garlic to choose from: soft neck and hard neck. There are a lot of nuances of difference between the two but, basically, soft neck garlic is considered milder in flavor and stores a bit longer (up to a year if conditions are just right). Hard neck garlic is rich in flavor and will store from 3-9 months in the proper conditions.

Hard neck garlic makes an interesting show piece in the garden, too, because it sends up a seed head called a scape. You want to clip scapes as soon as you see them shoot up because they will interfere with the garlic bulb development. However, while they're up, scapes look wicked cool; the stem of the scape curls back on itself, ending in a delicate bulbous shape. Once harvested, scapes provide a tasty addition to any stir fry, pesto or baked dish.

Pot Size: Pick a container that is at least 18" deep and 12" wide; you plant each garlic clove about 5 inches apart.

Sun: 6-8 hours or full sun.

Special Notes: Garlic can be susceptible to fungal root diseases. Be sure to add a light soil-less mix, like peat, to your pot, along with your favorite potting soil mix, so as not to bog down the growing garlic with too much moisture. Don't let it dry out, though; garlic likes even moisture. Remember, as with all perennials, if you live in extreme winter areas, plan to insulate your containers or move them indoors for protection from freezing.

5) Lamb's Quarter *(Chenopodium album)* – This is a reseeding annual that functions as a perennial. Considered a weed with many names, this weed is one of the very first things to start popping up in the garden and we love it. With silvery green leaves and taste like nutty-flavored spinach, lamb's quarter is a welcome sight come early spring when we're all starved for leafy greens. The baby shoots of lamb's quarter are wonderful in salad or added to stir fry, and the more mature leaves can be wilted with butter and lemon exactly as you would spinach. My kids think it's groovy when we cook weeds from the garden.

It reseeds very, very easily and prolifically and so some people do battle with it; believe me, it's not afraid of your winter and will be back in spring. However, I've known several wild foragers who deliberately planted the seed in their yard, so they could harvest it. In abundance, the seed can even be collected and cooked up like quinoa or amaranth. You can gather seed and re-plant it or, if seed is left in your pots, it pops up quickly in early spring and is usually the first food you can harvest from the garden. If you don't want to deal with saving the seeds yourself, cut back the seed stalks when they appear. You can find lamb's quarter seed for sale from seed and local food enthusiasts but it's not very common, so be sure to look around online.

If you want to treat this plant as an annual, just pull lamb's quarter from your pots and re-plant with spring peas, radishes or lettuce. To grow as a perennial, pull the stocks from your pots, but allow mature seeds to fall in the soil so that it will reseed.

Pot size: Really anything will do since these are weeds and they're biologically designed to grow anywhere. A standard pot that's 18" deep and 12" across will do just fine.

Sun: 6-8 hours, but it can take partial shade.

Special Notes: Lamb's Quarter is also known as 'Fat Hen', which indicates to you that it can be used to feed your poultry or really any livestock on the homestead. I use my wild crops of lamb's quarter around my property as an early spring forage crop for the animals. The goats eat it like it's candy. Bear in mind that the stalks are edible, too, but once they're thicker than a bamboo skewer, humans tend to find them a little tough; your livestock won't mind them at all, though.

6) New Zealand Spinach *(Tetragonia tetragoniodes)* - Also known as Botany Bay spinach, this tender perennial green is grown as an annual due to its sensitivity to frost. The plants grow vigorously and produce seed easily, meaning it will reseed quite often on its own. I think of any plant that does this as a perennial. If I don't have to mess with planting seed, that's good enough for me. If you empty your pots of soil each winter, you may not want crops that reseed, FYI.

New Zealand spinach isn't intimidated by heat and pests rarely bother it. The texture of New Zealand spinach is completely different from the spinach you're probably used to, *Spinacia oleracea*. New Zealand spinach is a bit rough on the surface and the dark, green leaves have a slight rubbery texture and a triangular shape.

You can go ahead and direct sow your seed right into a pot; be prepared for it to happily cascade over the sides of containers and troughs with its sprawling growth habit. You'll probably only need two to four plants for a family of four to have it produce throughout the season and keep you in spinach, depending on how much you eat each week. It will die back at the first severe frost but will reseed, producing that perennial effect. Like true spinach, New Zealand spinach is digested best after it's been rinsed and wilted under low heat. For the mildest flavor, pick young leaves. Dry the leaves and powder to add to smoothies, homemade pastas and soups.

Pot size: A five-gallon pot will hold one to two plants.
Sun: 6-8 hours, but it can take partial shade.
Special Notes: New Zealand Spinach is drought tolerant, although the leaves won't be as tender if the plants are dehydrated. For best results, keep the soil mulched and damp. Unlike true spinach, New Zealand spinach is not frost tolerant and must be planted when the temperature is at least 65°F/18°C. However, true spinach won't perform well in the warmest months of the year, unlike New Zealand spinach. To stay in spinach all year round, plant true spinach in the early spring and late fall; alternate with New Zealand spinach in the late spring and summer garden. In areas with intense summer heat, provide some late afternoon shade for your New Zealand spinach. If you let it reseed, it will come back on its own, like a perennial.

Can I Grow Veggies Indoors?

A typical vegetable requires six to eight hours of sunlight to produce food for both it and you. If you have a sun-room, enclosed patio or other indoor area that receives that much sun per day and it's the only space you have to grow food, I say, go for it! However, to grow indoors, you MUST follow the rules of container growing even indoors which include, but aren't limited to, the following:

- Each food producing plant must have 6-8 hours of light per day. For example: you may have one spot in your bedroom that gets that much light, so you place the plant there, turning it once a day to expose both sides to the same amount of sun.
- Each plant must be fed according to recommendations for that plant. For example: a tomato will usually need to be fed once a month during production, depending on what kind of food you choose to use. Food producing plants are typically heavy feeders themselves—think of how hungry a pregnant woman is. Since potted vegetables aren't planted in the garden, their roots are limited in food and water to what you provide in their container.
- Each plant must be given adequate water for its own growing requirements. For example: tomatoes like a long drink that dries out in between watering sessions, but beans like to stay evenly moist.

- Each plant must be provided the correct temperature. For example: tomatoes really do love some heat to ripen well and quickly, and do not tolerate frost.
- Air circulation must be maintained, and fresh air must be introduced from time to time. Don't overcrowd your plants in their pots. Overcrowding increases the chance that you might have fungus and bacteria problems. Be sure to open a window now and then.

RAPUNZEL, LET DOWN YOUR SALAD

Anna, of The Northern Homestead (www.northernhomestead.com) lives in Canada where the winter is longer than the summer and the only thing warm is the ice. Her homestead is tiny, too, weighing in at .8 of an acre, so she must diversify her growing methods. To grow as much food as possible, she uses the Back to Eden (www.backtoedenfilm.com) method in her front yard. In the back yard, she uses raised beds, grow bags in the greenhouse, and an aeroponic Tower Garden indoors. The Tower Garden is a vertical growing system allowing Anna's family to grow 28 pots in about 6 square foot of space which, at the moment, is a bright laundry room next to the kitchen. Talk about local food! Anna says, "Our salad only travels a few steps."

Anna likes the space savings the aeroponic tower provides. She also enjoys the controlled amounts of natural fertilizer and simple growing medium she uses (something called "rock wool"). Then there's the water. Anna says, "Container gardening requires a lot of water. I still struggle with watering the right amount (after a decade of container growing). A hydroponic system takes care of it. No watering or spilling water all over any longer."

To learn more about the Tower Garden, of you ask questions, you can visit Anna's business page at www.esau.towergarden.com or visit her blog where she writes a good deal about what's growing in her tower. Rapunzel would be jealous.

THE PLANTS ARE PRODUCING—NOW WHAT?

Enjoy them! Eat their fruits! Praise your God! If you're interested in learning to preserve some of that harvest, see the *Preserving What You Grow* section in the In the Homestead Kitchen chapter. Otherwise, eat hearty. When your plants are done for the year, you can give them to your city's green waste OR you can start a compost pile with them.

If you have backyard chickens or other livestock, they are usually quite happy to help you in your garden clean-up efforts. Be sure to read up on what's toxic and what's not; tomato plants, for example, should never be given to livestock. Be sure to get in the habit of cleaning and safely storing your pots once the growing season is over. If you're over-wintering seed, mulch over the tops of your pot and protect them from freezing during icy temperatures.

SEASICK CONTAINER GARDEN

Amber of The Coastal Homestead (www.thecoastalhomestead.com) lives on the coast of the eastern United States where she struggles to grow food in their garden against the dark forces of heat, humidity, poor soil and lots of shade on their small lot. Undaunted, though, her family has grown food in planters and "flower pots, hanging baskets, plastic bags, plastic box containers, five-gallon buckets, vertical gardening with planter boxes, border boxes, a Tower Garden (hydroponic growing system), and have even grown in a cardboard box."

Her most successful crops have been tomatoes, herbs, cucumbers and green and leafy vegetables like lettuce, collards, spinach, kale and chard. Although she also grows potatoes (sweet and regular), mushrooms, lemons, oranges, limes, squash, luffa sponges, beans, strawberries, edible flowers, peppers, avocado, and Brussels sprouts. Her methods are diverse, too, as she practices Organic gardening, companion planting, utilizing microclimates and any natural method that makes sense for her yard.

I asked her what five things she wished she'd known before she started growing food in containers and, as I read her answers, I kept saying, "Yep, me, too. Yep, me, too."

Here's her advice:

"Keep them watered. Smaller spaces dry out quicker, so they often require more water.

Remember radiant heat. If you're placing your container on concrete or a deck, you must pay attention to how hot the surface will get. I have cooked several plants by placing them directly on the surface. Use carts, shelves, a plant caddy or dolly, to get your plant up off the ground.

Use the right size container. Put a plant in the correct container, making sure they don't get root bound and have room to grow.

Be sure to plant for pollinators. We didn't have a lot of pollinators when we first started to garden, and it never occurred to me that I may have to 'assist' Mother Nature. Now we make sure we incorporate pollinator plants along with our produce.

Fertilize. A lot of produce plants are heavy feeders and when you put those feeders in a confined area, you must replenish the nutrients with organic matter as the plants demands increase."

The Children's Potted Garden

Veggies will grow in any container garden but I'm encouraging you to make this a family project for several reasons. I again bring your attention to the heart of the homestead—the home. What is it we're working so hard to build and preserve if not a haven for our family and a place of productivity for our community? Shouldn't something as important as a garden have room in it for children? Learning to grow your own is the goal of every self-sufficient person!

Teaching our children to grow their own food should run a close second. The vegetable suggestions in this section have been especially selected for their ease of cultivation and their crowd-pleasing attributes.

Gardening with children is a lot like cooking with children—a worthwhile investment in the future and an experiment in patience. You're tough, you can do it. As religious leader and author Gordon Hinckley said,

> "Without hard work, nothing grows but weeds."

Just keep it simple, keep it organized and do one thing at a time. As you teach a principle, practice it. For example, after you explain to the kids how deep the pot for the sorrel should be, go plant it. Give everyone a job so that they can all contribute to the planting. Even the toddler can dig holes off to the side while you teach the older kids to interplant your radishes with your beets.

If you want to start with favorites like peas, lettuce and beans then you'll be working with annual vegetables. Annuals are planted, grow and produce and complete their life cycle in one year. Perennials are planted only once and then they produce for many years after that. I think there's a place for both in the children's garden. Annuals are flashy and fast, and the rewards are easy to see and eat every year. Perennials are subtler; they're like a secret that we gardeners tell ourselves every spring as they begin to revive from winter's cold. My children have so much fun planting and eating peas but discovering the first runner bean shoots is something they can anticipate every year. Like searching for treasure.

When you're gardening with children, it's a fine line between encouraging successes and dealing with the realities of life in the garden which include weeds, work and wilt. We want our kids to enjoy prosperity in their gardening ventures without an abundance of failure (hey, it happens to everyone) and without adding too much hassle to our already busy gardening schedule. With that in mind, the suggestions made in this *Food in Pots* section will, most likely, ensure success without too much headache, even in the children's garden.

Some of these suggestions may seem obvious to the already seasoned vegetable grower and if that's you, then try picking a new variety of whatever vegetable you or your child have become proficient in growing. If your dear daughter is a tomato Ninja but she's never tried an heirloom, encourage her to choose one to try. Entice her with those amazing heirloom colors (hello, orange tomatoes with stripes!) and the promise of incredible flavors. If your son totally rocks it when it comes to spinach, but he's never tried starting it indoors in August to plant out in the fall garden, maybe this is his year.

Whatever choices you make, I encourage you to make them together. This is as much about your children's gardening efforts as it is about yours. Bear in mind, you know lots of garden wisdom that they're still learning, so take it slowly. What appears obvious to you will be brand new to them.

If you don't have children in your home, I encourage you to open your garden to kids in your extended family or neighborhood. My paternal grandmother invited me to be a part of her garden many times as I was growing up. Her garden was a magical place for me and I still remember her lessons on how to plant bulbs, how to grow in a clay pot, how to support delphinium stalks and why a creeping mint is a lovely choice for a shady spot.

Gardening has always been a passion of mine, but I sometimes wonder if my interest would have been sparked so early, or so deeply, if Grandma Gardiner hadn't taken the time to teach a kid how to transplant a Japanese maple. Don't let the next generation of gardeners and homesteaders enter their young adult years having to work so hard to learn gardening skills on their own, when you could easily teach them the basics by simply opening your garden and your heart.

> "Where you tend a rose my lad, a thistle cannot grow."
>
> — Francis Hodgson Burnett, *The Secret Garden*

ACTION ITEMS: Pick three crops to grow in pots this year, even if you have an in-ground garden. Pots are convenient, as we've already discussed, and if you don't want to do veggies, then try herbs. Write down your selections in your homestead journal and decide if you're going to grow them from seed or buy nursery plants. Sketch out where you're going to put them, what kind of set up you'll need and how they'll be watered. Make a materials list.

If you do have some yard space and would like to try the concept of container gardening in a bigger way, you may want to check out straw bale gardening. You read that correctly—growing a garden in bales of straw that have been conditioned properly for plant growth. Bales have many of the advantages of pots, namely excellent drainage and few weeds. *Straw Bale Gardens*, by Joel Karsten, is a good book to get your feet wet. Mr. Karsten also has a website with more information, including at blog a www.strawbalegardens.com/blog.

DIY: Convenient watering systems can be hard to create with container gardens, depending on your set up. If I'm failing at container gardening, it's usually because I haven't put enough effort into my watering schedule and system. People, being the geniuses they are, have come up with all kinds of ways to DIY self-watering pots for growing a container garden. What is a self-watering pot, you ask? It's a pot that uses gravity to evenly mete out water to the plants growing inside it. Below are some links to simple DIY self-watering pots. If you choose to make one or all, be sure to add the needed items to the materials list in your homestead journal. You will enjoy much more success in your container garden if you have simple, easy to use watering systems in place. Like I said, if I'm failing with growing food in pots it's because I've goofed up my watering plan.

To find a link for building self-watering containers out of tote boxes, and one that shows you how to use five-gallon buckets, please visit our links page under The Homestead Garden chapter: www.homesteadlady.com/homestead-links-page.

(If this is one thing you'd rather not DIY, there are plenty of high quality, commercial self-watering containers from which to choose, either locally or online.)

BUILD COMMUNITY: As we get older, bending and lifting become two activities that we'd just rather avoid. For grandmas and grandpas who love to garden this can be a tragedy. Ever heard of therapy animals? You know, the caregivers that bring sweet dogs, cats and bunnies into care facilities so that elderly patients can interact and cuddle with them? Well, you're going to create a therapy container garden for a garden-loving grandpa near you. Find a seasoned citizen whom you know loved to garden in their youth and surprise them with some of your self-watering containers (see the *DIY* section above) and their favorite flower or food producing plant. Don't give your elderly aunt one more florist fern, gift her a rosemary topiary instead!

It's important here to NOT create a burden for an ill or particularly elderly friend, or their care givers. Tailor the gift to the person and, if they're in a care facility, be sure to ask what the rules are for live plants. If your friend has access to a patio, then feel free to pick pretty much any variety of plant you think they'd like since that plant will have access to wind, sunshine and pollinators. If the plant will be living entirely indoors, try some hardy and obliging herbs. Most plants will need access to some light but there are those that will thrive very well indoors, like basil and oregano.

BOOKS TO READ: *Vertical Vegetable Gardening: A Living Free Guide*, by Chris McLaughlin for all kinds of practical, wholesome advice on growing food in small spaces, natural controls, good bugs and a run-down of plants to try.

Also, *McGee & Stuckey's Bountiful Container: Create Container Gardens of Vegetables, Herbs, Fruits, and Edible Flowers*, by Maggie Stuckey and Rose McGee is a delightful romp through the world of kitchen gardens grown up in pots. This book doesn't have any photos, just high-quality illustrations. *Small-Space Container Gardens*, by Fern Richardson is great for beginning gardeners.

For hands-on gardening information, organization and planning I recommend the simple gardening workbook called *The Gardening Journal* (TGN), created by Angi from Schneider Peeps (you find out how to acquire it on our links page under the Homestead Garden chapter). Note taking is an important skill for container gardeners to commit to because you need to keep on top of your watering and fertilizing schedules. There are a lot of educational resource pages to refer to and take notes upon, as well as plant profile pages for the most common fruits and veggies that allow you to record when you've planted them and what you plan to do with them. There are pages to fill out for pests and problems, calendar sheets to make plans for each month of the year, as well as suggested resources.

You simply print out the TGN pages you'll need for the year, put them in a three-ring binder and fill them out, printing new copies next year and keeping the old ones for reference in your binder. I'm embarrassed to admit how many times I've lost my pathetically scratched out notes or exposed my painstakingly outlined garden plans to the elements simply because I forgot I'd set them down in the garden. Using TGN will allow you to keep all your notes in one, safe place where you can make use of them for years to come.

WEBSITES: For specifically growing in pots, try Container Gardening at About.com (www.containergardening.about.com), and both Dave's Garden (www.davesgarden.com) and Hometalk (a garden and home improvement hub) under their container gardening section (www.hometalk.com/topics/container). These sites cover a wide array of ideas on what to grow and what kind of containers you can use. Especially seek out advice on watering systems and soil fertility, two concerns of particular interest for container gardeners.

To learn about the rain gutter gardening system that uses rain gutters, 5-gallon buckets and a float valve to create a self-watering garden, please visit our links page and look for Larry Hall's YouTube video under the Homestead Garden chapter.

HOMESTEAD LADY SPEAKS:
I once had an old-time, crusty gardener tell me that he thought container gardeners were wimps, because only *real* gardeners would work hard enough to plant *real* gardens in the dirt, in the yard. In his view, growing in pots was a cop-out; as if container gardeners were avoiding the commitment and responsibility of gardening. I happened to mention his comments on my blog's Facebook page and asked my readers what they thought about it. If you've ever been frustrated that your garden is relegated to pots as you long for more space, or if you get tired of garden references never mentioning you hard working containers gardeners, these comments might lift your spirits.

Deb said,

> "I'm a big container gardener. I have MS. By using containers, I can work around I can continue to garden. I raised 100 lbs. of tomatoes and 50 lbs. of peppers and over 100 lbs. of potatoes last year. So, I'm a happy 'wimpy' container gardener. I'm glad to help anyone garden anyway I can."

Roxanne said,

> "When one must move about, containers are easy to pack with you, easy to move inside when it gets cold and, nowadays, so many folks live where they are not allowed to dig up their rented yard area or they live in apartments. So, no, I do not think container

gardeners are wimps. Resourceful, is what I would call them, and determined to grow their own food—even a little of it."

Randi said,

"Container gardens and vertical gardens are my only option in an apartment, yet here I am still growing walls of food. If that makes me a wimp, then I'm a happy wimp."

Jennifer said:

"All gardens are good gardens."

For the record, I think container gardeners are super heroes.

"If you look the right way,
you can see that the whole world is a garden."

- Frances Hodgson Burnett, from *The Secret Garden*

For the Homesteadaholic: Plant a Wellness Herb Garden

With more of us opting out of the conventional this or that, there's been more interest in gardening in general and growing herbs specifically. Herbs are amazingly useful plants in the landscape even if you're not ready to use them for wellness. Most herbs are not very difficult to grow, many have lovely flowers and/or interesting foliage, and they can easily be integrated into perennial beds or any traditionally landscaped area. A lot of herbs grow well in pots, either indoors or outdoors, and many are very adaptable to different climates and types of soil. Another bonus of herbs is that many, many of them are pest resistant by their very nature.

Even people who aren't into herbs know the basics like basil, mint and garlic. All three of these are classified as culinary and wellness herbs, being both highly nutritive and flavorful as well as powerfully potent in supporting various health issues. My goal here is to cover a few basic principles on how to plan and plant a wellness herb garden—a garden planted with the goal of serving the needs of your general health support.

Always double check everything you read about herbs so that you can be sure of your information; accruing accurate information from reputable sources is the responsibility of the student. My intention is not to diagnose or prescribe anything, simply to present how a gardener can go about growing their own wellness herbs. These plants, in many instances, also have the added benefit of being useful in the kitchen and home.

Which Wellness Herbs to Grow

Most herbs are not terribly tricky to grow but they are plants and will require you to have a certain amount of gardening knowledge. Fortunately for all of us gardeners, nature is adaptable and resilient and whenever I have a garden failure I just say right out loud, "Well, that's why God invented next year!" My first piece of advice for effectively planning your wellness herb garden is to evaluate how much gardening experience and knowledge you realistically have. The best rule to follow for new gardeners is:

Aim Small, Miss Small

If you've never really grown much, try basil or calendula this year since both are easy to grow (from seed even, if you're feeling ambitious) and are very pleasing plants when they leaf and bloom. Not to mention, they're great wellness herbs for soothing tummy problems or skin imbalance.

Most quality, local nurseries will carry a selection of herbs. Walk through one and observe which plant speaks to you. If you're new to growing things, I'm limiting you to two purchases this year, because I don't want you to get overwhelmed and frustrated, suffer a loss and then figure you have a black thumb. You're going to be busy living your life AND tending your few new plants AND reading herbal gardening books from the library AND planning your larger herb garden AND looking for community gardening classes to join so that you can improve your garden Ninja skills. Two plants will be all you can handle.

If you've grown a garden before, I challenge you to pick up an herb you've never heard of or, at least, one you've never tried growing. Before you take it home, check out the label and make sure it's one that will survive the conditions of your climate and yard. Just because you want to grow something doesn't mean you can. Respect the plant's needs by knowing your growing zone. Let's discuss that a bit more.

WHICH WELLNESS HERBS DO I USE? CAN I GROW THEM?

There are literally thousands of very useful herbs you could grow but your climate, soil and other growing conditions will only successfully support so many of those varieties. Here's how you narrow down the list:

Sit down and go through your herb closet or shelf and note which herbs you use all the time. Is it Echinacea? Ginger? Garlic? What about Fennel? Mint? Licorice Root? More exotic? Are you always out of Ginseng? Myrrh? Black Walnut Hull?

Now, grab one of those herb books you've checked out from the library and start looking for information on each of your herbs' Cultural Requirements. These requirements are the conditions that each herb will need to grow, thrive and, hopefully, propagate itself in some way either by reseeding, producing seed for you to harvest, layering, cutting, and the like. Pay special attention to how many hours of sun your herb needs a day; if it says in your book that the plant needs 6-8 hours of sunlight, it probably means it. Further ask yourself what water requirements it has, what kind of soil it needs and, VERY important, what kind of winter and summer temperatures it can take.

Sometimes you can make exceptions with each individual plant here and there (a little less water, only 5 1/2 hours of sun, a soil that is only borderline quality) but winter temps, especially, are not forgiving. If you want to learn which growing zone you're in, look in your favorite gardening book for a hardiness zone map and/or look online for one. Growing zones are neat but be sure to ask the nurseryman what the temperature range is for the plant in which you're interested. This step alone will knock out a big chunk of your list of wellness herbs since some of the ones we've become accustomed to ordering for our favorite herbal proprietors are among those that will only grow in certain conditions.

For instance, in my climate, without a greenhouse, using the examples in step one I can only grow Echinacea, garlic, fennel, mint, licorice and walnut. Did I say only?! That's a pretty good list, all things considered. As I grow my own, I'll find herbs to raise at home that can serve as substitutes for the ones I can't grow. God wants us to be healthy and has provided all we need no matter where we live. I truly believe that, and I've bet my life on it, literally.

Stop here and make sure you have your list from step number one, edited and perfected by step two. Your list may change later but what you have now will suffice. Got it? Here we go...

Acquiring Wellness Herbs

Once you have a working list of wellness herb plants you know you'll use AND be able to plant, you need to decide whether you want to grow those herbs from seed or try to find a source for plants that have already been started from seed and are for sale, also known as "seedlings." There's no shame in buying an already established plant from your favorite local nursery—goodness knows I end up doing it a bit every year, as much as I grow from seed.

The only real roadblock to purchasing wellness herb plants is in the wellness or "medicinal" parts. Some wellness herbs are too obscure for your nursery to stock them for herb nerds like you and me. Echinacea or witch hazel may seem like normal plants to have in the garden, but most people don't grow either for their wellness benefits if they grow them at all. You may not be ready to grow all the wellness herb plants you need from seed this year. However, the reality is that you'll most likely need to learn this skill at some point simply because you'll have a hard time locating the plants you need otherwise.

For this reason, let's talk about finding a quality herb seed source. First, order an herb catalog from a quality seed house—in fact, order from two or three. My favorite seed house for wellness herb seeds is Strictly Medicinal (www.strictlymedicinal.com; formerly known as Horizon Seeds), which specializes in all kinds of herbs, but particularly those that can assist with supporting your health and well-being. Strictly Medicinal seeds are always viable, the packets have great information on them, the catalog is a wealth of knowledge and the people behind the seeds are some of the nicest with whom you'll ever do business. Whatever company you chose to begin with, when your catalog arrives, start dissecting it by reading the description of each herb and takes notes on herbs of interest. Keep your favorite herb book close by as a reference to answer any questions that arise about the plants which the catalog isn't answering. If you have further questions, call the company and ask. What you're really doing with this exercise is finding a vendor with which you want to work.

Ask yourself some questions like:

- Which company has the inventory you need?
- Will the company be a good educational resource for you?
- Is their website helpful?
- Is their ordering process easy and what does their customer service look like?
- What about ethics—are you trying to stay away from Seminis and De Ruiter (sister companies of Monsanto), genetically modified or even innocuous hybrid seed?

As I said, you may not be ready to start growing your herbs from seed this year since it is a step above keeping a plant alive in a pot on your deck. However, you will get there eventually and it's good to begin with the end in mind. Remember, if you want to create a wellness herb garden, the chances are that you'll exhaust the resources of your local nursery within a few years. You'll just be so herb savvy that you'll discover you've moved beyond the simple basil and sage options and need a wider variety from which to choose. So, go back to the library and get a book on seed starting, take a local class (try your university extension and/or your local seed exchange group), ask your gardening nerd friend if you can come see their seed-starting set up and pick their brains about what they do.

If you don't have time for seeds this year but want to begin your wellness herb garden, never fear because there are some online vendors who sell wellness herb plants. Just be sure to find one that's domestic to avoid complications with shipping. I'll be honest, unless you have a very small yard and a very big budget, stocking an entire herb garden with mature plants will be cost prohibitive, but a few herbs here and there shouldn't break the bank altogether. If you're in the States, Horizon Seeds mentioned above sells some potted herb and root cuttings that I've availed myself of in those times I just really didn't want to try to propagate something complicated.

Also, Crimson Sage (www.crimson-sage.com) out of California has an exquisite catalog of wellness herb plants. Full disclaimer, I didn't end up ordering from Crimson Sage when I started my first wellness herb garden, despite how often I visited the site, because of my limited finances. However, the proprietress Tina, was so helpful with information and advice that I felt empowered to try growing some of my own. She didn't make any money off me with her thoughtfulness, but she did gain my respect and appreciation. Herb gardeners are the nicest people you'll ever meet.

3 WELNESS PLANT PROFILES

Here are three wellness herb plant profiles from our book *Herbs in the Bathtub*, including tips on how to grow and propagate them. Are these the herbs for you?

#1) Mint - *Mentha Spicata* (Spearmint) and *Mentha Piperita* (peppermint) – The mint family is large and is a collection of herbs that many gardeners insist on keeping in a pot because of how invasive it can be in garden soil. By keeping mints contained in pots, you can enjoy their beauty and usefulness without having to worry about them becoming King Kong in your garden. If your garden is confined to a deck or patio, these herbs will be consistent performers for you. There are many more varieties than just the two listed above.

CULTURAL REQUIREMENTS:

Site – Partial shade or sun.

Soil – Moist, well drained, alkaline. It really will grow pretty much anywhere there's enough moisture. Go ahead, try to kill it – I dare you.

Propagation – Take root or stem cuttings or divide it in spring and autumn; in summer, root stem cuttings in water.

Popular uses – one of the most popular uses of mint is as a stomach preparation to support the body during nausea, gas or general gut malaise. It is also very cooling.

#2) Calendula - *Calendula officinalis* - Used in the Civil War to support open wounds because it simply loves to soothe imbalance on your skin. Sometimes called "pot marigold" because it does well in pots. The seeds are large and easy to handle, so they're great for kids. Even black thumbs can grow Calendula!

CULTURAL REQUIREMENTS:

Site – full sun to part shade; if growing outdoors, it will do well in zones 3-10.

Soil- tolerant of poor soils but a simple potting mix will do for your container.

Propagation – Direct sow in early spring through summer. Growing from seed is really the easiest.

Popular uses – Calendula is balancing for the skin – rashes, chapped skin, rashes, etc.

#3) Sage - *Salvia Officinalis* – As the proverb goes, how can a man grow old who has sage in his garden? As helpful to your sore throat as it is to your favorite chicken dish, sage is a great container herb for indoor or outdoor growing. It comes in a wide range of colors, too, from variegated to purple to gold.

CULTURAL REQUIREMENTS:

Sage is a perennial plant so make sure you add compost and fertilizer on a regular schedule.

Site – Full sun; may need to use supplemental light during the winter so that sage will get at least 6-8 hours of light each day. Protect from harsh winter winds if growing outdoors. If planting in the garden, sage will grow in zones 4-8 with protection in harsh winter areas.

Soil – light, dry, alkaline and well drained.

Propagation - Grow from seed or 4-inch cuttings, which is easier.

Popular uses – Sage makes a wonderful throat preparation for soothing sore, raw or dry throat imbalances.

DO YOU NEED ORGANIC SEED?

How do I know what kind of plant or seed to buy? Organic? GMO?
I'm not even sure what those labels mean…

Like you learn to do with your packaged food, paying attention to a label is important when you go to purchase a plant or seed. Doing research on terms like "GMO" and "Organic" will help you make informed decisions about your plant and seed purchases. Here are a few topics to ponder and some of my own considerations when I go to purchase a plant or seed, including a list of purveyors I prefer to patronize.

The Organic label is one given by the USDA to an item that has passed specific federal standards which are outlined at the USDA's website (www.usda.gov). At the time of publication, although it isn't specifically regulated to my knowledge, the Organic label usually means that the item is also non-GMO, that is, NOT a "genetically modified organism." Learning to ask good questions, even if they're hard to answer, is a healthy thing to do when it comes to our food and how it's produced. Whether or not buying Organic or Non-GMO is important to you is a question only you can answer for yourself with some research on current GM science. Is it important to me? In short, yes.

Though a bit dated (2011), we have an article on our links page under The Homestead Garden chapter by the GMO Awareness site that outlines information on GMOs and the Organic label and may get you thinking about the topics involved. You can find that link on our links page: www.homesteadlady.com/homestead-links-page.

I'm not as concerned about the official Organic label of fruits and veggies where pesticides, herbicides and fertilizers are concerned (although they do concern me greatly). As far as pesticides go, instead of having an outside agency tell me the food that arrived from across the country to my grocery store is safe, I'd rather talk to my local growers and simply get details on their growing methods. On an actual farm, at a farmer's market or a grocery store that features local growers, I can ask the producer if they use pesticides and commercial fertilizers and, if so, what kind and how often. I will say that I'm a lot pickier about pesticide use when buying food than when I'm buying a plant to go in my garden. The reason for this is because a plant can often outgrow its toxic-spray beginnings, just like a person can rise above a difficult childhood. Plants have a remarkable ability to adapt and thrive.

However, when you factor in the reality that the majority of 100% Organic products are most likely to have the smallest trace of GMOs available on the market, that label becomes a lot more appealing to me. When it comes to buying my vegetable plants, that fact is particularly important because I typically buy heirloom varieties so that I can save the seed myself to plant in the future. If I'm going to all the trouble of growing healthy food, I want to ensure that my seed saving plants are as "clean" and GM-free as they can be.

To avoid having to worry about the possibility of buying a genetically modified plant or seed, you may choose to do business with companies that have signed The Safe Seed Pledge. Those companies have pledged to not knowingly sell any GM plant or seed. (Incidentally, there are food producing companies that have signed similar pledges.) There are several places online to find a list of companies that have signed the Safe Seed Pledge, but you can access a link for one on our link page under The Homestead Garden chapter.

At the time of this publication, my favorite Safe Seed Pledge, non-GM producing seed and plant companies are: Baker Creek Seed, Peaceful Valley Farm Supply, Horizon Herbs, Territorial Seed, Seeds of Change, Botanical Interests, and Johnny's Select Seeds. There are several others, but these are a good place to start. You can usually expect to pay a bit more for seed from these companies than from their less-GM-conscious competitors. Like I said, I attempt to save seed from as many plants as I can so that I don't need to replace my seed stock every year, keeping my costs down.

If you have a place you already purchase seeds or plants from, it would be worth a phone call or email to discover if they sell non-GM products. My favorite seed house from which to buy in bulk (lots of seed for storage and for acres of garden) is E&R Seed—their prices are great, and they have a huge Organic selection. They do sell some GM seed and so I email them to find out which varieties I should stay away from, or I just purchase from their Organic selections. It doesn't have to be all or nothing when you to establish a personal relationship with your seed provider. Take some time and get to know your grower.

While we're on this topic, don't make the mistake of thinking that growers who do use and sell GM seed are all evil, earth-destroying hicks who are only in their trade to make a quick buck. Anyone who knows anything about farmers knows that, as a group, you'd be hard-pressed to find people who love and cherish the earth more. And the last thing they are is rich. I know of a lot of growers who truly feel that GM crops will be how we feed the world and that the science is safe and proven. The fact that I firmly disagree with that position in no way means I don't love farmers and the work they do. May God bless our growers!

All that to say that I can't emphasize enough how important it is to develop a relationship with the suppliers from whom you purchase. Get to know them and get a feel for their ethics and how they do business. A label will never tell you as much as a tone of voice or a look in the eye will; don't pass up the chance to get to know the farmers that feed you.

> **OTHER WAYS TO PROPOGATE**
>
> Some herbs can be buggers to grow from seed, but you don't have to learn everything all at once. If you are lucky enough to have a neighbor or friend who is already growing an herb you need, research the best method of propagation for that plant to determine if you can take a cutting or a rooting. For example, thyme can easily be propagated by a method called layering.
>
> *To Layer Thyme:*
>
> Take a supple but mature stem, lay it in the dirt and cover up a section of the stem with more dirt, weighing it down with a rock or garden pin.
>
> Keep it watered and wait for the point of contact with the soil to sprout roots.
>
> Once the roots appear, cut the stem from the mother plant with a sharp shovel, carefully dig up the stem with its roots, put the baby plant in a pot or in the ground and you have a new plant, no seed needed.
>
> Again, have a good book on hand to learn more about propagating by cutting and layering (see suggested resources at the end of this section).
>
> Also, where it's legal and the plants are available, consider learning how to wildcraft (harvest from native plants) the herbs that you need from your local environment. Please be sure to do this responsibly. Seed gathering from wild plants is usually straight forward—bring a paper bag and clip the seed heads into it. Only take what you need. Only harvest roots and cuttings if it's legal and never take more than you need, neither take more than the plant can do without. This will be species-specific so bring a field guide. For more information see the *Foraging* section in the Green the Homestead chapter.

Space For the Wellness Herb Garden

Be realistic about the space available to you when planning and planting your wellness herb garden. Are you in an apartment? Look at what you can grow in a sunny window or on a southern facing deck. Do you have a community or farm garden plot, or a friend who has extra space in their yard near you? You're into wellness herbs, right? So, you're used to thinking outside the box. Bottom line, find a decent amount of space to grow the herbs on your list.

What's a good size herb garden? Well, that depends. "Argh", you say, "it's impossible to get a straight answer from a gardener!" Sorry, but it really does depend on certain factors and I can't give you specifics for your garden because I don't know what herbs are on your list to grow. However, here are some things to think about:

- For how many people are you growing wellness herbs this year?
- How many different plants will be taking up space? (For example, fennel takes up a lot more space than thyme, both vertically and horizontally)
- How much of the area in your yard or plot is a good match for the plants you want to grow? Are you able to use your entire growing space or is there a lot of shade or unusable ground?

I grow plant medicine for seven people and it has taken years of experimentation to figure out what we use throughout the year, what I can successfully grow and how much I can harvest from each herb. I'm still constantly learning, especially what to grow, because I have yet to work with so many herbs. As an example, I have a spearmint patch (I inherited it with the house) that is about two feet wide and eight feet long. Over the years I've lived with it, I've learned to harvest at least twice, sometimes three times, a year by shearing the plant about six inches from the ground and letting it regrow. We dry all those cuttings, then use fresh harvests from the leftover plant throughout the growing season both in the house and in the barnyard.

With those two or three harvests (which equal at least three large, fresh bundles each) I have enough to last all winter for both the humans and the animals. I even have some left over most of the time. I typically use mint in recipes for pest control, desserts, teas and upset tummy tinctures. For my family, this 2' x 8' patch of mint is sufficient for our needs, but I only figured that out over five years of fiddling with it.

As an example of a successful experiment, this year I put a few sweet fennel seeds (not to be confused with bulb or Florence fennel) directly into the ground and grew up three patches with ease. Those three patches left me with a #10 can size (or 2 lbs. coffee can) harvest of fennel seed which was plenty for the livestock and the humans to be used in snacks, sweets, teas and tinctures. Remember, some herbs are also culinary (to be used in the kitchen) and you'll want to harvest enough from them during the growing season to provide for both your recipes and your medicines.

For those with limited space, you may need to confine your mint growing to a few pots. However, if you're diligent about harvesting throughout the season, the plants will continue to produce, leaving you with an abundant mint crop. If you don't have space for three seed fennel plants, grow one and observe how much of it you use. If you have a large garden, plant as many herbs as you wish and share your extra harvest with friends and family. I have no sense of proportion and plant way more basil every year than I technically need, but I'm always happy to pass on my surplus.

PLANNING THE WELLNESS HERB GARDEN ON PAPER

Are you still with me? So, you've done a good deal of thinking and studying, now get a large piece of paper and a pencil with a good eraser. Draw a sketch of your garden space and start plugging in plants—this will serve as a rough design for your new herb garden. Your design can be something as simple as a square foot garden bed devoted to herbs, or as complicated as an entire yard full of these great plants. You don't really need to get out a ruler and start measuring or counting off square feet yet, unless you want to do that. This is just a good time to start dreaming with your pencil in hand. I would suggest doodling a few different designs.

Whichever herbs you choose to grow, I encourage you to plant them together with great variety. Be bold and include your herbs in your showy perennial beds, put them in pots all over your house, plant them in every pocket of space you can find. Try to grow a new herb every year. Put them in unexpected places to experiment with where they'll grow best and to find new uses for previously unconsidered garden space. As the gardening duo of Joel Schwartz and his wife Colleen (www.growcookforageferment.com) remind us,

> "Everything that has a border, a size, a footprint, and existence has edges. Some species prefer these edge environments. Encourage these surprising and often unusual volunteers. Don't make perfect corners and straight lines, the edge effect dwells in curves and nooks and crannies. Don't plan everything, leave some area or aspect to chance. See what happens and encourage it!"

GET TO KNOW THE PLANTS AND YOUR PLAN

I don't mind the time it will take to mature my wellness herb garden as I continue to plant it since I'm using that time to learn more about herbal preparations, properties and functions. I'm also getting to know the plants themselves, as they grow and occasionally fail. Like any gardener, we often deal with horribly hot summers and terribly cold winters, sometimes shuddering to think what we might find in the garden in spring and fall. So, I guess my last piece of advice is to take your time and pace yourself. However, start this year and do something to plan and plant your herb garden, no matter how small the effort may seem. Just like growing a vegetable garden, the key is to:

Grow what you'll use and grow what will grow.

That is the unspoken truth of successful gardening.

ACTION ITEMS: Get out your homestead journal and do the assignments from this section, namely:

1. Honestly evaluate your gardening experience and decide how many new herbs you want to grow this year.
2. Sit down and go through your herb closet or shelf and list which herbs you use all the time.

3. Check cultural requirements for the herbs you use the most and answer the question, *Will they grow for me in my climate?*
4. Decide how you will procure your plants—grow yourself, take cuttings, purchase?
5. Begin to sketch your herb garden, including space requirements for each plant.
6. Herbs often play well with other plants and provide benefits for them so don't hesitate to mix up your herbs with your snooty annuals and showy perennials in any garden space you already have.

DIY: Set up an herb drying and processing space right now, even if you don't have an herb garden yet. This space doesn't have to be large, especially if your herb garden isn't. It's important to have a firm plan in place for where you will process and store your herbs before they start rolling in. Otherwise, you will end up wasting some or all your harvest. You want to choose a place that has good air circulation, the least amount of dust and no direct sunlight. Unless you have a dehydrator, you will most likely be air drying the bulk of your herbs. To use space efficiently you'll want several options for hanging your herb harvests, either from the ceiling or from hooks on the wall. You can build your own racks, or you can use repurposed materials to prepare your herb-drying areas. I've used a ceiling pot rack before with great success.

You also want to decide where you're going to store your herbs for the year once they're dry. A cool, dark place is recommended. You'll need containers to store the herbs in, so begin to gather plastic or glass, whichever you prefer. Labels, lids and other herbal medicine making supplies will come in handy, too. It is also highly advisable to put together an herbal binder for your family with your notes, recipes and ideas. You could use your homestead journal to begin with, but it will get messy with herb information, so I highly suggest you start a separate herbal notebook.

Strangers and Pilgrims on Earth (www.strangersandpilgrimsonearth.blogspot.com) has a wonderful series of posts called "Reinventing the Herbal" for modern homemakers and homesteaders which includes advice, recipes, free printables, simple instructions, inspiring photos and quotes and, in general, a great deal of what you might need to begin to set up your own home wellness center. You can find that series on our links page under The Homestead Garden chapter: www.homesteadlady.com/homestead-links-page. Jes, the author, in no way indicates that your herbal education should begin and end with her, but she can help you organize yourself and prepare you to be organized with your herbs. See the additional resources below on the topic of herbal education to further your study.

BUILD COMMUNITY: If you don't already do this with your homestead book group (see the *Homestead Book Club* section in The Homestead Community chapter), start an herbal discussion group with like-minded people. This group is NOT meant to be a replacement for quality herbal education, but there's just nothing like in-person dialog about a topic that opens your mind and sparks useful questions and personal reflection.

Your group doesn't have to be large, but it should be welcoming to all who are learning to grow and use herbs. If there are those who would like to share their tips for herbal preparations and crafts, invite them to do so. Always remind fellow group members that everyone is obligated to do their own homework on each herb and to be responsible for their own wellness. I have participated in several groups like this and they have all, without exception, been helpful to me as I self-educate on the topic of wellness and culinary herbs. As a bonus for the host, when there are demonstrations, your house ends up smelling wonderful.

BOOKS TO READ: *The Homesteader's Herbal Companion: The Ultimate Guide to Growing, Preserving, and Using Herbs*, by Amy Fewell is a go-to resource for both planning and planting your herb garden. She'll even get you preserving and using them. Also, *The Complete Book of Herbs*, by Lesley Bremness has solid general information, is very nicely produced, and covers a wide range of herbal topics from recipes to medicines to crafts. A very useful aspect of this book is the photographic layout of each herb—from seed to root to leaves, you can see every part of the plant. This is a very good book for the beginner and available used for very reasonable prices.

Medicinal Herbs: A Beginner's Guide: 33 Healing Herbs to Know, Grow and Use, by Rosemary Gladstar. Anything Rosemary Gladstar has written on herbs will be of use to you. However, you can start with this one because it's very basic and you'll learn something new each time you pick it up. *The Backyard Herbal Apothecary*, by Devon Young, is similarly helpful for figuring out what herbs will be worth stuffing into that overflowing herb garden you'll have.

The Herbal Medicine-Maker's Handbook, by James Green is primarily an herbal medicine manual but it has great suggestions on what kind of wellness herbs to grow in a garden setting, broken up by areas of the U.S. There are all kinds of charts in the back with suggested group plantings, herbs for water-wise gardening and much more. *Healing Herbal Infusions*, by Colleen Codekas, will give you simple recipes for health and well-being using the herbs you grow.

For learning to propagate plants, try *The Plant Propagator's Bible*, by Miranda Smith and/or *American Horticultural Society Plant Propagation*, by Alan Toogood.

WEBSITES:

For growing and using herbs, and good sense:

- Herbal Academy (www.theherbalacademy.com/blog/)
- Mountain Rose Herb (www.mountainroseblog.com)
- Herbal Prepper (www.herbalprepper.com)
- Nitty Gritty Life (www.nittgrittylife.com)
- The Fewell Homestead (www.thefewellhomestead.com)

For **herbal education**, please visit:

Herbal Academy (www.theherbalacademy.com) for online herbal classes, including beginner through professional herbalist.

Learning Herbs (www.learningherbs.com) also offers herbal education in various formats, including an herb fairy book series for kids and an awesome herbal board game called *Wildcraft*. We had to buy two of these games because the cousins loved it so much that when we all got together they fought over who got to play it first.

Speaking of teaching children about herbs, Herbal Academy also has a writer specifically for kids (http://herbalacademyofne.com/tag/introduction-to-herbs-for-kids/).

For **purchasing wellness herb seeds** try:

Strictly Medicinal (www.strtictlymedicinal.com; formerly known as Horizon Seeds). It is the premier source of *wellness* herbs seeds.

For great culinary and a few wellness herb seeds try also Baker Creek Seed (rareseeds.com) and Johnny's Seeds (www.johnnyseeds.com).

Most other seed houses will carry some herb seeds and even plants of interest. If you're looking for wellness herb *plants,* try Crimson Sage (www.crimsonsage.com)

HOMESTEAD LADY SPEAKS:
There's no shame in hiring a designer if this isn't your thing. Our herb garden is where our front lawn used to be. It seemed important, for the sake of dealing with those few in the neighborhood who are confused when you do something other than grass, that the herb garden be *designed*, not just thrown together. A designer can open your eyes to possibilities you might not have thought of before, particularly with the shape and contour of your garden, as well as varieties of plants that might be useful to you.

I ended up consulting with a designer and then taking her great plans and tweaking them the way I wanted. We also included a lot of edible plantings and even some ornamentals since the space was large and I wanted it to look full and rich all year round, especially for my bees. It will take me years to get in all the herbs I want and to grow up the edibles and ornamentals to a mature size. Illustrator and avid gardener Tasha Tudor says it takes over a decade for a garden to look like it's been there a lifetime—I think that will be just about right.

Incorporating a variety of plants, not just obvious herbs, also opened my eyes to the myriad of plants that have wellness actions associated with them. For example, we're growing rugosa roses to form a living fence at the front of the garden because they're lovely and will survive our winters. It turns out, though, that their hips are incredibly nutritious and powerfully soothing being full of Vitamin C, among other nutrients. We took our first harvest of those hips this season and, wow, did everyone from the children to the goats appreciate those plants!

The point is, don't hesitate to ask for and learn from other people's expert opinions and assistance; you don't have to do absolutely everything on your own, especially on the homestead.

"Of course, he will go on being educated every day of his life, same as father.
He says it is all rot about 'finishing' your education.
You never do. You learn more important things each day…"

— Gene Stratton-Porter, *Laddie: A True Blue Story*

For the Homesteaded: Growing Smarter With Permaculture

I remember the first time I read anything about permaculture; I didn't get it. I don't even remember what I was reading, but I do remember getting the idea that permaculture was all about hippies using gardens to achieve social justice. I'm all for people using gardens to help the world, but permaculture just didn't seem relevant to me. However, over the last ten years, as I've learned more about growing food, I've been drawn to less-toxic, more self-sustaining gardening methods.

When I lived in North Carolina I benefited enormously from NC State's incredible agricultural extension program and took class after class in new and intriguing topics like edible gardening, integrated pest management and ecological gardening. (See *The Homestead Community* chapter for ideas on how you can participate in similar opportunities.) Slowly, the wide world of gardening as nature gardens opened to me and I never looked back. I saw that the garden is all about cycles where producers, consumers and decomposers dance around each other in self-sufficient circles. In its simplest form, that's what permaculture is, so stay with me here.

We all start from somewhere on our gardening journey. For a lot of us, we learned to grow food the way our parents and grandparents showed us. Others of us have pulled what little we know out of books and magazines, desperately trying to recreate what we see in those glossy pictures. It seems, though, that the more we pour on the recommended fertilizers and pesticides, the more barren the land looks, the more depleted the soil, the more harmful insects rain down on our tomatoes and the more work we end up doing. In the end, we just run out of steam to expand the gardens and grow more. The garden has stolen our youth, so to speak, and we're disillusioned that we can grow anything but weeds. But, with our homesteader hearts, we yearn to produce, produce, and produce even if our homestead is an eighth of an urban-bound acre.

May I suggest you take a look at permaculture? Or, another look, if you're like me. One of the things that was hard for me when I began reading about permaculture was that there were too many details. It's not in my nature to retain minute specifics easily and there was just too much for me to remember; too many steps, too many rules, too much of a good thing. So, I stepped back. I analyzed what I was already doing and discovered that, apart from some design differences and an understanding of biological cycles, I was already a permaculturist!

Let's walk through a few permaculture principles that I've found to be of the most basic and immediate benefit to my homestead in the hopes that we can encourage each other to increase our yields in botanically intelligent ways.

> **ARE ROW CROPS FORBIDDEN?**
>
> Permaculture designer, Jared Stanley (www.jandjacres.com) says, "Row crops are not something shunned by Permaculture. Mono-cropping over a large area is. In other words, a 50' row of peas next to a 50' row of tomatoes next to a 50' row of lettuce isn't mono-cropping. To make it more 'permaculture' we are looking at how you are creating 'rows', where are we sourcing the nutrients for that soil, what we are doing with the crop we produce (recognizing that the 'food' isn't the only product from that plant), etc.
>
> "We also focus on scale. Some crops are suited for larger plantings in large rows that we'll only care about a few times a year: planting and harvesting. Others need more care and more frequent harvesting. This affects where we place the crops on a property. In other words, some items currently being 'row cropped' might be taken out of the 'main crop' area and brought closer to the home for the 'kitchen garden' where other methods besides row cropping can be used making the plants and the soil more productive.
>
> "I think the key [is to remember that] the food isn't the only crop you are producing. Roots are loosening and enriching soil. Stems and leaves could be compost, animal forage or even crafts (such as a grape vine). Then we can start to cycle things around—the cash from the craft can buy a new set of pruners, the animals can provide manure that mix with the plant residue for compost that will feed the soil for the next crop, seed can be saved from proper varieties for the next planting, etc."

Cycles of Smart

We're going to talk about power and energy, cycles and relationships between plants and their environment, which ultimately includes people and the animal kingdoms. Permaculture principles allow us to work smarter, not harder—or, rather, we cease to waste our energy in the garden. As Amy Stross, a permaculturist and author (www.tenthacrefarm.com) explained to me,

> "Since trees in the forest can't walk, they need their fertility to naturally be produced beneath them, in the form of other plants that grow, die back, and fertilize the soil as they decompose. Similarly, we want to minimize our walking distance in between elements of our homestead to maximize the use of our precious time.
>
> "A simple example of efficiency is the placement of elements on a homestead. Place elements that are visited frequently closer to the house, such as a vegetable garden or animal compound, and place elements that are visited less frequently, such as an orchard or timber stand, farther away. This design is a primary focus of a permaculture designer."

Placement of the gardens and their accessories is important, but so is creating living systems that can regulate themselves. In fact, in a permaculture garden you may find that you're doing less of what you'd call "work." That's not to say that permaculture is magic, and you will suddenly be able to produce food for the year without lifting a finger. Be wary of anyone, anywhere telling you that work is not required for success. I simply mean that you will most likely find yourself doing more of the work you enjoy in the garden (pruning, harvesting, tending the animals as they assist) and less of the work you don't (pulling weeds, hauling bags of fertilizer, spraying bugs, tilling up dirt).

I'm not a hyper-purist about tilling and spraying (Organic products, only). If you strongly feel you need to do it, especially in the beginning, go right ahead. I simply find that those activities are no longer necessary for me to produce the food I want on my homestead. I've learned the value of creating systems that mimic nature, using the power and energy of already available resources to thrive. Even when things appear to be "wrong," I can trust that nature will figure it out in the end. In short, I work smarter because I understand the natural relationships occurring in my garden and choose to nurture them. I guess you could call permaculture the love language of the garden.

> **THE FRUIT TREE GUILD EXAMPLE**
>
> Amy Stross says,
>
> "Let's compare planting a fruit tree in a conventional garden versus planting a fruit tree in a permaculture garden. With conventional gardening, we might dig up a ring of grass, plant the fruit tree with some store-bought fertilizer, water it, and then mulch it with some store-bought mulch. If the tree became infested with a pest or disease, we would look for a spray of some kind to kill it.
>
> "In permaculture, however, we try to model what nature would do. Underneath the trees of a forest are usually a random collection of a diversity of shrubs, herbs, ground covers, and sometimes, even vines. Ironically, this plant diversity can improve the tree's vigor.

> "In permaculture we want to under-plant a fruit or nut tree using this forest model. A fruit tree planted in this way is called a guild—a central element (fruit tree) under-planted with support plants.
>
> "To do so, we look for plants that do five things: they naturally fertilize, attract beneficial insects, resist pests and disease, naturally provide mulch and grass suppression, have edible or wellness qualities
>
> "We look for plants that are multi-functional, i.e., plants that do more than one of these important tasks. For example, Russian comfrey will fertilize, attract beneficial insects, and provide mulch. Then we plant those plants underneath our fruit trees. We gather as many multi-functional plants as it takes to meet all of the five needs of a tree. In this way, we don't have to purchase fertilizer, mulch, or insecticide; the plants do this for us, which saves us money. And since we don't have to haul in mulch or treat pests, it saves us time, too.
>
> "We are regenerating fertility and biodiversity because even if we leave the property, those fruit tree guilds, with their naturally fertilizing underplantings, will continue to enrich the soil, the flowers will feed beneficial insects and pollinators, and the fruit will continue to feed wildlife."

*As with every section in this book, but especially with this one, do NOT stop your search for information on permaculture with just this information! The following is not meant to be a comprehensive explanation of the whole philosophy of permaculture—there are several worthy books that can do that for you. This will be more like taking a university textbook and turning it into a child's board book. We're sticking to the very basics so that you can feel empowered to learn more!

The Power of Mulch

Every gardener worth their salt knows that what you're really growing in a successful garden is good dirt. I think one of the easiest places for conventional growers to begin with permaculture is with mulch. Mulch is everybody's friend, making soil rich and healthy over time. Even casual gardeners can appreciate it, if only for the basic reason that mulch is attractive once you lay it out on top of your garden beds. Mulch, made up of material like shredded leaves, wood chips, shavings and pine needles, acts to retain water in your garden, encourages the presence of beneficial microbes, provides nutrition for your plants over time, cools the soil and, in general, provides a cozy place for your plants to thrive. If you leave it long enough and it breaks down, it becomes a kind of compost.

Manure and compost are helpful soil amendments made from various types of organic matter that can be applied as a top dressing to work their way down into the soil over time. As Phil Nauta reminds us in his book *Building Soils Naturally*,

"We've seen that organic matter takes center stage when it comes to water-holding capacity and drainage, promotion of air in the soil and resistance to compaction, holding onto cations and anions, and providing fertility because it is made of nutrients and other substances.... Not only is it a source of nitrogen, but nitrogen-fixing bacteria actually need it for energy, so they can do their job."

If you've ever read about going back to Eden in your yard, lasagna gardening or sheet mulching, then you may have already learned about the benefits of having layers and layers of mulch and other organic materials in your growing space. Mulch can be turned into your soil to add organic content but my favorite way to use it is for sheet mulching. With sheet mulching you to take a patch of unused or abused ground, cover it with a layer of cardboard, and then add 4"-6" layers each of compost, manure and mulch. You can put down these layers a season in advance of your planting; put them down in fall to plant in spring. Or, if the materials are suitably composted, you can plant directly into these layers whenever you're ready. You do NOT have to pull up grass to create a new growing space on your homestead; simply smother it and cover it with sheet mulching. See, smarter.

If you have an ugly space in your homestead garden or you just can't get the weeds clear in that one patch, please try this method. Even if you must wait over several growing seasons to observe all the results you're looking for, I promise that nature will find a way to even everything out. I've seen it repeatedly. Toby Hemenway, author of *Gaia's Garden*, tells of a time when he accidentally brought field bindweed into his garden with a big haul of wood chips. Bindweed would grow on the dark side of the moon and can re-plant itself in new places with just the tiniest spit of a root. He was bemoaning his fate as he ripped it out vigorously but, over the next few years, as he added more and more mulch, he noticed that the field bindweed was lessening until it disappeared completely.

Have faith, keep adding organic content in several inch layers each year and the garden will do the rest. See the *DIY section* below for more thoughts on compost and mulch.

The Power of Bugs

Most of us are trained, from a young age, to think of bugs in the garden as being evil. Except for butterflies, maybe. And honey bees. And lady bugs. Well, maybe bugs aren't so bad after all. Just like with people there are "good" bugs and "bad" bugs — bugs that do what you want them to do and bugs that do the opposite. They're both present in pretty much any garden but the real question we need to be asking ourselves is, what proportion of Good vs. Evil do I have going on in my garden? You're going to have bad bugs, even if you use every pesticide marketed today, so just get over it. The key is to keep your good and bad balanced. It's very Zen.

Amy contributes the following to help us correct our way of thinking about the bug issue:

"In a conventional garden, if pests are found, we usually look for a spray to get rid of the pest. In permaculture, however, we think about all the reasons why pests might be attracted to a particular crop. We look for ways that the crop isn't getting what it needs to thrive and therefore has a weakened immune system. We ask questions like: Does the soil lack organic matter, fertility, proper drainage, or is the pH wrong for that crop? Is the plant not getting watered or is it located in too much or too little sun? How about air circulation?

"Nature grows what it wants and doesn't grow what it doesn't want.
[emphasis added]

"If a particular plant (fruit tree or vegetable crop) just won't grow without a ton of pest care and coddling, it is probably not well-suited for that site. In this case, nature would choose to grow something else, and that is what a permaculturist usually decides to do. For example, I realized after four years of trying to grow winter squash and failing (squash bugs) that my yard is not an appropriate site for this crop.

"Our land is bowl-shaped with little air circulation and is perfect for attracting squash diseases, which in turn, attract squash bugs. There's nothing I can do to improve the air circulation of my property, so squash plants here will always have an ailing immune system that makes them susceptible to pest and disease. I could keep trying to grow it and be disappointed, or I can grow any number of other crops that seem to thrive here without coddling."

What a great perspective and it's so true that the first rule of successful gardening is to grow what will grow, and bugs can help! The best way to ensure that you have as many beneficial insects (good bugs) in your garden as possible is to invite them to visit and stay for as long as possible. Think of yourself as a gracious host. What does a good host always provide for his guests? Why, a comfortable place to rest, refreshment and the safety of his protection, of course. Permaculture encourages the gardener to plant a wide range of plants, all of which perform several functions (we'll talk more on that in a minute) and one of those functions is to encourage beneficial insects to come and stay.

By planting a wide variety of flowering plants, native grasses, herbs, vines, trees and shrubs, we create a place of comfort, refreshment and safety for the myriad of insects and arachnids that can act as sentinels and warriors fighting for the survival of our garden. When you see a tomato hornworm in your garden, don't freak out. If you've done your job as a host, there will be parasitic wasps laying eggs in that hornworm that then feed on his flesh as they mature. Once they hatch, they'll spin cocoons on his back, further weakening him until they hatch out leaving him, quite literally, deflated...and dead.

Believe me, whatever minimal amount of damage that one hornworm inflicts on your tomato plant, the number of beneficial wasps (don't worry, they're so tiny you hardly notice them, and they don't hurt people) that will then be living in your garden will make it all worth it. If your bug numbers have gotten out of balance (important word) and you have too many hornworms, remove the ones without cocoons and feed them to your chickens, who will love them.

This is just one example of the dance that is going on in a healthy, balanced garden where bugs serve a vital purpose. Before you try to eradicate that one pest, study it. Figure out why it's there and what it's doing. Do you have aphids swarming one rose that's yellowing but hardly any at all on the rose that has dark, green foliage? This tells you that you most likely have a soil problem that needs to be addressed—yellow leaves tell the bad bugs that the diner is open, and lunch is on the house. If they can easily feed off a weak plant, they'll often leave a healthy plant in peace. The point is, the bugs tell a story and that story can be a powerful tool in helping you fortify the weak places and duplicate the strong areas in your garden.

Just a side note, it's been my experience that nearly every time my garden gets out of balance, the problem can be traced back to the soil. So, don't neglect the power of mulch and manure. Grow good dirt before anything else and learn to pay attention to what the garden is trying to tell you. As Jared Stanley reminds us,

> "It may well be a life-long effort to look at one acre of land and try to understand every single connection that takes place inside of it. How many migratory birds fly through and deposit nitrogen rich manure? How many cubic feet of organic material falls to the ground each fall? As summer came on how did the bumper crop of ants affect the soil structure and did that help water and nutrients reach further into the soil?
>
> "Questions like those typically overwhelm people, as they should. We don't need to understand those numbers, we just need to understand that those associations DO happen. Once we can understand that then we can start to think about our own yards in a new way."

The Power of Water

An argument could be made that I should have made this the first principle we talked about since water is so basic to life—no water, no life. Many of us take water for granted, thinking of it as a never-ending resource that comes from a hose. That may be where many of us access water, but it is not, in fact, it's point of origin. Water is in the land and the air all at once, surrounding us and sharing its energy with us so that we may live. God be praised for the water! Permaculture helps us establish a working relationship with the water that's available to us and the principles of water management are sound whether you live in highland deserts or humid forests. The key to managing water is to pay attention. As Mark Shepard explains in his book *Restoration Agriculture*,

> "Since water is of such critical importance for plant life and growth, the very first step for a restoration agriculture farmer, no matter where the farm is located, is to optimize the land's relationship with water."

A homestead designed with permaculture principles will first figure out where it stands with water that is already present in the ground and what typically falls from the sky. If the area is a dry one, steps will be taken to build up soil fertility, so the dirt holds what water does come in, this includes preventing erosion on hillsides. Varieties of ground covers, herbs, native flowers, small shrubs and even trees will be selected for their ability to thrive on less water as the garden becomes established. Mulch can be layered on top of barren soil to increase fertility over time and to hold water where it needs to be. If there is an abundance of water, then strategically dug ditches can direct the water to plants that need it, thereby avoiding useless run-off. These ditches are called "swales." If appropriate, ponds can be dug and used to slowly fill swales in dry times. Developing a relationship with the water on your homestead is about analyzing where it wants to go and where you need it to go. If the water's path is destructive, we re-direct it with soil manipulation and plantings until the garden soil can absorb the water on its own to be used later.

Mark Shephard also teaches us that,

> "Water is held in the pore spaces between soil grains. All life in the soil is 50-80 percent water and represents its own type of water storage. In short, the soil and the soil life represent a massive water storage system. In restoration agriculture systems we want to start with storing more water in the soil...By slowing down fallen rainwater and spreading it out you then allow it to have an increased residence time in the landscape. This gives the rain time to soak in rather than run away."

You may find that rain barrels are especially useful for you with water collection and redirection, especially if you're on a small lot. Water gardens, cisterns, ponds and marshes all can play a part in water retention and provision on your land. The key is to figure out which of these options is best suited to your space. Try a few ideas and see which function best for you—keep going until you find a system that works! There are countless permaculture success stories of people taking literally dead land, like doornail-dead, and turning it into a lush garden just by using these principles. Water is its own kind of energy and energy is precious. Like water itself we need to be energetic in our efforts, strong in our vision and wear away at the barren until life springs forth. One man's garden is another man's metaphor.

> "Energy is very important in a Permaculture system. Think of it like this: If a cup of water, and the nutrients it transports, were pouring down a ramp and off to nowhere it would be a total waste. Nothing is gained from it. However, if we can capture that energy then we can employ its use—over and over again. Energy could be human, wind, solar, water, animal, etc. Generally, we are looking for ways to expend as little energy as possible while retaining as much as possible."
>
> - Jared Stanley

The Power of One

I like to be efficient. I'm also very task-oriented and if I can get two duties done by performing one job, I'm very happy. The point I enjoy probably the most in permaculture is that, ideally, each plant is performing multiple functions in the garden. Every small space homesteader should prick up their ears at that. I'll say it again, ideally, in a permaculture garden, every plant is doing multiple jobs in one place. Let me illustrate with a personal favorite, the Guomi bush.

I originally planted Guomi bushes simply because they produce edible fruit and are zoned for my cold climate (they're hardy to -20°F/-29°C). The year I planted them I had annihilated my front lawn and put in an edible/herbal garden. Guomi was on my purchase list from my favorite online plant source, Raintree Nursery. I wouldn't read about sheet mulching for at least another year, so we dutifully rented a sod cutter and removed our grass via that method. We gave the sod away on our local, online classifieds, which was nice, but I could have prevented strain on my poor back if I'd known simply to layer on the compost and mulch over the top of the grass and weeds. I did add some of those items, as well as manure, but not nearly enough. Ah, well, live and learn.

Anyway, I planted the Guomi and after a few years noticed how well they were doing and how well other things grew around them. I read up and, sure enough, Guomi fix nitrogen in the soil. Certain plants have the capacity to take nitrogen out of the air and convert it to a usable form in the soil (there's way more chemistry happening there, but that's basically what they do). Also, Guomi are lovely in the garden with their demure berries, their lovely green and silver foliage and their pleasing shape. Their prolific leafing can be pruned back to provide compost fodder or mulch. The bush is drought tolerant and not particularly picky about what kind of soil it's forced to grow in. Because of its medium height, there was plenty of room for it near a full-sized tree and with the afternoon shade it provided in our hot, dry summers, lower-lying plants were able to thrive around it. So, with one plant I had nitrogen in the soil, compost components, edible fruit, an attractive bedding plant, and drought and frost tolerance and, oh, did I mention how much the bees and other pollinators loved the sweet, spring flowers of the Guomi?

Just like with people, you should never underestimate the power of one plant to make a large impact on your productivity in the garden. Mostly, I hope that you and I will find success in our gardens as we step back long enough to observe how the garden is humming along. When problems do occur, strive to regain balance by thinking of which system might be out of whack—is it the soil, the water, a vital plant that's missing? How can you help the collections of plants on your homestead behave like a loving family instead of a mess of botanicals at war with their environment? How can you work your land less while having it produce more? I submit that permaculture will probably have an answer to all those questions.

> "Please also remember that permaculture is not just about growing plants. It's about how buildings should be constructed and where animals, structures and plants should be placed. It is about community development and regeneration of the planet. It is about being a positive force for the earth. Gardening is a great way to be introduced to Permaculture, but it is so much more."
>
> - Jared Stanley

ACTION ITEMS: Get online and search your city's name and the word "permaculture" together—note what pops up. Did you get any hits? Write them down in your homestead journal as potential contacts for learning more as you try to figure out how permaculture applies to your homestead. Now do the same thing with the name of your state just in case you didn't get any hits with your city's name. You may not end up with local permaculture mentors in your area (at least, not ones that can be found online) but there are a lot of resources on the web. Start searching and take notes of what looks interesting or any source that generates questions.

Check out *Gaia's Garden* from the library and start to read it. Do you have any of the plants mentioned there already planted in your garden? If so, make a list of plants you might use to flesh out the areas surrounding those already present. Are there other resources already in your garden like fruit trees, water collection devices and beneficial animal habitats? Is there a framework already in the garden upon which you can build, or should you start doodling design notes for your new permaculture garden?

DIY: Rainwater catchment is a simple place to start as you begin to adapt permaculture principles to your existing homestead. If you have land and can begin to place swales in your garden areas, do that. Swales direct rainwater exactly where we want it to go by using the soil to create pathways for it. If your garden is very small, or if you simply want to begin with catching rain in containers to pour out via hose or watering can at your leisure, then building your own rain barrel may be the best place to start for you. If you want to learn how to make some simple rain barrels for your property, please see the *Water - Storage and Conservation* section in The Prepared Homestead chapter.

Another great permaculture practice to implement, besides water direction, is the production and use of compost. At this homesteaded level you may already have an established compost system. If not, I encourage you to immediately do some reading on the topic and create one—even if it's just a cool compost bed (basically one you don't bother to turn or tend much). For a great read, try *Let it Rot: The Gardener's Guide to Composting*, by Stu Campbell, a Storey's Down-to-Earth Guide that will give you the run down on the basics of composting.

For me, one of the hardest parts of an active compost pile is getting it turned, which helps the decomposition process and produces high quality compost faster. There are two solutions to this problem that I've found to be the most useful. One, is to use my chickens and pigs to turn my compost. The other is to use a compost tumbling machine—more on that later. If chickens are allowed access to the compost pile, they will not only eat up much of what is there, pooping out valuable contributions to the compost, but they will also incessantly scratch and move the contents of the pile. They're extremely effective at this—sometimes I wish I could set them to cleaning up my house.

Pigs will work into the ground for the most part, constantly tilling up the dregs of the compost as they root around looking for edible morsels. Usually heritage breeds, with their typically stronger snouts and disposition to root, are more naturally adept at rooting than others, but really any pig will be useful to you in this endeavor. A chicken will eat pretty much anything, even chicken. Ew. Pigs are a lot more discerning and won't eat something they distrust, being highly intelligent little critters. Even so, there are a few materials that I don't put into my compost heap such as egg shells (which I wash, dry and crush right into my garden soil). I also preclude items like plastic, meat and anything poisonous.

There are several ways to set up a compost area so that it's accessible to animals. If your space is large and you have a lot of animals, piling manure, kitchen scraps and homestead debris into one massive pile and letting your animals have at it is simple and effective. Be sure to fence the area, especially if you're using pigs and/or use a livestock guardian animal to oversee the process to prevent animals from wandering off. If space is limited, creating a three-sided box with some pallets wherein you dump all your compostable materials is a good solution. Your livestock can access the materials from the open side or from the top and it keeps the resulting compost somewhat contained. During the day, the chickens especially may disperse the materials as they scratch, but a few swipes with a rake can usually get things back in order. I never worry too much about mess with my compost and just let the animals do their thing. However, for those homesteaders in municipalities or who simply like things tidy, it may be an issue.

If you don't have chickens and pigs to assist you in turning the compost, you might consider building a compost tumbler. A compost tumbler is a drum fitted with a handle that connects to an internal agitator. You put your compostable materials into the drum, add water to keep things damp and then turn the tumbler every day for a few weeks. This constant turning, coupled with the water, produces compost in a matter of weeks, as opposed to months with a traditional pile. You can certainly buy commercial versions of a compost tumbler if you don't mind spending the money. If you would prefer to DIY this one, go to our links page (www.homesteadlady.com/homestead-links-page)under The Homestead Garden chapter for a few suggestions on building compost tumblers with five-gallon buckets and a fifty-five-gallon drum.

You can also check your local, online classifieds to purchase used tumblers.

BUILD COMMUNITY: Get together with your homestead group or book club and purchase the necessary plants to create a simple apple guild. Find a local school (a charter or private school would probably be easiest), care facility, halfway house or rehabilitation center that already has a watering system in place and offer to donate the plants and your labor to get the guild established. Make sure they understand that this will be a *fruiting* tree and they'll need to plan to harvest from it.

Make a schedule for the next six to twelve months for various members of your group to stop by and do simple maintenance on the guild—prune the apple tree, do chop and drop with your green manure plant, dead-head the pollinator flowers once they're spent, etc. While you're there, train a staff member or teacher at the school to take care of those few chores so that they can enjoy an apple harvest every year. These guilds are meant to be self-sustaining and, in the end, you get apples. Sweet rewards.

Similarly, find out if your local prison or youth detention center has any kind of garden to which you could donate the supplies and labor to set up a few plant guilds. If no prison near you has a garden project, and you have lots of energy, you can investigate starting a program yourself.

When I was still in my master gardener classes in central Utah, we toured Utah County's prison garden that worked in part through the agricultural extension of Utah State. That was one of the most wonderful garden tours I've ever been on! The garden was growing beautifully under the care of the men who enjoyed the privilege of working with the earth in the fresh air, and all the inmates got the benefit of fresh veggies from their harvests. They even sold their crops at local farmers markets. The real benefit to the program wasn't the money, it was the difference that the opportunity to nourish something in the earth brought these men. To read more about the program and get inspiration, you can find the article list on our links page under The Homestead Garden chapter.

BOOKS TO READ: Amy Stoss's own book, *The Suburban Microfarm: Modern Solutions for Busy People* is a fantastic read for smart permaculture principles in normal human-speak. *Gaia's Garden*, by Toby Hemenway is equally useful. Every gardener should probably own both books, even if they don't employ every single principle right away. This is permaculture explained simply and thoroughly for the everyday gardener. Once you master the principles in these books, you can go off and explore other permaculture titles because they probably all have something to teach. Amy and Toby will explain, if nothing else, the "why" of permaculture. It's an easy sell after that, in my opinion.

For those looking for more in-depth information, *Restoration Agriculture*, by Mark Shephard is so thorough on a variety of real-life permaculture topics. Mr. Shephard walks the walk and really knows from experience what he's talking about. Apart from the other useful material, his chapters on integrating animals into the permaculture homestead alone are worth reading the book. It's a great resource, but don't read it if you're completely new to permaculture as it might be too much information that will short out your brain for a time. Ask me how I know.

Amy Stross says,

> "If you're interested in diving deep into permaculture plants and design concepts, Edible Forest Gardens, Vol. 2, by Dave Jacke has been indispensable to me. It isn't cheap, but the plant lists of nutrient accumulators, mulch plants, and pest repellents, etc., etc. are simply amazing."

Remember, for basic composting information, *Let it Rot: The Gardener's Guide to Composting*, by Stu Campbell, or really any publication or post on the topic. I guarantee that the more you read up on compost, the better your methods will become.

WEBSITES: The official site of the Permaculture Research Institute is the home of permaculture online (www.permaculturenews.org). Sun Calc (www.suncalc.net) helps you to determine the sun's path on your property.

Says permaculture designer, Jared Stanley,

> "A good resource for our temperate climate in the USA is Temperate Climate Permaculture (www.tcpermaculture.com). The author, John Kitsteiner, does a good job of quantifying all the data he puts on that website. He doesn't just list a plant because it is 'common knowledge' that it is good for a particular use, he bases his listings on data that has been researched."

I checked it out on Jared's recommendation and, he's right, it's helpful.

Both Amy's (www.tenthacrefarm.com) and Jared's (www.jandjacres.com) sites are quality permaculture blogs that will help you learn basic principles. Jared is very active and effective on YouTube. Grow, Forage, Cook, Ferment (www.growforagecookferment.com) will be helpful to you as you learn to think of plants as multi-functional in the garden, the kitchen and the home. You may also find value in Organic Life Guru's site (www.organiclifeguru.com) that offers courses from experts like Toby Hemenway and Ross Conrad.

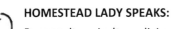 **HOMESTEAD LADY SPEAKS:**
Permaculture isn't a religion, so relax. As often happens in the real foods movement, and the natural parenting community when you mention topics like Cheetos or disposable diapers, if you start talking about tilling your garden, some rabid permaculturist may verbally pounce on you. Don't worry, for every crazy there are at least 100 normal gardeners out there who simply enjoy the increased soil fertility and symbiotic relationships they can create on their homesteads by employing permaculture principles. Talk to lots of gardeners, read several permaculture books and just dig into learning. There's so much more for you and I to explore on this topic and we shouldn't let anyone tell us that there's only one way to grow a garden.

Don't let a fear of doing it wrong or doing it differently keep you from doing something. Begin from where you are and start with just one of our simplified principles. As Toby Hemenway writes in his book, *Gaia's Garden*,

> "Overall, doing an imperfect something is better than doing a perfect nothing."

Having said that, however, it is important to learn to make the most of our permaculture assets. I asked Jared Stanley to weigh in on a question I had. I asked him:

For the homesteader who's never even heard of permaculture, what are five action items you can give them to focus on right now that will head them in the permaculture direction?

Here was his wise reply:

1. "**Identify what you already have**: Learn about wild plants and what is already growing that you can use for yourself, your animals, etc.

2. "**Learn why certain plants grow in certain areas**: Plants can be key in determining what is going right or wrong with your soil—and if your soil isn't ready for the plants you want to grow, then you are putting the cart before the horse.

3. "**Make Compost:** If that amounts to a small 'cold compost' pile that won't be ready until next year or heaps of wood chips and horse manure—make compost! It is by far the best thing you can do for your soil and therefore your crop production.

4. "**Don't Fight Nature:** Nature has been working just fine before humans got involved and it will keep on working after we're all gone. Start fighting for the winning team instead of against it. If your climate is too hot and humid for your favorite store-bought apple variety then it is time to find a new variety, it is not time tell nature that you're going to grow it anyway. She'll just laugh at you.

5. "**Start Small:** There is a reason this advice has been repeated since the beginning of time. It's true. Don't go out and try to "permaculture" your whole property at one time or in one season. You'll spend more resources than it's worth. Plan it out and phase it in over time. You and your property will be much happier that way."

CHAPTER 3

Green the Homestead

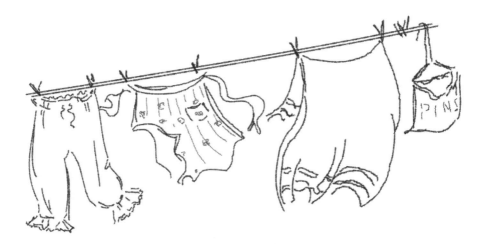

There many valid ways you and I can be good stewards of the resources we have, while providing for our needs in an economically and ethically sound way. As homesteaders, the number and kind of "seeds" that we plant in many different areas brings us rewards in similarly diverse ways. If the term *Going Green* is familiar to you, then you've probably already thought of some changes you can make to your lifestyle that will make it more renewable. Revised habits can reduce the amount of input you require, reduce the amount of waste you produce, and teach you to re-use everything you can. These ideas have found their way into the political arenas of our time, so try to separate yourself from that for a bit, as politics are not our focus here.

Recycling and *upcycling* are two fairly new terms, but both can assist us as we begin to improve our stewardship over the little plot of earth that is ours. Have you ever thought that by deep cleaning your home you might be going green? The subject is broad, and we'll be covering some topics that might appear unrelated. What's shared in this chapter are projects we've tackled on our homestead that have increased our appreciation for the resources we have, while helping us to use them better. I am so grateful for this earth and the bounty and beauty it provides. That sounds like an overly simplistic statement of a sentiment that is so large that it swells and edifies my soul. I am so *grateful* for this earth.

Here's what we'll be discussing in this chapter about going green on the modern homestead:

HOMESTARTER level: We'll be talking trash in this section—upcycling, recycling and waste! Homesteaders don't like to waste anything, ever. However, our culture has produced in most of us a natural tendency to discard when we might instead repurpose. Here we'll cover a few basic ideas on how to look at waste in a whole new light.

HOMESTEADISH level: In this section, we'll help you ask yourself some hard questions about downsizing and de-cluttering your home so that you have more mental and emotional energy for the homestead. Ever felt like your "stuff" rules your life? Learn to gratefully pass it on to someone else as you clear the cobwebs and start a new chapter on your homestead.

HOMESTEADAHOLIC level: Green up your homestead by, literally, bringing in the green! Learn to forage for wild food and wellness plants to make good use of the natural resources all around you. Here we'll cover some practical advice on the basics of foraging from a seasoned forager and plant-lover.

HOMESTEADED level: Energy is a topic that every homesteader thinks about at whatever level of homesteading they're living. Here we'll be talking about practical ways to save energy all over the homestead, as well as some ways to generate our own. Everything takes energy to run, including you; let's learn to conserve and expand the energy available to us.

As with every chapter in this book, be sure to follow along at the end of each section as you reach the Action Items, DIY Projects, ways to Build Community, resource recommendations and parting advice.

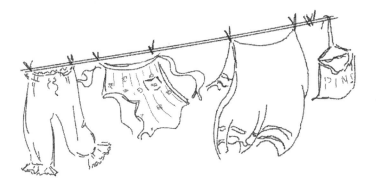

"Use it up, wear it out, and make it do or do without."

- Attributed to American Pioneers of the early West
and Homemakers of the Great Depression

For the Homestarter: Who Put the "Up" in Recycle?

First, what's the difference between *recycling* and *upcycling*? Briefly, *recycling*, in our modern culture, is taking an item after it's been used for one purpose and chemically altering it to form it into another item that can be used anew. For example, when you're done with your juice bottle, you rinse it and put it in your recycling bin. Your local recycling truck picks it up and takes it to the recycling facility where it's broken down and turned into a new plastic item. *Upcycling* is when you take the same plastic bottle, clean it out and turn it into a candle mold for your candle making project. That bottle spends the rest of its life in your candle making box being useful.

The point of either exercise is to retrain our brains to look at items we might otherwise discard and give them a second chance at being useful. I'm not going to deliver a lecture on the ecological repercussions of throwing stuff away because it shouldn't be necessary. If you're reading this book then, at some level, you're seeking to be more self-reliant. You are becoming more aware of what you consume and how you can reduce it and/or produce it yourself. So, all I have to say to you is that throwing stuff away is a waste. An absolute waste. Ack! Homesteaders don't do waste!!

Wasted Food

Let's start with food and waste. I've read some online studies that claim Americans waste as much as 40% of their food—meaning, they throw it out to rot in the landfill. The solution to this problem in almost all these articles is governmental; to form a committee to explore the issue and educate the public on food awareness and waste reduction. I suppose that would be one way to tackle the problem. But homesteaders don't need a committee to tell them how to stop wasting food. We just need to inspire each other with ideas we haven't thought of on our own or didn't learn from our moms.

For example, here are a few kitchen and homestead tips to reduce food waste (some of these you may recognize from the *Use All the Parts* section of The Homestead Kitchen chapter):

1. When you roast a chicken, save the extra meat for chicken tacos and use the bones to make broth or stock. You can use the bones an extra one, or even two times; just add a tablespoon of vinegar to the new batch of bone broth to draw more flavor and minerals out of the bones. Freeze or can any leftover chicken meat or broth that you can't consume right away. Any meaty bits or scraps of skin can be fed to your cat or dog.
2. Don't discard corn cobs, shrimp shells or vegetable tops for the same reason—they make great stock. Stock doesn't have to be saved for just big batches of soup. A cup of stock can be used to sauté vegetables, create a sauce, or be added to the rice pot for richer flavor.
3. Instead of throwing out bread scraps, save them to make bread pudding, bread crumbs and croutons. Keep a bag of scraps in the freezer until you're ready to use them.
4. Raise a few chickens. Chickens make quick work of any leftovers or produce that, despite your best efforts, you still end up not being able to use in time. Chickens will, literally, eat anything. Chickens will even eat chicken. Bleh. (I never feed our poultry with scraps of poultry dishes because it seems supremely inhumane to turn an animal into a cannibal without its knowledge. A chicken sees food and it eats it. Feeding it chicken or turkey isn't fair or just, in my opinion.) Very little of your food need EVER go in the trash because the chickens will take care of it.
5. Also, get a compost bin as soon as possible, especially if you can't have chickens where you live. A compost bin will absorb all your rotting veggies and fruits, coffee grounds, tea bags, shredded junk mail, ripped up paper plates, paper shopping bags and any other organic material that can break down with water and heat. If you don't have room in your garden, or if you don't have a garden, you can still compost all that stuff with worms (see our *Vermicomposting* section in the Livestock Wherever You Are chapter). If you can't use the compost yourself, I am sure that you can find a gardening friend somewhere who can. Better yet, join a community garden in your area and use it yourself. You do NOT need to throw those things away ever again. Turn them into something useful—compost is the ultimate upcycled material!
6. Save for and purchase a glass or stainless-steel water bottle so that you don't need to buy plastic water bottles ever again. Besides, have you ever taken a drink on a

hot day from a plastic water bottle that's been sitting in your equally hot car? What does it taste like? Plastic, right? Purchase a high quality, reusable bottle and that will never happen again. Just FYI, I like the glass bottles with the silicone sleeves. They prevent breakage and the water always tastes cleaner. Get stainless steel cups for hiking and park days because they're not as heavy as the glass bottles.

7. Instead of going to the grocery store for specific ingredients for that special recipe, have one or two nights a week where you use accumulated leftovers to create a new meal. This takes practice if you're not used to preparing meals this way—so, practice!

For example: take that roasted chicken mentioned in step one to prepare dinner.

- **Sunday** night – prepare roast chicken with rice and carrots.
- **Monday** night - have chicken tacos and make your rice Spanish by adding powdered or canned tomatoes and a little spice.
- **Tuesday** night - make a Mexican lasagna with the leftover tortillas and beans (and reserve some for later in the week). Add rice and any leftover chicken.
- **Wednesday** night - take out some of the broth you made in number one. Prepare a lovely minestrone with the stock, reserved beans from Tuesday night, a bit of pasta and some veggies.
- **Thursday** night - have any leftover minestrone served in homemade, sourdough bread bowls (make some extra dough).
- **Friday** night – make it homemade pizza night with the extra dough you refrigerated and whatever veggies and cheese you have hanging out in your fridge. Splurge and fry up some bacon as a topping because you can use it as a garnish on the quiche you'll make...
- **Saturday** to use up your stockpile of eggs.

Without even breaking a sweat, you have a week's worth of meals planned and it all started with a chicken!

A Word on Recycling

Recycling facilities have an environmental impact just like any industrial facility. But since plastics, glass, paper and metals are not going to fall out of use any time soon, it seems like a logical thing to simply recycle whenever and wherever we can. Most municipalities in the U.S. have curb-side recycling program, while others, especially for more rural areas, allow drop-offs at the recycling centers themselves. Does it take some work to rinse and set aside recyclable material? Sure, but since when did work scare a homesteader? You'll want to recycle materials wherever you can because useless waste makes the hairs rise along the back of your neck with the pure wrongness of throwing something away that could reasonably be turned into something useful.

If you're new to collecting recyclable materials, set up a simple recycling station in your home. Pick a spot where both the garbage and recycling containers can be placed together. If your collection service allows you to combine recyclable materials, make your recycling bin the same size as your garbage bin to show that these two have equal importance in your home. Pick a recycling bin that's a different color from your garbage can and write "RECYCLING" on the top, so that guests and small children know what to put into it. If you have curbside pick-up of your recycling, you should be able to simply dump the contents of your indoor recycling bin into your outdoor pail, without taking out the plastic bag lining your bin. Recycling must be rinsed and clean to be accepted, so your bin won't really get dirty. We are usually able to use the same bag lining our recycling bin for many months.

Some towns require residents to sort recycling into paper, plastic and metal. If so, there are several stacked bins you can purchase, often designed for sorting children's toys or shoes, at various home improvement stores. For others, your city will provide them. Overseeing sorting the recyclables is a great job for kids because it can be fun to put things in their proper place, and it's an important accomplishment in the home. Just be sure to double check that the food containers have been cleaned out and that the baby hasn't added pet food to the bin while your back was turned. Your children can also be on the watch for any materials that might be repurposed in your home instead of recycled. My kids are so good about pulling various kinds of paper out of the recycling and putting it into the kindling pile for the wood stove or outdoor fire pit.

> **IT'S SIMPLE, REALLY**
>
> Of all the many ways we can live more mindfully, most municipalities see to it that recycling is one of the simplest ways to get started. If you live outside city limits, recycling will require a bit more work since you'll most likely need to bring your items to a recycling center. However, for those who enjoy city-wide recycling programs, you shouldn't have to go too much out of your way to participate. Don't let the prospect of setting up a bin or two and keeping a list of what your city recycles stop you.
>
> The first place to visit is your city's website, which will tell you when your recycling is picked up and what can be properly placed in your containers. Be sure to thoroughly wash all your recyclable containers or they just might end up back in the landfill after all your trouble. I saw a presentation during a city council meeting from a gentleman who worked for the city's recycling center. He reluctantly reported that his facility had to send about half of what they received on to the dump because the items either weren't recyclable or were too soiled to qualify for recycling. Wash out your milk cartons, rinse your cans and keep your cardboard grease free.
>
> Don't forget your city's yard waste program, especially if you don't have a lot large enough for a compost bin. (Although, even on a small lot, you can still recycle things like grass

clippings and leaves into useful soil amendments for your garden.) Often your city will take your green waste and turn it into compost themselves, selling it back to residents at prices much cheaper than they could purchase it at garden centers.

Some items like magazines can be given away on sites like freecycle.org or donated to schools and libraries. For more ideas, visit sites like Earth 911 (www.earth911.com; check out their "Recycle Guide" tab at the top of their homepage).

A Word on Upcycling

Do you know what I read the other day? An entire home decor book consisting entirely of items made from upcycled materials. Home decor! No joke. Some people just naturally have a gift for looking at an object and seeing its infinite possibilities. Hence the phrase, "one man's junk is another man's treasure." Often also called "re-purposing," this is a must-have skill for our times, I believe. Our culture creates and creates and creates, and it's marvelous! However, we have so much stuff. SO. MUCH. STUFF. (We'll address that further in the next section.) Often, perfectly useful items wear out and are no longer usable as originally designed. However, we can redesign them and make them useful and beautiful again.

I am the first to admit that making cool home decor "thingies" is not my forte, however much I admire it. I'm much better with utilitarian upcycles, like turning a table and some benches into a chicken brooder. Or making a towel hook out of a horseshoe. I geek out over upcycles like that.

My favorite surprise in upcycling occurred during a recent energy drill where we turned off all lights in the house for a day and night to learn to do without. I kept my stove on (this was just a light drill) but soon discovered that there was no way to position a flashlight, lantern or candle so that I could see what I was cooking. I'd just finished using up the contents of a glass avocado oil bottle and, in desperation, I shoved a short taper candle in it and lit it up. The height of the bottle was perfect to diffuse the light over what I was doing. It's not like I was the first person to think of such a simple thing. Historically, the reason candlesticks were rather tall was so that diners could see their plates in an era of no electric overhead lighting. It was just a neat experience to realize a need and be required to look around for a non-obvious solution.

SOME SIMPLE UPCYCLE IDEAS FOR YOUR HOME:

1. Convert that milk jug or spaghetti sauce jar into the perfect waterer for your fragile, indoor-started seeds by drilling tiny holes in the lid, filling the container with water, securing the lid and sprinkling the water on like salt and pepper. I use these all the time! At the end of the season I recycle the container.
2. On that note, yogurt and sour cream containers make great seed starting pots if you just pop a few holes in the bottom for drainage. Keep the lids to cover the

seeds for the first few days to keep them moist and warm. These also can be washed and recycled after the season.

3. Ever end up with those cheap-o, made-in-China baskets lying around? (Where do they come from because I don't buy them?!) Instead of tossing them or donating them, spray paint them to give them new life. If you decorate for holidays, use holiday-themed paint colors and create catchalls for kid's crafts and instant containers for holiday gift baskets. The homemade bread for your friend will look much better in one of these baskets.

4. Save all the clothes and sheets that you just can't repair anymore and cut them into 1/2-inch strips. Connect the strips by sewing a 1/4" seam across the ends. Wait until you have long strips of about the same length, braid them together, and stitch them side by side in an oval shape to create a rag rug. This is not brain surgery, don't be intimidated! The Internet is your friend if you want to learn more about the specifics but making a rug like this is simple.

Remember, these ideas are only a place to start. The main point is to begin retraining our brain on what we think of as "waste."

> "Everything that is produced on a homestead is recyclable on that homestead in one way or another. No homeowner owns a plastic making machine. Plastic is imported onto the homestead by the homesteader. Reduce the amount of plastic that is brought to your homestead. All veggie scraps go in the garden to compost. Glass is 100% recyclable and reusable, paper and cardboard can be tinder for your woodstove, weeds are compost, leaves are mulch, and scrap materials have future uses."
>
> - Joel Schwartz (www.growcookforageferment.com)

ACTION ITEMS: If you have a recycling bin, dump it out on your lawn atop a large tarp or into a big box. Sort through the items and make a list of them in your homestead journal, along with some detailed notes. What is it you're consistently buying and, consequently, which containers are you always recycling? Of what material are the containers made? My guess is plastic—am I right? Plastic and paper are common materials to consistently find in your recycling bin.

Start asking yourself some tough questions like, "What can I learn to make myself so that I don't have these plastic containers in my bin?" One of the things I learned to make right quick after this exercise was yogurt. Yogurt is super simple to make at home and doesn't require a plastic container to produce. Paper is easy to repurpose if you have a fire place or an outdoor fire pit because it can be saved to start fires. I keep all my cardboard to put down in the garden to repress weeds, covering it with a nice layer of wood chips or compost.

If you don't recycle yet, duplicate this process with your trash bag. Gross, I know, but you'll survive. If you start vermicomposting or get a small flock of backyard chickens, they can take care of the food scraps you're no doubt mucking through for this exercise.

DIY: Pick one of the package remnants from your bin for a product that you'd like to learn to make yourself. What is it? Bread, yogurt, honey, crackers, potatoes, underwear? You read the last one correctly. Clothes have their own kind of packaging, too. What will it be? What do you use the most often? That might be a good way to decide where to begin.

Let's say you pick bread; here's what you do:
1. Look up a recipe online or in your favorite cookbook.
2. Make a list of ingredients and make sure you have them all.
3. You'll also need certain equipment, like bread pans (if you choose to use them), an oven and mixing bowl.
4. Try that recipe three times.
5. If you don't like it or can't get the hang of it, try another recipe for bread. Try that recipe three times.
6. If you must keep buying bread in between attempts, go buy it.

If you get sick or go on vacation, buy your bread from the store and don't worry about it. Keep going with your baking experiments until you've perfected your method and found the recipe you love. Look what you did! You learned to make bread—that's the most amazing and wonderful thing!

Now, go pick the next plastic-packaged item on your list. *Rinse, lather, repeat...*

BUILD COMMUNITY: If you've joined us on the DIY Homestead Facebook Group (https://www.facebook.com/groups/TDIYH), please feel free to chime in and share with us your favorite upcycling tips. People are so creative, and I'm prepared to be amazed at your ideas. If you're not on social media, look around your home and find unused magazines, books and other literary media to donate to your local shelter, library or school. The garbage can will not be edified by these items you no longer use, but some deserving dad or hard-working student sure will be.

BOOKS TO READ: *One Plastic Bag: Isatou Ceesay and the Recycling Women of the Gambia*, by Miranda Paul. This is a children's picture book that tells the story of one woman living in Gambia who decided to put the plastic bags that were littering her town to good use by crocheting them into fashionable handbags and then selling them. An inspirational story simply told that is a perfect illustration of our discussion here. Read it with your kids or all on your own. Does anyone else just love children's picture books?! I get stuck in that section of the library so often, cross legged and enraptured while my kids are finished and waiting to leave.

If you're looking for ways to keep a non-toxic house and garden, you can try the delightfully illustrated, amusingly helpful and only occasionally pedantic *Slug Bread and Beheaded Thistles*, by Ellen Sandbeck. Also useful in its own simple and practical way is *The Newman's Own Organics Guide to a Good Life*, by Nell Newman.

WEBSITES: To jump start your brain, try Upcycle That (www.upcyclethat.com) which is overflowing with ideas on upcycling and recycling every day materials. Attainable Sustainable (www.attainablesustainable.net) and Healthy Green Savvy (www.healthygreensavvy.com) are full of information on many green topics, from upcycling to gardening to general common-sense sustainability. There are a lot of green products you can purchase for your home; a simple one is beeswax wraps in place of plastic wrap or even plastic storage containers. Visit Bee's Wrap (www.beeswrap.com) to get you started but browse the Internet for more ideas. You can learn to make your own beeswax wrap with a link located on our link page under the *Green The Homestead* chapter: www.homesteadlady.com/homestead-links-page.

If you're into frugal but fun upcycled home décor then My Repurposed Life (www.myrepurposedlife.com) has many upcycled projects, with a heavy emphasis on furniture. If you have kids or just enjoy a good creative project, a lot of the craft websites use repurposed materials for their projects. Try 30 Minute Crafts (www.30minutecrafts.com) for buckets of ideas.

HOMESTEAD LADY SPEAKS:
Don't get overwhelmed. The first time I overturned my recycling bin onto a big table in my backyard and started looking through it to figure out what was in there, I nearly gave myself a migraine. I'm big on doing things myself and working hard; it's just part of my nature. As I sat there staring at the mound of mostly plastic in front of me, I realized that the bulk of these materials represented items that I *could*, technically, learn to make myself. The truth is, the only solution to not having anything in my recycling is to learn to completely produce everything I need on my homestead. Alas, I was failing at that—woe is me!

As I sat there looking at the butcher paper that had surrounded my plastic-wrapped, grass fed beef I saw that, while I could recycle this clean paper (though not the plastic), it represented the need I had to produce all my own meat. Then, I saw the olive oil container and noted that I hadn't done that research I'd meant to do on growing black oiled sunflower seeds and getting an oil press to produce my own oil. Next to that lay a cardboard egg carton and with dismay I remembered that I hadn't figured out how I was going to build back up my laying flock this year. On top of that was the raw milk gallon jug I was having to purchase this year because my milk goats had been relocated for our impending move and I was without backyard dairy. I was faced with a pile of discarded packaging that all screamed the same thing at me, "You're not doing enough! You're not truly self-sufficient!"

It took me a minute to chill out. After a deep breath and a reminder to myself that this was supposed to be a learning exercise, I started to make some notes about what I could start doing today to diminish this pile of recycling. Then, I wrote down a few long-term goals and gave myself a time line to come up with a viable plan for each one if I had one. None of us can realistically make the jump from a typical, modern lifestyle into a completely self-sufficient one overnight. In fact, 100% self-sufficiency is not a very realistic goal for most of us, though you can get close with a lot of hard work. As Jared Stanley reminds us,

> "Make no mistake: You will have waste! The goal is to minimize it, not stop it.

> "So, let's make sure we understand waste—waste is the energy, nutrients or products that are not used by another element on the site. To take it to an extreme, you cannot harvest all the wind or sun that crosses your property. However, you can make the most use of the water, soil and nutrients and let as little of it escape off the property as possible.

> "There are several approaches you can take on an existing homestead to think about how to reduce waste. One key example is to go outside in a rainstorm and see what water escapes off the property and then think about how you can put it to use. How can you get the water into the soil and off the surface?

> "All that said, I would encourage a homeowner to think more about the connections between elements rather than just focusing on waste. If you can understand the connections, then the waste will decrease to the point where it is not noticeable."

However, even the most capable and sustainable modern homesteader is not a complete island unto himself. There are certain products and equipment that we will always need to acquire from an outside source. None of us can do it all. Most of us don't have the skill sets required to even know where to begin. Besides, there will always be products that someone else can produce with their expertise that I will be willing and happy to pay for because I lack the ability or the time to do it myself.

Instead of expecting ourselves to jump right into not purchasing anything from a store, maybe we could take Kris's advice (see below). Let's just start with eliminating plastic-wrapped products as much as we're able to this year. Let's figure out what we're buying that has plastic packaging and determine if we can purchase something in a paper package instead, or, if we can learn to make the product ourselves. Just see. Give it a whirl. What have you got to lose except some yogurt containers and a headache?

> "If I had to choose one thing that a family could focus on it would be this: Remove single use plastic from your household. Plastic shopping bags, take out containers, plastic fruit boxes, water bottles—they've become so prevalent in our lifestyle that often we

overlook these! Step back and look around—especially in the kitchen. Are there lots of plastic containers? What came in those? Maybe there's an option with less packaging or you can figure out how to make that particular item at home.

"Where do the items in your home come from? Are you buying food locally when possible? Other household items might be harder to source locally, but consider checking second hand shops before forking out for a product manufactured in another country… "

Kris Bordessa, Attainable Sustainable (www.attainable-sustainable.net)

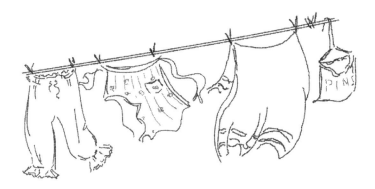

"Out of clutter, find simplicity."

- Albert Einstein, Physicist

For the Homesteadish: Downsize and De-Clutter

I know how challenging it can be, believe me. With five little kids, food storage, homeschool stuff and just the normal items that families collect, we were fit to bursting. We started to realize, though—as we filled every nook and cranny of our 4,000+ square foot house—that something was terribly wrong with the "stuff." Not all of it was ours. We lived in the house with my mother (and the "stuff" from her own home), as well as her parents (with their 60 years' worth of house "stuff"). To that was added our family's stuff, which included homesteading "stuff" perched like a cherry on top. Stuff was coming out of our ears!

When we decided to sell our urban homestead, my grandparents and my mother purchased another home not far away and, of course, took their belongings with them. We stayed behind to manage the property until it sold. But the dynamic of our "stuff" had changed quite a bit in our time in that house. I've never been a pack-rat, but in anticipating another move something in my husband and me just snapped. We looked around at our possessions and began to realize just how much energy we had to put into cleaning, storing, using, repairing, packing and moving them. We are both extremely practical, most of the time, and so we decided we'd had enough.

We carefully selected only those items that had been used in the last year, and then we donated everything else. We went through movies, tools, clothes, books and supplies. Then we went through them again. Finally, we waited a few months and went through them once more, each time finding something we didn't truly need.

THE LIFE CHANGING MAGIC OF TIDYING UP

It was about this time that I was sent a book to review called, *The Life Changing Magic of Tidying Up*, by Marie Kondo. It's a small book but, wow, does it pack a punch. Marie teaches the art of de-cluttering and organizing with her own method called KonMari. I must say that Marie and I have different energy profiles (think personality types) and so not all of what she wrote resonated with me. Plus, we're at very different phases of life—I have five homeschooled kids and a homestead, so I do NOT have time to fold my socks. Period.

Having said that, though, there is so much in this book for anyone who lives in a home with anything other than yourself and your underwear. The truth is, we all have too much stuff. Let me share some ideas with you and see what you think.

De-Clutter All At Once

Marie writes,

"People cannot change their habits without first changing their way of thinking."

She teaches that you need to de-clutter in one shot, despite the oft-held notion that tidying here and there is a better approach. Why? Because until you feel the results of getting rid of unnecessary items in your home, the desire to de-clutter won't become part of your way of thinking.

Take my kitchen counters. Everyone, including me, uses our kitchen counters as a place to deposit various items on our way in and out the door. It's a dumping ground for school work, cookie cutters and fermenting foods. I used to poke at it here and there when I had time and would ineffectually mention to people that maybe we should deal with some of the items residing there. However, it wasn't until we had to really clean the counters (once we started showing the homestead, after listing it for sale) that I realized it was possible to keep it clean—or, at least, clean*er*. I cook from scratch and we eat all our meals at home, so there will always be some trace of that reality on my counters. What is now gone, however, is our prior belief that clutter is part of our "normal" when we look at the counter.

Today, cleaning forty to fifty things off the counter—from dry dishes to school work—is an almost daily assignment on the chore list and, everyone expects that the counters *should* be tidy. We all just needed to realize that it was possible, by getting it truly clean and clear of misplaced and unneeded stuff!

Exercise Self-Control

Marie Kondo points out that putting things away, "creates the illusion that the clutter problem has been solved." Which is why it's imperative that when we start cleaning, we also start discarding. Being tidy is not the end-goal; being free of clutter, is.

I love, love, love the connection that Marie draws between our stuff and our state-of-being. She connects the tangible symbols of who we are (the things we choose to possess) to the intangible inner person. She makes the point that what we're really trying to create is the lifestyle we want—we want to be free in our own space to be and live in a way that is true to ourselves. We can't do that if we're surrounded by possessions taking up space in our homes, as well as our minds and souls.

As an illustration of this point, her criteria for deciding whether you should keep something is quite simple. You pick up the object and ask yourself, "Does this spark joy?" What an awesome way to figure out what you truly need! For example, take me and dishes. I'm such a "girl" when it comes to dishes and linens; I love them all. I have a good set of sturdy dishes that are for every day and my nice, trousseau dishes that we use on special occasions (we do use them, and they do bring me joy). Added to that, I had a knock off set of Blue Willow. I love the Blue Willow pattern and this cheap-o version was all I could afford. I had them for years and we used them all the time. But, I realized that I was repeatedly washing way too many dishes. We would let the dishes pile up simply because we could—we had different sets, after all.

One day, while laboring to wash the blasted dirty dishes—a thing I do NOT enjoy in the least—I picked up one of my faux Blue Willows and had an almost irresistible urge to throw it against the wall. I suddenly realized that this lovely dish had become a burden, where once it had been a blessing. With Marie's advice fresh in my head, I washed each dish and placed it in a box. I sent out an email to some ladies from church and asked if anyone needed a nice set of dishes, tea set included. I got an email from someone who was so excited to get them since she, too, was a Blue Willow lover. I had a moment of doubt – I was so sure I was going to miss those dishes, even if it was only a little. However, we brought them over to my friend and wished her well.

Happily, until this moment, I haven't thought of those dishes. Why? Because they were no longer bringing me joy. We can thank the things we own for the service they've rendered and the blessing they've been to us, and then, peacefully, pass them on to someone else. In doing so, they bring new joy to another.

As an aside, Marie talks a lot about *throwing away* items—so much so that I started cringing. If that starts to bother you, too, as you're reading, just insert the word *donate*. Gather up those things that no longer spark joy, thank them, and then donate them to someone in need. You will feel so free, I promise.

Spare Nothing This Scrutiny

Marie has you go through each type of object in your home to analyze it for keeping or passing on. She suggests a particular order—clothes, books, papers and so on. She insists her clients pile everything in the center of a room to inspect and handle each item. I do this twice a year as I change out our clothes by season. To reiterate, Marie and I have very different natures and purposes, as I do have some items that I save solely against a day of need. It is acceptable to me to store extra food, tools, school supplies and clothing. Why? Because it is far less expensive for me to slowly accumulate frugally and store these items in a measured way. I probably store a lot more than Marie would, but I understand completely what she's saying about analyzing each item you own.

Getting back to our clothes, I store the kids clothing in plastic tote boxes that I keep organized by age, season and gender. Periodically, I send out an email request to my homeschool and church groups asking for any clothing in the sizes I need that they're no longer using for their kids. In doing so, I often offer a trade of our clothing, or items that the homestead produces. Consequently, we hardly ever buy an article of clothing new. Buying new clothing for seven people each season is cost prohibitive for us. But I also don't like the idea of buying something new when a used item, that still has its positive energy to share with us, is a perfectly acceptable alternative. The items we can't obtain from people we know we usually purchase at the thrift store.

Every spring and fall I pull out each clothing box from the shed and I search the house for every stitch of underwear and socks. I have the kids bring me their shoes, their jammies, and their jackets. Then we put everything in one big pile in the middle of the floor. The first thing I do is sort through it all looking for items that have had their day—anything full of holes, with too many stains or that's just looking too thin. I put these into the rag bag to be cut up into cleaning cloths or cut into longer strips to become rag rugs (our newest upcycle hobby). I then sort the clothes into piles by gender, then by season, then by age of the child. Each item gets folded and placed into a box with a label for the next year's season. Then, and only then, do the kids get to go through the boxes for the coming season to find what they'd like to wear.

My children don't know any different, so their reaction is just like anyone going into a department store with a gift card! They're so excited to be able to pick out what they want to wear. I put a limit on how many pieces of clothing they can choose, though, because I have other hobbies besides laundry. Each person gets roughly seven shirts, seven pairs of pants, a few sweaters and socks. The girls get one or two dresses for church and I pretend I don't notice when it turns into four or five. (A little girl should be able to twirl and feel beautiful as often as possible.)

Each piece of clothing must pass the Mommy Test, too. I ask them repeatedly, "Are you going to wear that?" It takes me several days to get the clothes all sorted again so the kids have time to vet their choices and make some changes before I put everything away again. I do set aside some extras in each size, to account for changing tastes and personality. In this way, they can enjoy a variety of styles in a way that pleases them, and they're comfortable not pulling everything out all at once. I've discovered, however, that I'm often the one guiltiest of holding onto clothing that nobody wears—usually just because I think someone should. There's this lovely blouse that a friend gave us years ago and that I adore. Even though none of my girls have wanted to wear it, I've kept it for...well...for too long, let's just say. I finally donated it this year because no one was going to wear it, (Mom!)! I'm slow but I eventually get there.

So, go through each item—yes, *every* item—in your home and do it in an organized way. If today is clothes day, make sure you go through every closet, every box, and every drawer. Don't skip any place in your house (including storage) looking for clothing. Make a big pile and touch everything with your hands. If you keep a deliberate cache of items for future use, great! Just make sure that each item is needed. And don't hang onto a blouse just because Mom thinks it's cute and someday, someone in this house just might agree to put it on. There really is a life changing magic that occurs when you clear out and clean up!

TOSS AND LOSS

When I was young, we moved quite a bit. In an eighteen-month period, we moved nine times—and we weren't even a military family! I got used to going through my possessions and figuring out what I really needed and what I could live without. Truly downsizing and de-cluttering for me has become a familiar process, one I've even missed in some measure as my adult life has increased with children and other blessings. My problem with downsizing lies in how I personally weigh the importance of something. Some things require they be kept, whether they bring us joy. Financial documents come to mind. I am too apt to toss something in this category only to need it later. I am sometimes too casual in my connection to items of sentimental value, as well. Be careful with your assets, if you're like me.

On the other side of the spectrum and those for whom this process of downsizing will be unexpectedly painful. Sometimes, without really realizing it, we attach a level of importance to our possessions that becomes a part of our very identity. It can be a physically painful process to excuse a thing from our home that is seemingly a part of it. If this is you, don't despair! Begin from where you are and sift repeatedly until you find yourself out from under all that rubble. A homesteader is so busy that we don't have the time or energy to carry the weight of the "things" that cry out for our attention, but for which we have no meaningful use. Thank those items for how they've blessed you in the past, and then let them go.

Whichever type of person you end up being in this process, don't be surprised if the journey is an exhausting, though liberating, one.

ACTION ITEMS: Make notes as to which clutter pile has been flashing into your mind as you've read this. I'll bet there's something that immediately sprang into your thoughts when you read the word *de-clutter*. Was it clothes or media or knick knacks? If nothing came to mind right away, let's start with clothes.

In your homestead journal, write down a day this week to gather all your clothes together in a pile. For this first experiment with de-cluttering, let's just stick with your personal clothes. Gather boxes, bags and a dark marker for organizing. Make a spot for what will go to charity, what will stay with you, and what you still need to think about. On the appointed day, make a pile on your bed of all clothes, scarves, underwear, shoes and ties for every season. Follow the process described in this section. As you sort, organize what you have into either charitable donation, items to keep, or those you're still pondering. Be sure to take breaks, if it ends up taking a while. If letting go of possessions is hard for you, plan a big reward for when you're done. I am deeply motivated by chocolate and reading and will often use them as rewards for when I've accomplished some big goal. For example, when I finish writing this book, I get a jar of Trader Joe's® Cookie and Cocoa Swirl Butter (no judging) and my favorite Jane Austen. Whatever reward you choose, I promise the whole process will feel worth it when you're done.

Once your clothes are seen to, move on to your family's wardrobe, or the hall closet, or your sewing supplies. The best area to begin de-cluttering is the place in your home that seems to be sucking the life out of you.

DIY: This is a caution to not make this project too DIY. What do I mean? Don't make all the de-cluttering decisions for the other people that live in your home. If you live alone, then you're safe. But for those of us who live with other people, we can often assess their possessions as being of less value. My bric-a-brac may be completely useful and lovely, but someone else's needs to go. Right?

Plus, as a parent, we can get so desensitized to our kids' stuff because they appear to have so much of it, and it's always on display on the floor of their room. As stewards of our home, it's true that our job involves keeping the house tidy and making sure the space is used appropriately. In large measure, we're also responsible for the spirit and overall feeling of contentment in our home. However, that does not give us the right to toss other people's belongings out, without thought and without consulting them. Involve your family in ways you feel are appropriate in this process and be respectful of their needs and wants. Be willing to make de-cluttering the home a process that completes itself over time.

BUILD COMMUNITY: There are many charitable groups in your area that could use some help providing for people in need. All they ask for are your gently used items. I once saw a billboard ad for a local thrift store that said, "Give stuff a second chance." This fits perfectly with our discussion about items bringing us joy because it suggests that our possessions might go on and start fresh with a new owner. There are entities that are looking for your used shoes, books, clothes and even cars. There are others that would love any extra canned or boxed foods you might have.

To remember to do something like a donation round up, I need to attach it to a holiday or some fun event that we're already involved in, so that it gets done and done with flair. May I suggest you do the same? One family donation drive takes place on Martinmas, in November. Even though we're not Catholic we love this simple holiday celebrating the life of a generous man. We've turned our celebration into a reason to get friends and family together for a winter clothing drive. Our other big effort is connected to Easter and Passover. Not only do we learn good lessons with these holiday celebrations we also eat fabulous food, meet with great friends and always remember our family donation drives. (For more ideas on family building activities, see the *Family Times* on the Homestead chapter.)

BOOKS TO READ: *The Life Changing Magic of Tidying Up*, by Marie Kondo. Also, *Sink Reflections*, by Marla Ciley aka The Fly Lady (if you prefer her information in a book format; if not, her blog is listed below). For simple tips, especially for young adults and newly marrieds, *Apartment Therapy - Complete and Happy Home*, by Ryan and Laban. *Art of Home*, by Hannah Keeley, was written to help you manage your home according to your personality type. So, if Kondo drives you bonkers, try Keeley.

WEBSITES: Depending on your personality type, the famous Fly Lady (www.flylady.net), may be of benefit to you. Also of interest:
- Becoming Minimalist (www.becomingminimalist.com)
- The Minimalist Mom (www.theminimalistmom.com)
- Get Green Be Well (www.getgreenbewell.com)

HOMESTEAD LADY SPEAKS:
One last thought on downsizing with the family. Our last move took us from a very large home (the 4,000 sq. ft. one) to a comparatively small one. My kids have always shared rooms, but they've been feeling the pinch of less storage and living space. Before we made this move, though, we prepared them a year beforehand by verbalizing the anticipation that the next house would be smaller, so that we could afford to buy more land. We asked them to help us as we went through our personal belongings, so that they saw my husband and me getting rid of items we liked but no longer needed. We talked about the average living space of other people around the world and how ours compared.

It took some time, but we eventually got each child down to one box of "personal stuff." These were items that were really, truly theirs and weren't part of our homeschool or homestead family life. For some of my kids, this was a really, really difficult task to accomplish. I'm so proud of them for being willing to work with me as we prepared for a downsized life. (Don't get me wrong, our current house would not qualify as a "tiny house," but the seven of us are using every inch.)

In general, the de-cluttering journey will go much more smoothly if we share the "why" of what we're doing with our family. We don't need to ask our children's permission to de-junk our house, but it is incumbent upon us to explain why it's necessary. If we can inspire them by cleaning up a communal area first, it's possible they'll catch onto the feeling of freedom that follows. In turn, they may start going through their own belongings. It's been my experience that whether they willingly participate or not is going to depend a lot on their nature and my approach. I have a few children who enjoy items while they have them but never miss them once they're gone. However, I have one that acts like I'm asking for blood during our semi-annual donation drives.

We've made a tradition of donating toys, books and clothes to our local charities twice a year. It's an activity that they anticipate, but I still get days of grief from this one kid. I'm always tempted to go into her room and bag up half of the contents, just to get the process over with. But the stuff is not the point. What I'm really trying to teach her is a certain way of being, not an acceptable number of things to possess.

My point is, know yourself and plan to include your family accordingly. If you naturally throw things out and don't often have a sentimental connection to items in your home, be aware that not everyone is like you. The de-cluttering process might be hard for your spouse or your children. If you are always squirreling items away that are difficult to part with, be prepared to let your family toss things out freely, if that's in keeping with their nature. Don't try to enforce your de-cluttering philosophies with anyone but yourself. Although it's perfectly appropriate to expect children to keep a tidy living space, it's not necessarily the right move to go into their room armed with a trash bag but not having their consent. You are the best judge. Be the smart spouse and parent you are and pave the way with your personal example.

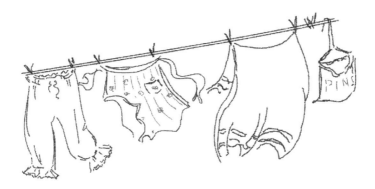

"When weeds go to heaven, I suppose they will be flowers."

— L.M. Montgomery, from *The Story Girl*

For the Homesteadaholic: Foraging

A favorite way to go green is by purchasing local food, particularly fruits and veggies grown on farms near you. Buying local reduces the overall amount of shipping and fuel required to keep your family fed throughout the year. It also supports the farms that feed you and your local economy benefits. A person may also choose to start a home garden to keep their family supplied with fresh foods. You can't get more local than your backyard, after all. However, without purchasing seed or plants, you may already have a lush garden full of food just outside your door. Let me introduce you to the wonder of weeds!

Did you know weeds have virtues? Well, they do and a lot of them.

Weeds are Free

One thing every homesteader knows, whether they live in a city, or the country, is that free stuff is good! Free food and herbs are even better and foraging from the weeds already in your yard and neighborhood can provide wholesome nutrition and medicine without any financial investment. When we go on walks we always look to see how many useful plants we can identify. Knowing which plants are edible, and which aren't, is a great skill to have.

As I write this, our homestead is for sale. The spring weeds are emerging and I'm having to pull them out, so that the yard looks pristine. In doing so, I'm realizing just how many of them I use every year. It's painful to have to pull the weeds before they've even had a chance to reach their potential. You know you're a homesteader if you think of weeds as having potential! I deliberately let my lamb's quarter grow up every year so that I can feed it to my goats. Having to remove these plants from my garden this year has made me more grateful for them—GRATEFUL for the weeds!

Let's face it. Weeds survive because they forage successfully. Their roots have superpower abilities to find and store nutrients in the soil. This is the reason behind the almost disturbing capacity that weeds have to grow out of sidewalks. What this also means is that weeds are powerhouses of nutrition. And, seeing as they grow everywhere, that translates into a free meal for you and me.

Dandelion greens and flowers go into salads; you can even use the blossoms to make cookies or jam. Lamb's Quarter leaves cook up like spinach. Shepherd's Purse leaves work as a styptic. Mallow seeds pickle like capers. Nettle leaves make a nourishing tea or even pesto. These weeds are so versatile and useful. "Weeds" are just herbs that haven't learned to grow in tidy rows. Embrace the wild-child in your garden and let the weeds grow up so they can nourish you. And be proud of yourself for being so plant savvy that you cultivate these selections. You're very smart about plants.

> **FRIEND OF THE BLACK THUMB GARDENER**
>
> Not everyone with a homesteader's heart is going to be a great gardener—some don't even like that part of homesteading. I've met a few who plunk some seeds into the ground every year and call it good so that they can go work the chickens or the pigs and not be bothered with cultivating corn. I get it, I really do. I feel that way about the value of higher math in my everyday life.
>
> Many things on the homestead are there to be done, though, whether we like doing them or not. And growing at least some of your own fresh fruits and vegetables is a wise endeavor. For those who struggle to enjoy gardening, I would first remind you that no homesteader's garden looks like the Monticello estate unless you have a team of gardeners and volunteers at your disposal. As author Sharon Astyk reminds us,
>
> "The perfect gardens only live in my head."
>
> Secondly, I turn your attention to the value of learning to forage food that is already growing in areas where the plants naturally thrive. Foraging food and plant medicine is a way to reap what God has sown, naturally, and without your sometimes-fumbling efforts of cultivation.

The How-To's of Foraging

Wild animals forage for their meals, and you can often look to them to determine what is in season and edible. What you're able to forage for, and what end up being your favorite foraged foods, depends upon where you live. After interviewing several homestead-foragers I learned that their favorites ranged from beach asparagus to dew berries to mushrooms. A lot of foragers go out in search of food and others in search of medicine in the form of herbs. Still others hunt for pure protein sources like nuts, fish, and small game. Even children can learn to forage alongside their parents, and it's a great preparedness skill to have. What I know is limited mostly to herbs, so that's what my kids know best. But we've been studying our foraging books lately to get more comfortable with gathering berries, leaves and roots.

Speaking of books, every forager should have a great foraging manual for their area and some other important tools on hand.

TOOLS:

1. A reliable guide book is a must and there are several listed in the resources section below. If you can find some specific to your locale, those will be perfect; otherwise, get a book that covers your area of the country.
2. It's also a good idea to have quality boots and a bag or basket for carting your treasures home.
3. Practical tools include a spade, sharp shovel, utility knife, hat, gloves (get the correct size for the kids) and a first aid kit (especially if you have the kids with you).
4. If you can bring along a foraging mentor, especially for your first few forays into wild areas, it would be wonderful. A live person can explain what you're looking at, and answer questions as you go. While a guide book is useful, a live mentor can be priceless

WHEN AND HOW MUCH

Plan to learn how to forage year-round because, even in places with harsh winters, you may still be surprised at what you can find. For example, pines are evergreens and the needles are a wonderful source of Vitamin C. Foraging dead wood for tinder is a worthwhile pursuit, too, unless you've waited until after a storm drops three feet of snow.

Being ecologically aware is important, regardless of the season. If it's been a drought year and the pickings are slim, be sure to leave the bulk of the harvest for the animals. Being responsible about the amount you take is a must, regardless of the growing conditions that year. I've had foragers tell me to take no more than one third or as little as 10% of the plant. Others have advised me to know the plant well enough to know its reproductive capability; annual weed seeds stay viable in the ground for years, so it's nearly impossible to over-forage them. Mushrooms should be harvested and carried in a way to spread their spores as you return home (e.g., in a mesh bag), thereby helping to "plant" more as you go.

Either way, commit to harvesting only what you'll use. Fire up your dehydrator if it looks like you won't get to processing all your harvest; once dry, you can come back to your foraged foods as you have time.

NOT JUST FOR FOOD

Lest you think that food alone is what can be found in the wild, read on! Food is awesome, don't get me wrong, and I love contemplating all the wild berry jellies and mushroom sandwiches I'm going to enjoy. However, the wild can also provide me with craft products and medicines. Here are just a few ideas:

- Moss, stones, berries, vines and wild flowers can become gorgeous centerpieces for any holiday gathering.
- Snake skins, fossils, bug shells, and scat can all become fodder for rockin' school projects.
- Berry juice and walnut hulls can magically transform into usable ink, and a dried fern frond can become the perfect addition to homemade paper.
- Wild herbs can be tinctured for medicine, brewed for tea, and stepped in oils, which can then become salves and lotions.

You and your children can fall down a veritable rabbit hole with foraged items, never becoming bored and always learning something new and useful to do with your wild fare.

> **ONE LAST NOTE ON LOCATION:**
>
> You really can forage anywhere you are, for the most part. Your backyard, the woods, park trails and out-of-the-way spots all may have something to offer your foraging basket. Be sure, however, that you've checked ahead of time what, if any, herbicides or pesticides have been sprayed nearby. Just as we're careful with how our garden produce has been sprayed, keeping your foraging limited to non-chemically treated areas will better ensure that what you're ingesting is good for your body.

Foraging Interview with a Forest Homesteader

The following is the text of my foraging interview with Chris of Joybilee Farm. The interview was simply delightful and so I include it here as both informational and inspirational fodder for aspiring foragers. Chris lives on a 140-acre farm in Canada, in growing Zone 3. Cultivating crops can be a challenge but her mountain home provides an abundance of foraged items.

WHAT ARE YOUR FAVORITE FORAGED ITEMS?

We forage wild herbs like rose hips, Oregon grape roots, St. John's Wort flowers, yarrow flowers, mullein leaves, nettles, and flowers. We have an abundance of wellness herbs growing untouched on our land. We also forage for wild blueberries, huckleberries, strawberries, juniper berries, and Saskatoon in season, if we can beat the birds to them. Wood for [lathe] turning is also plentiful where we are— there's no need to buy turning blanks with the amount of fallen trees around us. [Chris's husband is a wood worker.] Mushrooms are a little more difficult to identify but there are a few that are easy to find like morels, Shaggy Manes, and puff-balls that we pick if we can find them. In a damp season they are plentiful but short lived.

WHAT ARE SOME QUALITY TOOLS A FORAGER SHOULD HAVE?

I carry a basket or just use a hat as a basket when we go for a walk. It's nice if I remember to have a pocket knife, too. But often my only tool is my hands and the sticks I find lying on the ground. Hands are amazing tools. If I had a set of tools for foraging I'd want a tightly woven basket with a handle, a set of light-weight, good quality pruning shears, and a French pocket knife for trimming branches. But really you don't NEED these things, they are just nice to have. And I have them, but I forget to grab them when I walk outside.

And it's always good to carry a stout hazel walking stick to whack the black bears on the nose if they get too close while you are picking their huckleberry bushes. At least that's what I tell myself when we go huckleberry picking. So far though, I haven't met the bear.

CAN YOU FORAGE ALL YEAR ROUND?

I forage for herbs beginning in April, picking the cottonwood buds for salve, and end with the first snow-fall, picking the last of the rose hips. I could grab pine pitch and spruce needles in the deep snow, if I needed them. However, by snowfall I'm ready to rest for the winter and enjoy the fruits of my harvest. We are in the mountains and our first snowfall comes at the beginning of November and spring comes in April.

HOW DO YOU USUALLY CONSUME FORAGED FOODS?

When we pick berries, we eat a lot of them while we are picking, but the herbs need to be dried or infused in alcohol or oil to make them fit for use. I have a commercial dehydrator to help with drying at low temperatures, but most often I just bundle the herbs and dry them on hooks around my living room.

WHAT PRECAUTIONS DO YOU TAKE AGAINST CONTAMINANTS?

I don't pick near road-ways, and other than chemtrails [from planes flying overhead], our land is free of pollution. If I'm going to make a tincture or an infused oil I will just take the herbs 'as is' and macerate them in the oil. If we are going to eat the berries, though, I usually rinse them in spring water before we eat them.

WHAT LAST FORAGING ADVICE CAN YOU GIVE?

I think it's super important to correctly identify the plants that you are picking, in the field, before you bring them home. A little bit of effort to learn some botany before you venture out could save you a belly ache. There are so many look-alikes and you don't want to gamble with your family's health and well-being. Don't pick anything that you can't positively ID. And don't put anything near your lips to 'see' if it's edible. There are a few plants that can kill an adult with a single drop of sap.

CAN YOU FORAGE WITH CHILDREN?

Puff-ball mushrooms and Shaggy Manes are easy to ID, and you can't go wrong if you cook them as soon as you get them home. I taught my daughter, Sarah, to know all the wild plants on our acreage as a 5-year-old. We would go for nature walks, and ID the plants, in various seasons, and learn their benefits, whether they were edible, and if they had health benefits. It's so important to pass on these skills to our children when they are young.

When Sarah was in senior high (homeschool) we invited the government weed control rep to come to the farm and instruct our homeschool group on noxious weeds in our area to teach them about the biological controls that they had. Sarah was able to inform the group of the wellness benefit of some of those same noxious weeds and many of the native plants, during the hike with the other students and their families.

Another fun project to do with children is foraging for natural dye plants. Woad blue, cleavers (bedstraw) red, and many, many yellows can be obtained from a morning spent looking for leaves, roots, and flowers. There is no eating involved and kids are thrilled to see the color developing on cloth. My daughter Sarah grew a science fair project from dyer's woad. It took her to the national science fairs 3 years in a row and won her 4 medals for her foraging and growing the indigo plant. But it began when she was 5 and we were gathering yellow lichens off the neighbor's broken fence for dyeing fluffs of wool pink and blue. We use [foraged plants] for crafts, for medicine, and for food. Natural dyes are a passion of ours, along with the fiber arts like spinning and weaving. We've even foraged nettles for fiber to spin. You must use mature plants and harvest them once new growth is visible coming up from the ground at the base of the plant. The stems must be 'retted' like linen by letting them lie on the ground in the dew for a few days, turning them often so they don't rot.

There was a tornado here a few years ago that devastated a children's camp. Pines Christian Camp lost 90% of their Ponderosa pine trees and the life of one camper in 20 minutes. It was a tragic beginning to the summer camping season. Sarah and I foraged for the longest pine needles from the fallen trees and bundled up a laundry basket full of pine needles. I wove a 12-inch basket from those needles and presented it to the camp—beauty from ashes. And we did a pine needle basket weaving workshop with the camp leaders to help them process the very sad event before the summer was over.

During the summer all the trees were removed, and the camp was transformed with brand new cabins and walking paths. There was very little to mark the tragedy, beyond a plaque to remember the young camper. The Pine Needle basket sits in the foyer of the dining hall as a memorial to God's protection over the 170 campers and staff that were at the camp on that fateful day.

****

I include Chris's personal note here because I think it will resonate with you, as it did with me. Don't we all get a little overwhelmed by this homesteading lifestyle sometimes? It's good to be reminded of why we do what we do.

> "Thanks, Tessa, for asking these questions. You reminded me of all the reasons I love my farm—the whole 140 acres. I was feeling discouraged with the winter and feeling overwhelmed and thinking maybe it was time to sell and move to a smaller property with less stewardship responsibility. You just reminded me of why this place, at this time, is the best place for me.
>
> "We truly have so much bounty here, despite living in Zone 3 and struggling to grow tomatoes, melons, and squash. But I can buy tomatoes, melons, and squash at the farmer's market every summer, or at the grocery store year-round. We are totally blessed with other things that you can't buy in the store or at the market—like fresh wellness herbs, berries, and natural dye plants. Not to mention mushrooms, birch syrup, pine needles, and so many other renewable treasures that we haven't even tapped into yet."

> "Country children once were familiar with, and enjoyed the edible qualities of many native products, which today are scarcely known, since tropical fruits and manufactured sweets have found their way to the smallest village stores. The young wintergreen leaves, the artichokes, the sweet flag ground nuts, seeds of the sweet fern, and many others have been delights of country children from time immemorial. Almost anywhere along the highway or the railroad tracks, just now, we may see Italian women and those from other countries, who know the value of such foods better than many Americans, gathering the first dandelions for salads and greens."
>
> - from "American Cookery Magazine," published May of 1919,
> as quoted in *A Garden Supper Tonight*, by Barbara Swell

ACTION ITEMS: Check out a field guide for wild, edible foods for your area from your library, or order a copy to own. Bring your guide and your homestead journal along with you to your backyard, local park, or waterway. You may want your camera, too. Start identifying some edible plants and take notes in your homestead journal. Figure out which plant you have in the most abundance and find at least one recipe that includes that plant. The usable part you find could be the leaves, stems, roots, seeds or berries. So, don't limit yourself by just looking for one plant part.

Make sure you note what time of year it is and exactly where the plant is growing in abundance. The flavor of wild food can fluctuate from season to season. So, you'll want to pay attention to how it tastes in spring, as opposed to how it tastes in fall. You certainly want to be able to relocate it easily each year so take good notes!

If your neighbor thinks you're crazy for picking great buckets of dandelion blooms from your lawn, just gift her with some dandelion jelly this spring. Who knows? She may come out and help you pick dandelions next year.

DIY: If your homestead journal is getting too stuffed with notes, magazine clippings and recipes, create a field journal. This journal will be specific to your foraging efforts. This won't be a garden journal. This notebook is only to chronicle your education as a burgeoning forager. I like the 1" three-ring binders for ease of use, but your foraging notebook could be made from anything.

You may want to invest in a small flower press to carry with you. This will enable you to take specimens of what you're finding, press and dry them, and include in your journal with a label. There may be plants that take a while to identify, so keeping a sample can come in very handy. If you're an artist, you can make sketches, too. Include notes on location, weather, season and other plants growing in the area. Throw your notebook, press and a bottle of water in a bag, and you'll be ready to do some exploring.

BUILD COMMUNITY: This is the time for an herb walk! Herb walks are guided tours of various local sites that have an abundance of wellness and edible plants. See if you can find an herb walk already taking place in your community—several would be even better. Not only will you learn about herbs and other edible plants, but you will learn so much about your community and the people in it. If you don't have a community herb walk, organize one by asking your herbal-foraging-smarty-pants friend to head one up. You get the group together and your friend can pick the location. Enjoy being together and sharing information and experience. If you need help locating an herb walk mentor, try contacting your local health food store, quality plant nursery or university extension agent for suggestions.

If you're on Facebook, there are several good foraging groups you can join to share experience and ask questions. Be sure to read the rules of each group carefully and be ready to comply. If you're on Pinterest, pinners have created whole boards on the topic of foraging that you can easily follow and utilize. To get the link for the Homestead Lady foraging board, as well as other social media information, see the Resources section in the back of the book.

BOOKS TO READ: If you can find books specific to your region, certainly investigate those titles. *The Forager's Harvest*, by Samuel Thayer and *Edible Wild Plants*, by Elias and Dykeman are both great for North American foragers. For mushrooms, a good book with quality pictures is *All That the Rain Promises and More*, by David Arora.

WEBSITES: There are several Facebook groups dedicated to the topic of foraging. Also, of value are Grow, Forage, Cook, Ferment (www.growforagecookferment.com) and Edible Wild Food (www.ediblewildfood.com) which both cover various foraging topics.

HOMESTEAD LADY SPEAKS:

It's important to make and keep friendships with the weeds in your garden. To do that, you must control their growth by cutting back their seed heads and thinning thick clumps of roots, so that the weeds don't take over your cultivated space. Everything in the garden is about balance. For example, I love borage and grow it on purpose for my animals to eat, to use as green manure in the garden, as a forage plant for the pollinators, as well as a salad green and flower. And simply because it's lovely. HOWEVER, borage reseeds rapidly and easily in my climate. Reseeding is when a plant produces seed after it flowers, dropping the seed to produce new plants. Sometimes this process can occur multiple times in one year, as is the case with borage.

I don't want my borage taking over my garden, bullying other plants and making it impossible for them to grow because the borage has sucked all the nutrients from the soil. So, I need to be vigilant about either cutting back the flowering stalks before they drop their seed or pulling the mature plants out by the roots and feeding them to my goats. Borage leaves have a healthy amount of nitrogen, so composting them is a good idea, too. Nothing goes to waste. Plus, I do have enough seed fall while I'm busy with other tasks in the garden that I get more plants all season long and reseeding into the next year.

Remember that weeds are vigorous, and you must keep a firm hand with them. If you're willing to do that, they can be some of your biggest allies in the garden and the home.

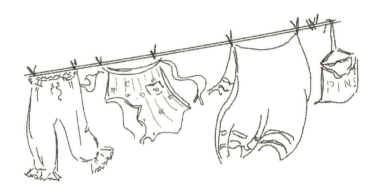

"These times are too progressive. Everything has changed too fast. Railroads and telegraphs and kerosene and coal stoves —
they're good to have but the trouble is, folks get to depend on 'em."

- Laura Ingalls Wilder, from *The Long Winter*

For the Homesteaded: Energy on the Homestead

It's a phrase you hear a lot anymore: "Going off grid." It can mean different things to different people but typically the phrase refers to the energy independence many people are seeking as they disconnect from public utilities. I remember when I started identifying myself as a homesteader to people, I fielded all kinds of different questions about what I did. People would ask about small farm livestock and gardens and canning, but hardly anyone ever asked about energy. Once, during the break of a Master Gardener class, a fellow student and I struck up a conversation during which I mentioned that I had a homestead. She immediately began peppering me with questions about energy production and savings. She asked if I used solar or wind power, if I used rain barrels, and if I line-dried my clothes after washing them by hand. For her, homesteading was all about energy independence.

For you, maybe it has always meant produce independence or dairy independence; maybe even complete independence from the grocery store, or the public-school system, or your job. We're all striving to be more self-sufficient, including me. My friend's questions really got me thinking about what it means to be independent in energy production. I also started to wonder how I might learn to conserve the energy required to run my homestead, including my personal energy. The longer I pondered, the more definitions the word "energy" seemed to have!

This section, much like each of the previous sections, is not meant to be an all-inclusive treatise on how to conserve and produce energy on your homestead. One reason is because this is something I'm still exploring with my family as inspiration strikes. It's also not entirely about going off-grid. In fact, the bulk of that information is reserved for the very end of this segment. Rather, I'm going to throw out a bunch of ideas to get us thinking about energy in a serious way. I'm not just talking about electricity, although that certainly is an important part of our lives. Foremost, I'm really talking about work.

How much work are we doing that we could stop doing poorly and start doing better? Are we being thoughtful about the amount of work that is being done on the homestead by everyone and everything? How are we using the energy being produced and is there a way to make or capture more? Also, if we would like to go off-grid, how can we do that on our homesteads—especially if we already live on them?

WAYS TO CONSERVE ENERGY AND RESOURCES ON THE HOMESTEAD

Here we can brainstorm together, finding new ways to train ourselves to use less energy to produce the outcomes we want. This is a behavior most of us will have to learn and learning new habits can be hard. It's a truth that our culture has become very used to living with what our great-grandparents would have considered luxuries. (I'm in the United States but it happens in lots of other places, too). Flipping a switch to produce light in a dark room is a relatively new miracle in the history of the world. Driving a car instead of riding a horse or walking to get somewhere is a similarly recent convenience.

Some of us may pursue energy conservation because we anticipate the electrical grid going down, or the economy fluctuating in a way that makes commercial energy cost prohibitive. Some of us just hate spending our money on product we don't really need. Whatever your motivation, let's begin to re-train our brains to become aware of how we manage our energy and resources.

Appliances and Such

Let's start with a few basic areas of energy conservation in the home.

Lights out. Your dad was right. He wasn't made of money and neither are you. So, turn off the light switch when you leave a room. While we're at it, don't stand in front of the open fridge for ten minutes deciding what you want to eat. I thought I'd get the obvious out of the way first.

Think before you set. If you use an automatic thermostat in your home, be smart about what temperature you maintain. For people with or without central heat and air, you can use various methods and tools to efficiently regulate the temperature of your home. Some of these include: whole house fans, shades on south facing windows, insulated windows, strategically planted shade trees, as well as heavy draperies and a fire place.

The thing about our indoor lifestyle is that we aren't used to being uncomfortable. When it's cold outside, we're often still wearing our t-shirts indoors because the heat is cranked up.

Apart from saving money, being cold enough indoors during the winter to put on a sweater and socks does several different things for our bodies. Winter is a time to hibernate, to move slower, to draw in and to rest from our labors of planting and harvesting. Our modern lives don't operate seasonally anymore but, time was, winter was a chance for the busy agriculturalist to rest a bit before the new cycles of planting and harvest began. I believe our bodies still crave this.

Most people's lives are no longer agrarian-based, but a homesteader's is, at least in some measure. Allow the changing of the seasons to touch you by being in tune with them. If it's hot outside in July, go swimming instead of going indoors where the air conditioning blasting all day. The height of summer is also time for a brief rest. It's typically too hot to plant and the harvest isn't ready yet.

Even how we eat is affected by our core temperature. In the middle of summer, after being hot all day from working and playing, the last thing you're going to find me eating is a heavy hamburger. All I want to consume is salad and water. Perhaps what we need to do to lose a few pounds is eat with the seasons and allow our bodies to move with them as well. I'm not suggesting you turn off the thermostat altogether if you're not ready for that. However, I am saying that we are too disconnected from nature because of the artificial environments we create inside our homes. If we aren't willing to learn to be a bit uncomfortable from time to time, we're never going to have what it takes to be truly self-sufficient through our labor. (Incidentally, I hate being hot. So, keeping the air conditioning down or off is hard for me in summer, but I try to do it anyway. In some measure, my body adjusts, as bodies usually do. I'm still a baby about it, though.)

Use energy efficient appliances and fix ones that are broken. That's kind of like your mom always telling you to buckle your seat belt. You know you need to do it, I'm simply compelled to say it. If you're already off-grid with solar power, or will be soon, consider using a compatible solar refrigerator and freezer. These units usually look like chest freezers, but they use DC (or direct current) power, instead of AC (alternating current), or propane. Try not to buy anything with a digital screen, like a clock, that will draw power to keep you informed of the time, whether you're looking at it or not.

If something is plugged into a plug, but not in use, unplug it. These appliances are still pulling power, just not as much. This is called "phantom load," which makes it sound spooky, and older appliances are the worst (read below regarding the One-Watt Initiative*). Off-grid homesteaders that produce their own energy (like solar) are accustomed to monitoring their power usage closely. They can see that phantom load is a waste of energy. Not the biggest one on the homestead but a waste nonetheless. It's sort of like staying awake while you sleep, which doesn't make any sense at all.

Jaimie, who writes about her off-grid family life on the blog An American Homestead (www.anamericanhomestead.com), advises against any appliance with a heating element, too. She says:

> "Practically speaking, when you choose what electrical appliances to use (if you plan on having solar and/or wind power), remember that anything with a heating element is a power drain. The off-grid families that I know who use some electrical appliances in their homes, [use appliances without] heating elements. For example, no refrigerator, Crockpot, clothes dryer, etc.
>
> "Many off-grid families can generate enough power to use electronics like laptops, ceiling fans, and a washing machine. Sometimes supplementation with a gas-powered generator is necessary on cloudy days. Many off-grid families use propane for things like refrigeration, hot water heaters, and stoves."

The One-Watt Initiative was begun by the IEA (International Energy Agency) in 1999 to ensure that by 2010 all appliances produced in the world would only use one watt when in standby mode. That initiative being mostly successful by their standards (although be aware if you have an electronic device that was manufactured before 2010), their new concern is networked devices, or those appliances that are connected to the Internet. TVs, phones, computers and more are constantly connected to the Net, and are using energy to remain so. Whatever you may think about an international energy agency having so much influence over our policies, their point is valid about how connected we are and how much energy we're consuming. For anyone desiring to reduce their energy consumption, and especially if we desire to go off-grid, these are things we're going to have to think about. And for which we'll need to find personal solutions.

Consider alternative cooking methods. A conventional oven takes a considerable amount of energy to heat, but a toaster oven or crock pot use far less. A Wonder Box only requires a little bit of energy to initially heat your pot, and the insulation of the box does the cooking after that. Open flame cooking requires a bit of wood. A solar oven requires no energy to heat except the suns. The kitchen is where so much of our energy gets used up, both personal and appliance. Examine how many appliances you have and pick one to live without for a few days. Then, figure out if you can make the change permanent. We did that experiment with our microwave a few years back and ended up getting rid of it because we just got out of the habit of using it. For more information on alternative cooking methods, especially Wonder Box use, please see the *Off-Grid Cooking* section in The Prepared Homestead chapter.

Ditch electronic appliances and replace with more manual methods. Remember, we're trying to form better habits, and that means old habits will have to die off to make room for new ones. Don't worry, you can do it with a little ingenuity and some research. For example, Krystyna, of UP Pastured Farms (www.uppastured.com), runs her kitchen without a refrigerator. It's the coolest thing to get her talking about how she does it, even though I'm not ready for that. Kendra, formally of the blog New Life on a Homestead (www.newlifeonahomestead.com), chronicles her family's journey from a typical homestead to an off-grid lifestyle. She tells me you can buy some simple parts to change your energy efficient chest freezer into a refrigerator easily. This idea is on my list to figure out because, as grateful as I am for it, I'm tired of paying for my energy-sucking fridge. For more off-grid and energy saving ideas from other homesteaders, see the end of this section.

Sew it up. First, by way of confession, I hate fashion sewing. I'd rather walk on my lips than must sew a shirt. I like quilting and repurposing fabric pieces into a usable product (at least, I did before I had kids, which was the last time I really quilted anything). But fashion sewing requires patterns, and patterns require black magic to interpret. I still sew, but I don't enjoy it.

My oldest daughter is especially gifted with hand-sewing and has been waiting for me to service the sewing machine, so she can start making skirts and all kinds of fun stuff. She finds the fashions in the stores to be too tight and too short for her taste and standards and wants to simply make her own clothes. What an awesome kid! The problem is, sewing your own clothes isn't cheaper anymore. The fabrics are so pricey these days that making your own clothing isn't going to save you money, when compared to shopping at Walmart. If you're shopping at Saks, you might save a few pennies by sewing your own, though. As homesteaders, we're all trying to save money and time.

I don't say that to discourage you from sewing your own clothes because, like my daughter, you may have motivations beyond cost effectiveness. Go for it! Use coupons, thrift stores and your grandma's stash to acquire fabric and supplies and unleash your creativity. Something else you're saving by sewing your own is the labor of others, possibly children, abroad. You're also negating the need for your clothing to be shipped or trucked from who knows how far away. Energy touches everyone, not just our immediate family.

One place you can really save some money (which represents both time and energy) with clothing is in learning how to mend rips, tears and worn out spots. My kids have pants that are perfectly good except for a few places in the butt and knees that have torn out from climbing trees and sliding down rocks. I've learned how to patch from the inside and even make it look cute. I've also learned how to repair the inseam of jeans, which I didn't think I could do. YouTube and blog posts have been such a help to me in learning how to mend. If you don't know how to hand sew, I encourage you to learn. You can mend many things perfectly well using a sewing machine, too. If an article of clothing is beyond repair, repurpose it into cleaning cloths, mama cloth, diaper liners, craft batting or rag rugs.

Each of those ideas represents yet another skill for you to tackle. Don't worry or get overwhelmed; you have your whole life to learn these things. By the time you're ready to meet your Maker, you'll be completely ready to live the life you want—isn't that always the way?

Laundry and Dishes

Hang-dry your laundry. You know what's funny about this one? Granted I hang out with a lot of homestead-type people (like you), but almost across the board people tell me how much they love to line-dry their clothes. Even though it takes longer. Even when you must work around rainstorms and winter. Even when you have drying clothes hanging from every high point in your apartment and you keep knocking them down. Even though... There's just something calming and wholesome about it. I don't think I can adequately explain the magic, but I can honestly tell you that if you've never done it before, there's a good chance you're going to love it.

Your clothes will, at any rate since dryers can exhaust and wear out clothes over time. Line-dry clothes to make them last longer, so you don't have to purchase or produce as many articles of clothing. Not to mention that, as an appliance, dryers are huge energy consumers—typically even more than your refrigerator.

Hand-wash your laundry and dishes…or don't. Honestly, I'm on the fence with both tasks. I've read some studies that conclude hand-washing dishes and laundry saves water. I've read other studies that say they don't, especially as compared with modern, energy-efficient machines. I washed my clothes by hand or by simple machine (a water-filled metal drum filled with an agitating plate) the whole time I lived in Russia, and it still took a lot of water. And that was when I was single and had only one person's clothes to wash. I've tried a High Efficiency washer, too, and as little water as it used, it never got the clothes clean (although I'm sure different brands perform differently).

I really love having an electric clothes washing machine – Ah, the time savings! We purchased a plunger-washer and a hand-wringer and have learned to operate them, just in case. They're a simple to use, off-grid option, if a lot slower than their electric counterparts. However, they require an enormous amount of physical labor to utilize well. Hand washing clothes for my family by hand is a several days process that takes up huge chunks of time.

I like that I can clean my laundry in one day with an electric washing machine. However, during our humid summers I can't get all the clothes to line-dry in only a day. That means that I must learn to keep pace with the weather and be willing to do laundry more often during the week to give everything time to dry. I have so much else to do that I get a little obsessive about getting the laundry done in one day. However, if I were willing to wash all my laundry by hand, one load at a time, I would be forced to do only one load of laundry a day, because there's no way I could hand-wash all the laundry for seven people in one day. Hmmm…still working this one through in my head. What do you think I should do?

My dishwasher just doesn't work that well and is a waste of energy and water, in my opinion. So, the kids and I made a goal to hand-wash our dishes more often until we finally stopped using our dishwasher at all. Even when using a dishwasher, there are a few things we can do to reduce energy consumption. First thing, look for a machine that isn't computerized (and that goes for clothes washers, too), because they draw power every minute of every day. Finding a dishwasher with mechanical buttons and dials is harder than you might think. You will have to look hard, but you can expect to pay less since computerized models usually cost more. The unit should also have an "air dry" or "energy saver" cycle that avoids using the heating elements in the washer. Heating elements basically cook your dishes until they're dry, much like an oven would. An "air dry" option prevents that from happening.

Also, as with a clothes washer, only run a load at 100% capacity to fully utilize the water and power involved. For those still on the public utilities grid, you can choose whether you want to practice these money/power saving techniques. For those who've already switched to alternative energy, these steps are vital. Even when following them, you're only going to use a washer on days when your electrical systems' batteries are fully charged.

Reduce the number of clothes you wear each week. This is related to the last point and will reduce the amount of laundry you're doing. Holey Moley! Whether we're washing by hand or by machine, we need to stop changing our clothes every five seconds! Look, if you're still working in the professional world, you're just not getting that dirty. Our society is nowhere near as dirty as it was 100 years ago. If you're homesteading full time, working with animals and dirt and blood and sweat, that's a different story. Only you can be the judge of what is dirty. But if the volume of your laundry isn't something you've thought much about before, I encourage you to do so now. Whoever oversees the laundry in your household every week will certainly thank you.

We have generous friends who give our children wonderful hand-me-downs, sometimes to the tune of several large bags. My kids had so many clothes floating around their shelves and closets that it became a game, for the girls, anyway, to determine how many new outfits they could try on every day. Where did all those discarded clothes go? Yep, into the laundry hamper. Finally, this mama said, no!

I keep all those clothes as part of our family resource management program. Now, however, I have plastic tote boxes of clothes sorted by age and gender that I store year-round. When we need to go "shopping," we pull out the boxes at the change of seasons and the children find what they need in there. Even so, they are only allowed seven shirts, seven pants/skirts, several jammies and a few church dresses/neck ties. They can have all the socks and underwear they want. They are also only allowed a few pairs of shoes—church, every day, muck and just for fun (only the girls use that last category). For more details see the *Downsize and De-Cluttering* section of this chapter. I was once spending my life doing the laundry and now I don't.

Incidentally, I follow this same quantity rule myself, but I follow it differently. I only have so many shirts, skirts and pants, but I re-wear my clothes for as long as possible each week. I wear an apron all day long; literally all day. Sometimes I fall asleep in it. The apron protects my clothes from surface dirt and so I'm able to re-use my clothes for several days, depending on the season. A century ago, girls used to do this with pinafores they wore daily to protect the dress beneath during planting season, house chores and school recess. Just call me "Anne of Green Gables" (who, thanks to her adopted mother Marilla, always had a clean pinafore).

> **JUST A NOTE:**
>
> If you find your body scent is just too malodorous to re-wear your clothes, please see the section, *Ferment all the Things*, in the Homestead Kitchen chapter. There you can research balancing your gut by cleansing it of excess yeast.
>
> Every body creates odor as toxins are pushed out of our cells. But there's normal body odor that a simple, natural deodorant can usually handle, and then there's knock-down, nasty body odor that even commercial deodorants can't cover. And ladies, (seriously, guys, don't read this next part), if your underarms are that foul, your southerly parts are going to be aromatic, too. You don't need a stronger deodorant, you just need a balanced gut. Really, it might just be that simple for you.

Our culture has become super-saturated in perfumes—deodorants, soaps, wipes, lotions, even foods. We're used to things being stripped "clean" and soaked in scents. The trade-off of this process is to be found in our wellness and I encourage you to research what those stripping and scented chemicals can do to your body, especially your skin (your largest and most exposed organ). Slowly, as you make small changes, you'll find yourself a lot less concerned about what other people consider clean or balanced. You'll form your own opinions.

Yes, I know, you now have a picture of me being one of those B.O.-ridden hippies with garlic breath and dreadlocks. Hey, some of my most wonderful friends have been B.O.-ridden hippies with garlic breath and dreadlocks! That's not my style, though. I assure you, I shower every day and wash my hair and clothes with soap. They're just short showers and laundry sessions, accomplished with homemade or natural soaps and with an eye to work and water conservation.

Mix it up. Make as many of your own cleaning products as you can to keep your gray water cleaner (and, hopefully, usable), and to keep your body balanced. If you can produce all your own ingredients for those products, cool! But I don't know how to make baking soda. I purchase most of the ingredients I use for homemade soaps, cleaners and beauty products. However, making them myself accomplishes several things that have become important to me. First, I can control the quality of the ingredients. If I want certified Organic ingredients, I can get those. If I want minerals, oils or clays, I can get those, too.

Second, making these items myself teaches me a new skill. Yes, it is one more thing for me to have to do, and my days are already busy. So, I've learned to make large batches of soaps and cleaners all at once to better use my time. It takes just as long to make a gallon of laundry detergent as it does to make a quart. If, however, I have days or weeks or even years where

I'm just *not going to* make my own, I know lots of cool homesteaders and crafters who make and sell their products. I can support my local economy by buying from them. It's a win-win for us all.

Third, having the kids participate in the process of making our own products is a wonderful learning experience. These activities are simple to incorporate into our homeschool curriculum, too. Just the other day we learned about heterogeneous mixtures and whipped up a batch of homemade tooth powder. See what I did there? Sneaky mom. Having the children help also keeps us all more accountable about the quantity of each item we use. We are far less prone to waste something we have produced ourselves.

I'll never forget the first loaf of bread I made as a grown up. I had one or two children by then and we all sat around eating the warm, buttered loaf with delight. As I was cleaning up, though, I discovered that someone had not eaten their crusts and someone else had let half a piece fall to the floor. To say that I was upset doesn't really convey the feeling that flashed through me at all that wasted work and effort. I had ground the grain into flour, mixed every ingredient, kneaded by hand, waited through rise cycles and lovingly baked that bread. To see all that work so unappreciated was clearly unacceptable. That same feeling accompanies every hand-made item in my home. The best way I've found to re-train myself and my children not to waste anything, is by making whatever it is ourselves—from bread to soap to toothpaste. As author and lecturer Mary Ellen Edmunds said,

"It's wrong to waste no matter how much of something you have."

In the Garden and Barnyard

Let it Rot. Learn to compost and turn your organic garbage into soil amendments. Or, keep chickens so that no food scraps go to waste but, instead, get converted into eggs. Create various cycles that are self-sustaining, especially with the animals; instead of expending your own energy to turn compost, employ chickens and pigs to do it. Grow plants that feed and tend your soil by producing nitrogen and other supplements.

Similarly, stop tilling the land, and use layers of organic matter to keep the soil friable instead. Research low or no-till methods of gardening and implement them as you learn them. Not only will this save you fuel costs from running a tiller, but it will also save you time and energy overall. You won't need to take time to till, but you also won't need to spend all season trying to manage weeds and water. You will, at least, significantly reduce the amount of time you spend on those "lovely" pursuits. (For more information see our *Growing Smarter* section under In The Homestead Garden chapter.)

Let it lie. If you mow your lawn, leave the clippings on top of the grass instead of throwing them away to provide a layer of mulch.

Maintain it. If you're using a tractor, or any other piece of equipment with a motor, make sure all the pumps and parts are working well. And keep tires inflated. Investigate bio-fuel to determine if that's something with which you want to get involved. An ounce of prevention is worth a pound of cure when it comes to using and maintaining homestead labor-saving tools!

Plant toward the sun. Always plant the gardens, orchards and pastures with the path of the sun in mind to maximize its use. So, too, plan to position your barns, greenhouses and outbuildings to best make use of the sun throughout the seasons. (Visit www.suncalc.net to calculate the path of the sun across your property.)

Conserve water. Save water everywhere you can on the homestead. Rain barrels are marvelous tools for nearly effortless water collection and they can be set up underneath any gutter pipe in the front or back yard. Be sure to check if they're legal where you live, though—especially if you have a HOA. Cisterns are also very useful, above or below ground. Utilize drip-watering systems instead of overhead watering in the garden. Also, use heavy mulch in your growing areas to retain water as much as possible.

Save gray-water from the laundry, and even from the sinks and shower, to use in the garden. This will require some hoses, buckets and a little ingenuity but this can really go a long way towards living in a water-wise way—especially if you live where annual rain fall is low. The least you can do is turn the water off while you brush your teeth and wash your hands.

Also, store water for your family to drink in times of emergency (see the *Water - Storage and Conservation* section in The Prepared Homestead chapter for more details). Further, find the closest body of water and map the quickest route there in case your public water is interrupted or your well pump stops working. Buy several water filters to keep in your grab-and-go-bag, your home, your office, your car—in short, everywhere you might find yourself stuck. Water is fundamental to life and maintaining your energy.

Cut the cord. Consider getting rid of the long extension cords going across the homestead providing light to goat sheds and hay barns. You know the ones I'm talking about. Replace them with solar light kits for your chicken coop, outbuildings and barns, as you can afford them. If you aren't ready for solar panel arrays, buy solar garden-path light kits at your local big-box hardware store. They can be very useful as a cost-efficient indoor lighting, even though they're small. Charge them during the day and bring them into the coop or barn at night. Don't expect to light the whole building, but they help in areas that need just a bit of light for you or the animals to see. Incidentally, they'll provide similarly useful light in your home, if you'd like to not use electric lights but would like to save money on candles and oil lamps.

Directing and Generating Energy (Going Off-Grid)

We are blessed to live in a time of wonderful thinkers and innovators and knowledge that spreads around the globe with the touch of a button. A great many people have been working on energy production for off-grid living, but the most practical choice for most us is solar power, in my opinion. To learn more about the realities of solar power on the homestead, I suggest Internet searches where you can read real-life experiences from people who use it. I don't have much time for reading blogs, but I make an exception for the practical advice on every day, off-grid living that I find on two of my favorite blogs: An American Homestead (www.anamericanhomestead.com) and Homestead Honey (www.homestead-honey.com). Jaimie, Zach and Teri, the voices behind these sites, always inspire me as they talk about energy.

I suggest you also keep your eye on companies like Tesla Motors, which recently developed the affordable and efficient Powerwall solar battery, whose release we are hopefully awaiting. (To learn more, visit our links page under the Green the Homestead chapter: www.homestead-lady.com/homestead-links-page.) Solar technology, and other off-grid options, keep improving and we're living in amazing times. I can't give you specific, name-brand suggestions simply because the technology is developing every second and constantly changing. (In fact, by the time you read this, Tesla Motor's invention will most likely have been surpassed by something else!)

Whichever technology you chose to produce energy on your homestead, remember that going off-grid is not an all or nothing venture. You can pick and choose which alternative energy equipment appeals to you. You can decide where and how you might like to generate energy and experiment with a method to make sure it's a right fit for you. Maybe you're not really that into the idea of a composting toilet yet but putting your well pump on a solar panel doesn't seem that daunting.

The following are some practical tips from experienced off-grid homesteaders, answering specific questions that I'm sure you've asked yourself.

HOW DO I CONVERT MY HOMESTEAD TO OFF-GRID?

Kendra (formerly of the blog New Life on a Homestead) is doing just that with her current homestead and home and has this to say:

> "It can definitely be tricky converting a modern home to be off-grid. Many floorplans aren't designed to be energy efficient anymore. You just have to do the best you can with what you have.

> "To keep your home cooler during the summer months, plant shade trees strategically around your home. Build trellises for climbing vines to help absorb some of the heat around your home. Install awnings over windows and doors and put up window shades and heavy window treatments to block out hot sunlight during the day. Cook outdoors as much as possible or eat raw foods more often.

"To stay warmer during the cold months, install a wood stove if possible. Also, wear more clothing, keep the blinds open so the sunlight can heat your home, close off unused rooms, and pile more blankets on at night.

"Replace electric appliances with non-electric alternatives. There's so much that can be done by hand if you just have the patience. Keep track of your energy usage and make goals to use less electricity. Buy a Kill-O-Watt meter and test all of your appliances to make records of where your energy consumption is going. Repair or replace any faulty equipment, such as the seal on your fridge, to make it more efficient.

"I'd say our biggest mistake here on our homestead was not going off grid right away. Of course, hindsight is 20/20. When we first moved here, we had no idea that was the route we'd end up wanting to go. Honestly, we never even considered it as an option. So, we paid a pretty price to have power lines dug out to our property, a heavy-duty air conditioning system installed, and an expensive submersible well pump put in. If we could go back, we'd definitely do things differently. We could have saved a lot of money. "

WHICH ATTITUDES WILL HAVE TO CHANGE AS I GO OFF-GRID?

Jamie (www.americanhomestead.com) shared this:

"You have to decide what is important to you. But first and most importantly, make sure that you really want to make a lifestyle change and why. Define your reasons for doing so because when things get difficult, it's easy to forget why you started this off-grid thing in the first place. You'll be tempted to give up and I've seen people do that. You have to make sure that this is what you really want. Secondly, there is an adjustment period. I would say give yourself at least a year or maybe two. I feel that things start to get easier the 3rd year, but that was my experience.

"I equate a lot of my initial adjustment to moving off-grid with a newborn baby. In your adjustment period, don't allow yourself to dwell on all the conveniences that you used to have. It's not productive to be washing dishes and thinking about that dishwasher that you could just load and push the button. I believe the most "work" you have to do in moving off-grid is to forget the conveniences you left behind.

"Again, remember why you chose to do this in the first place!"

IS THERE SOMETHING THAT WORKS BETTER FOR YOU OFF-GRID?

Tammy (www.trayerwilderness.com) really prefers using hand-powered equipment, believe it or not!

"Well, we always have power even when the local power is out! When we use [hand-powered] tools, although they may take more muscle power and more work, the life using them is more rewarding and we stay in much better shape without the need for a gym. Our lifestyle was a choice for us and one that we dreamed of. It is not for everyone, but I feel the freedom that comes with such a life is so very worth it. We will never look back!"

Jaimie's praise of her humanure toilet system compared to an on-grid toilet might surprise you!

"I love our composting toilet/humanure system. Yes, you have to carry out the bucket and dump it regularly, but the rewards are so worth it. The bathroom definitely stays cleaner without the regular flushing of a toilet. I don't have to deal with the nasty film that collects all around the toilet and surrounding floor. A simple wipe down of the toilet box is all I need to do. I always hated cleaning a flushing toilet!"

WHAT DO YOU MISS?

So, I asked my off-grid homesteading friends to share with me what they missed from their on-grid living times, or what they don't enjoy about their modern-convenienceless life. Their responses echoed the exact the things I've been pondering myself. What they shared with me made me feel so much more confident about my potential to learn how to live the energy-savings lifestyle that is the goal of my family.

Jaimie, with her multi-generational family of six, said:

"I'm not gonna lie—I'm so happy my youngest is out of diapers! Washing diapers adds another level to washing clothes by hand. It's not hard, it's just time consuming. It generates about twice as much laundry for me...family cloth is currently out of the question. My husband cannot wrap is head around it. Ha! In all seriousness, I'm thankful. If it ever came down to that without using a washing machine, he would be washing his own.

"I really don't miss electricity that much. Once in a while, I think a refrigerator would be nice, but I know now that a certain amount of what most people have in their fridges doesn't need to be there in the first place. Other refrigerated items, we simply do without. A washing machine would be nice when I have been too busy with other things and the clothes are piling up. It's in the plan one day, but we would need to increase our solar power. I miss running water the most. But hopefully this winter, we will finally get our system installed. I've learned that going off-grid is more like a marathon. Things get easier over time. You learn how to do without things. You learn how to do things in other ways. But you also add back conveniences as you can afford them and have time to build up more of your infrastructure. It all takes time. It doesn't happen overnight!"

The advice to take it slowly and allow yourself a learning curve was echoed by the others I interviewed. Kendra, with her family of six, advised you make sure that everyone is on board with energy savings so that, even if you pine for a few things while you're making the adjustment, it all evens out.

"When you make such a big lifestyle change, the whole family has to be on-board. Of course, the younger kids don't really get it yet. But the older ones were excited about being able to produce our own electricity. And all of them have been really good about paying attention to not wasting power or water, though we've always tried to be conscious of conservation—albeit not to this degree."

Tammy and her family of three lived in a canvas tent for over eight months while they built their house on their off-grid, Idaho homestead. That time meant no electricity, no appliances, and no walls except those of the tent during that frigid winter. When I asked her if she missed anything, she had to say no because of her experience in that tent. The doing without conditioned her to simplify and so, once their house was built, even without appliances and conveniences, it felt luxurious. She says:

> "First let me say that the time we spent in the tent for those 8 1/2 months is something I yearn to go back to. That was the simplest living I had ever experienced, and it was beyond rewarding and extremely adventurous. There is really something to be said about simple living! While we were in the tent all our belongings were in storage and, you know, I did not miss ANYTHING. Now that we are done building for the most part and working daily, I miss those days of just living.
>
> "We still worked hard on a daily basis, but it was different—I felt like a pioneer. When we incorporated the Internet it honestly took away from that simple life we lived. There was quite a freedom in living a fully disconnected life from modern conveniences. I have always said that I was born in the wrong era, so I really don't miss anything; the only time I think about it is when we have guests looking for the microwave. Everything else we do is really the same."

So, my last question is really for you, the reader, to ponder on your own: what would *you* miss?

> "To find the universal elements enough; to find the air and the water exhilarating; to be refreshed by a morning walk or an evening saunter... to be thrilled by the stars at night; to be elated over a bird's nest or a wildflower in spring — these are some of the rewards of the simple life."
>
> — John Burroughs, *Leaf and Tendril*

ACTION ITEMS: In your homestead journal, make a list of ten things you can do right now to save energy in your home or on your homestead. Visit any of the websites above or talk to your energy conscious friends to pick up more ideas. Present your list to your family and agree to begin working on at least five items for one month. Consciously pace yourself and your family because, if you go overboard, you're going to have a mutiny in the ranks and people will want to give up.

DIY: Run an energy drill for a family night this week. This does NOT have to be complicated. Pick a day, then pick an energy source. For example, decide to do without electricity this Sunday.
- Try not to alter your normal routine much but plan ahead for obvious problems.
- If you have an electric stove, you'll need to prepare meals using alternative cooking methods.

- If you have electric lights, you'll need to have something else on hand once the sun goes down. Or plan to go to bed early.
- Is it winter, and do you have an electric furnace?
- What will you do instead?

Without endangering yourselves, find the holes in your energy plan if the energy source fails. Really ask yourself what habits you'd need to alter and what equipment you'd need to have on hand for an extended public utility power loss. Take notes during your drill, especially on equipment you'd like to buy, even if it's only for emergency power outages. Also, make a note of each family member's emotional reactions. How did they feel during your drill; how did they react to the simulated situation? A lot of what we're striving for in our preparedness/self-sufficient lifestyle is emotional stability in difficult times.

BUILD COMMUNITY: Off-grid experimentation is way more fun with friends! Plan an off-grid camping trip with family or close friends. Keep pregnant women, nursing mothers, small children and the elderly in mind when making your plans but make it as off-grid as you can so that you really learn together. Plan menus, including meal preparation methods and equipment. Figure out how you'll provide water and make hygiene preparations for washing and using the bathroom. What will you use for fuel, shelter and, most importantly, fun?

Yes, don't forget the fun! Plan games, fishing trips, songs, competitions and even a talent show for your off-grid adventure. Add a few special desserts, too. Remember that a lot of what we do as parents is to impart our belief systems to our children. A well-placed S'more may just mean the difference between getting your child to willingly cut back on energy consumption, and constantly trying to wheedle them into shutting off the light when they leave the room. Think I'm over-simplifying? Just try the off-grid camping trip with lots of fun and practical experience, and then get back to me.

BOOKS TO READ: Most off-grid homesteading books that I've read seem to be either too preachy, too technical or too vague. So far, the only one that I connect with and enjoy is an E-book by Teri Page called *Creating Your Off-Grid Homestead*. At about 100 pages, Teri covers everyday aspects of creating an off-grid homestead from the ground up (which is what her family did), along with suggestions for those converting an already existing homestead. I couldn't put this book down, so to speak, and found it extremely inspiring and informative. To find the book for yourself, visit our links page in the Green the Homestead chapter: www.home-steadlady.com/homestead-links-page.

If you want to download a free book on humanure (making compost with human waste), which also includes plans to build your own composting toilet, visit this link from www.humanurehandbook.com to download each chapter of *The Humanure Handbook*, by Joe Jenkins, as a PDF. (The book is also available for sale in published format.) In my limited reading on composting toilets one thing I've noticed in most customer reviews is that all the commercial options available for purchase have big design flaws, despite their high price tags. Several people I know who use composting toilets have recommend that we build our own, instead of buying a manufactured one. According to their advice, the pre-made units are too expensive for how many problems they have, and Mr. Jenkins design is not only sufficient, but adaptable, as needed. I share that for what it's worth, as a composting toilet is something I have yet to build on my homestead.

Read the *Little House on the Prairie* books to not only improve character, but they also teach how life was lived when energy was produced only by raw labor. Similar books, like the *Little Britches* series, will also be beneficial. Don't be fooled into thinking that you're too good to learn from a bunch of children's books.

WEBSITES:

- An American Homestead (www.anamericanhomestead.com)
- Homestead Honey (www.homestead-honey.com)
- Practical Self-Reliance (www.practicalselfreliance.com)
- Trayer Wilderness (www.trayerwilderness.com)

HOMESTEAD LADY SPEAKS:

Even if you're not ready to go off-grid completely, or even if you've never even thought about it before, there's still a great deal of value for you in reading the stories of off-grid homesteaders. Everything about becoming self-sufficient is learned precept by precept. No one wakes up one morning and decides they're going to eat only the food they can produce when they've never even grown a green bean. Similarly, learning to step back from dependent, modern resources, like electricity, can be something we work into over time as we read and practice, read and experiment.

Chatting with Jaimie about her humanure toilet was particularly interesting to me since I've been researching the idea for a while. It still appeared like a too foreign idea for some of us, though. Jaimie made the whole notion feel not only attainable, but preferable! It was surprising to me that she preferred cleaning the bucket of her humanure toilet over cleaning the bowl of her old conventional toilet. Jaimie is smart and practical, and there's no way she'd waste her time with a system that didn't work. Having her experience tucked away in my mind helps inspire and motivate me to learn more. If she can do it, maybe I can really do it, too!

As I was reading about Teri's efforts creating her outdoor kitchen (see that chapter in *Creating Your Off-Grid Homestead*) it made me seriously stop and consider the benefits of having an outdoor kitchen. My home isn't off-grid, but my kitchen is west facing and hot in the summer. There's a nice, shady spot in my backyard, in a corner with level ground not too far from my house. I keep looking at it and thinking about all the benefits Teri outlines of cooking outside. This was the system she used exclusively while the indoor kitchen of her off-grid cabin was being built and it's still one she happily uses in the summer time. Her advice and observations come from real life experience in using an outdoor kitchen to cook for her family. She shares her tips as one busy, homesteading parent to another.

I soak in all these ideas and experiences as I read about them and, suddenly, they don't seem so foreign, or even so impossible to consider. I feel like I really could make a change. Isn't this why we share our stories, after all?

CHAPTER 4

Livestock Wherever You Are

    Can I be a "real" homesteader without animals? Is this a question you've asked yourself? Do you have visions of strolling out to toss grain to the chickens as the sun sets over your lovely garden? What about those equally idyllic daydreams of milking the dairy cow as your children frolic, carefree in the distance? It seems like animal husbandry is a watershed issue for homesteaders. We feel as if we've truly made it if we tend an animal or two. There is a reason why we feel so connected to the idea of achieving some self-sufficiency through animals, and that reason, which is one we'll discuss in this chapter, goes clear back to our first farming ancestors.

Not all homesteaders will seek to produce their own protein (milk and meat) for whatever reason, but I do encourage every homesteader to still consider raising some small farm livestock. If for no other reason than that the land will manage itself much easier with the animals than without them. In my opinion, livestock, in whatever form you choose to work with, is essential to a healthy homestead because of what they have to offer in the way of manure production, brush control and light tilling. Unless you're a vegan homesteader, the idea of raising some kind of animal that produces some kind of protein has probably already crossed your mind anyway.

There are several animals from which to choose, and you can suit your decision to your family's needs and the capacity of your land to support them in a healthy manner. If you have less space, the noble rabbit (meat), sturdy poultry (meat and eggs) or the worthy Nigerian dwarf goat (milk and meat) are all quality choices for homestead protein production. If you have a bit more space, larger breeds of goat (milk and meat), cattle (milk and meat), swine (meat), sheep (meat and wool, or milk) and others can be considered. (Be sure to research the breed of any animal in which you're interested since some are better suited to a specific product.) Even honey has a measurable amount of protein, though not much, if you're looking for an excuse to keep bees! Even if you're an apartment homesteader, animals can still be part of your homesteading ventures, as we'll cover in the following material.

Here's what we'll be discussing in this chapter on livestock for the modern homesteader:

HOMESTARTER level: Ever heard of vermicomposting? This is the perfect solution for the small-space or beginning homesteader who wants to raise livestock, especially to contribute to the garden soil's health. You might even turn this into a small business, if you'd care to. This section will get you started.

HOMESTEADISH level: You may have, or will soon be acquiring, that flock of egg-layers you've dreamed about. Have you thought what you might do when it's time to replace the older hens? Where will you find new chicks in the quantities you require to serve your needs each year if you decide to raise a batch of meat birds? Here we'll talk about finding a solution to this problem yourself and "making" more chickens on your homestead.

HOMESTEADAHOLIC level: You've finally made the decision to get a dairy animal, or you already have a goat and are wondering what a cow might be like. Join us in this section for an exploration of both options, as we chat with some homesteaders who've worked with each. Here's the real scoop on these two very different dairy animals.

HOMESTEADED level: Planning how much livestock to raise for meat production every year can be a real chore because there are so many things to consider! Factoring in hunting and preserving methods, we discuss here some suggestions for calculating what type, and how much of each animal, you might need. We've also thrown in a few harvesting tips and tricks.

As with every chapter in this book, be sure to follow along at the end of each section as you reach the Action Items, DIY Projects, ways to Build Community, resource recommendations and parting advice.

"My whole life has been spent waiting for an epiphany, a manifestation of God's presence, the kind of transcendent, magical experience that lets you see your place in the big picture.
And that is what I had with my first compost heap."

- Bette Midler, Actress and Vocalist

For the Homestarter: Vermicomposting

Like I said, livestock is a common goal for most homesteaders but it's not possible for all of us. Or it's not possible yet. For many of us at the Homestarter level and/or simply because of where we live, we don't have enough land, or the proper zoning required for livestock. Sometimes we just don't have the know-how or the desire. To get your feet wet with the idea of "livestock" that will fit in any space, may I suggest you take up worms?

Vermicomposting or vermiculture is, simply put, using worms to maintain an active compost pile in a small space. It can be done on a larger scale, of course, but worm composting is an ideal solution for the small space or even "no" space gardener. If you've ever read anything about composting, you know that traditional methods involve a certain amount of yard space devoted to the endeavor, a great deal of back-breaking turning of the compost materials and lifting of ready-compost with a pitchfork. The older I get, the more appealing the small size and maintenance requirements of a vermicompost bin are looking. Even if you already have a backyard compost bin for garden waste and kitchen scraps, mixing worm manure into it will make it that much healthier for your garden.

There are several different ways of doing vermicomposting, but it basically boils down to a tote-sized box system indoors, or an in-ground worm tower in your garden. I will briefly cover each as an introduction and encourage you to research further with the suggested reading materials at the end of this section.

The Small Space Livestock Solution

First, if you are in an apartment and have no land available, compost still has value for you. You might have a potted garden for which you'd like to produce soil amendment, or you may just not want to throw out your compostable materials as waste. If this is true, then the first option of creating a tote-sized worm home to be stored inside or on your deck is probably the best one for you. This is a simple DIY project if you care to make your own, but there are also worm condos you can buy. The basic idea for either is that you have one shallow (12-18" deep) bin with a series of holes drilled out in the bottom that holds a bedding material like damp, ripped newspaper, along with the worms themselves, and then your kitchen scraps. (See the *DIY section* at the end for some useful links.)

After a time, usually 4-6 months, the worm castings (which is their poop and the compost they make) builds up and you can harvest it. The easiest thing to do when it's time to get your compost out is to have another tote box* to place underneath the first with fresh bedding and new food for the wormies to enjoy. When they run out of food in their upper quarters, they move through the holes in the bottom of the bin to their new digs in the tote box below. Commercial worm towers have a series of trays that can be moved around to accomplish the same thing. The point is, you get the worms to move and they leave their castings behind after they've consumed the materials, including the bedding, so that you can dump their compost into a pot and use it right away. All you need is a small counter or space underneath a sink to maintain a worm bin.

*If you're trying to eliminate plastic in your home, these systems can be set up using galvanized steel or wood if you prefer.

Worm Towers in the Outdoor Garden

If you do have some garden space, then a worm tower in a grow bed might be easier for you. The bin system is simple, but it's one more thing for which you need to find space and one more thing to keep alive inside your home. People, parakeets and sourdough culture may all be organisms in your home that you're already keeping alive and you feel filled to the brim!

A worm tower is nothing fancier than a 2 foot in length, 4-8" in diameter PVC or HDPE pipe with a series of ¼" + holes drilled into it. PVC can leach chloride when it gets hot, so you may choose to use a food grade pipe, like the HDPE, instead. Your pipe is then buried, standing up, with about 4-6" poking above the soil line. Into this you put some damp bedding material, a bit of water and some kitchen scraps. On top, you place a piece of screen and secure it over the pipe. You'll want to cover that with an upside-down clay pot or something similar to keep out the rodents.

The composting worms will come from your garden soil to your tower for the food. Or, you can add a handful of composting worms to your tower when you put in the food. They'll stay and reproduce because you're an excellent host. Every few days you'll toss some more scraps down the pipe and a bit of water, too. The worms will take it in and give you back castings which simply stay in the soil with no need to be harvested. They'll also make more composting worms for your garden as they reproduce. These worm towers can be placed in your grow area 5'- 6' apart to facilitate the spreading of helpful worms in your garden. Don't go crazy putting in towers; too much of any good thing is still too much. You don't want to crowd out other worm species who have their place in your soil, albeit much deeper, and should be welcomed, also.

Here's a little graphic from my vermicomposting friends at Homestead Chronicles (www.homesteadchronicles.com).

Method	Pros	Cons
Bin	Castings can be applied in a controlled manner. (See *How To Apply The Castings*) Visual confirmation of colony health.	Must harvest and apply the castings yourself (yuck factor). Worms do not aerate the soil (because they are not *in* the soil). Requires regular maintenance, cleaning, fresh bedding, etc. Can be stinky especially if not maintained regularly or properly. Must be temperature controlled (between 50°F-86°F at the extremes but best between 59°F - 77°F). If temps exceed this range in either direction, the bins must come indoors. Worm sitters required if you need be away for an extended period.
Tower	Castings are spread by the worms and deposited throughout the garden so there is no need to harvest or apply castings.* Worms aerate the soil in the process. Other than feeding, no maintenance required. Temperature control is unnecessary. ** No worm bins in the house. Worm sitters not required. Just top off the pipe with extra kitchen scraps and go on vacation.	Castings cannot be applied in a controlled manner. Must trust nature to ensure the health of the colony and can only spot check by turning up some soil or using a trowel in the pipe to pull some up.

* The worms are free to move about the garden, typically stay within a 35 foot radius of their food source, and will distribute their castings where it needs to go (at the roots of your plants) with no involvement from you.

** When temps are extreme, worms instinctively go below the frost line for warmth or cooling.

© 2014 Homestead Chronicles
Used With Permission

The Benefits of Vermicomposting

Here are just a few of the benefits of composting with worms:
- Worm castings, like all compost, provide nutrients for your dirt and can even be used to clean up volatile organic compounds (chemicals like fuel) in contaminated soil.
- After an initial investment to set up, vermicomposting can run itself with minimal input from you, which means that you won't be spending a lot of money to make your new project function. Just some food scraps, some water (worms breathe

through their skin and must stay hydrated), some ripped up newspaper and you're good to go.
- Vermicomposting doesn't take up as much space or time as a more conventional farm animal. They may not give you milk, but you also don't need to milk them, and they poop for free!
- A lot of materials you might currently be throwing away are consumable by composting worms. I read an account of one man using lint and random food material he'd swept out of his kitchen to feed his worms and they ate it up. I recommend you be attentive to their food but, really, they'll eat just about anything besides meat.
- If you already keep rabbits, worm bins can successfully be kept underneath your furry friends to compost their waste.
- Worm bin maintenance is a great chore for any child and can become a truly rewarding experience for your young, budding homesteader.

Successful Worm Composting

You will do your own research, but here are some suggestions for successful vermicomposting to get you started:
- To prevent fruit flies in your worm bins you can first freeze your kitchen scraps to kill any larvae. Also, be sure to bury your scraps completely each time you add them to your worm bin.
- To a new worm bin, add a bit of grit and native soil so that the worms have a way to break down the scraps you'll give them in their gullets. They're kind of like chickens in that they don't have teeth and require sandy bits in the soil to help them break down their food.
- Do a squeeze test on your paper bedding for proper dampness. If you squeeze your damp paper and water runs down your arm, it's TOO wet.
- Egg shells won't always be consumed by the worms, but they often enjoy laying their babies inside them.
- Never overfeed your worms or they won't be able to keep up with the volume of scraps and your bin will start to stink with the decomposing materials.
- On that note, don't feed them citrus, meat or oil because they won't eat it and you'll get the same stink problem.
- The kind of worms you want are Eisenia Fetida or Eisenia Andrei. These two varieties adapt well to shallow living and eat decomposing material.
- You can harvest your own composting worms at the base of someone else's outdoor compost (the right kind of worms will congregate there), a friend's vermicomposting set up, or you can order them online.

- To deter ants from invading your worm bin, keep it up off the ground if it's outside and put down a cayenne powder barrier on the ground that the ants will be unlikely to cross.
- You can make a compost tea from your castings with which to water your plants, providing them with a nutritive boost.
- Many people will pay a good price for worm castings—want to start a small business selling worm poop? Bet you'll be the only one in your office doing that!

> "The message is not so much that the worms will inherit the Earth, but that all things play a role in nature, even the lowly worm."
>
> - Gary Larson, Cartoonist

ACTION ITEMS: In your homestead journal, write down all the pros and cons you can think of about vermicomposting. How many do you have? Time to do some more research. I'd go to a website first, just to skim some basic information and then check out a book from the library. Start taking notes about potential cost and decide if you want to make your own bin or buy one. You may decide after your research that you don't want to do it all. That's OK, too. Sometimes, the most important thing we learn from reading up on homesteading topics is what we do NOT want to try right now. Worms are a living thing and you will be responsible for their care, so be sure you're ready.

DIY: If you decide to give worm composting a try, build your own worm bin or tower. To learn to build your own **worm bin**, visit our links page under the Livestock Wherever You Are chapter: www.homesteadlady.com/homestead-links-page. To construct your worm bin, you will need two 8-10-gallon storage boxes, a drill with various size bits, newspaper and about one pound of worms.

To learn to build your own **worm tower**, visit our links page. To construct your own worm tower, you will need one 2 feet in length, 4-8" in diameter PVC or HDPE pipe, a drill with a 1/4" bit, piece of screen cut to fit the mouth of the pipe and a heavy pot or rock to place over the screen.

If you already have a vermicomposting set up, try harvesting your own composting worms instead of purchasing them. You will need a shovel and a bucket to harvest your own worms. Harvesting your own composting worms is nothing more complicated than digging around in the bottom of a seasoned, outdoor compost bin and putting the worms you find into a bucket. Identifying the proper worms is the part that requires skill because there are several different variety of worms you're bound to run into. To learn more about the possible worms you might find, how to properly identify them and to visit with a gentleman who really loves worms, have some fun with the link found on our links page under Livestock Wherever You Are.

If you still feel uncertain, take a class at your local agricultural extension or plant nursery. You can also use YouTube to learn a great deal, but I always favor learning from a teacher in person so that I can interact with them and ask questions.

BUILD COMMUNITY: Post on local homesteading groups when you have extra worms available to give away for free every now and then. Even if you decide to run a vermicompost business and want to sell worms and castings, be generous with your worm stock. This will not only create good will but will serve your higher purpose of being of use and giving back to the homesteaders where you live.

BOOKS TO READ: I enjoyed *The Complete Guide to Working with Worms*, by Wendy Vincent. I found this book to be so useful because it was full of quality information, personal experience and completely practical advice. The topics covered are broad inside the vermicomposting umbrella and include a lot of very insightful information on building your own vermiculture business. This information covers creating a business plan, practical tips for dealing with worms and using the Internet (from search engine optimization to social media to websites) to establish yourself in the trade.

Also worthwhile is *How to Start a Worm Bin: Your Guide to Getting Started with Worm Composting*, by Henry Owen. This book is simple and straightforward and makes starting your own worm bin so easy that you'll take longer reading the book than you will setting up the bins. There's a lot of useful information on the biology of worms, too, written in normal person speak.

WEBSITES: The internet in general, and YouTube specifically, can be helpful for examining how different people stet up their vermicomposting systems. The websites listed below all have useful posts on this topic so be sure to peruse their site menus as you're doing your research.
- Red Worm Composting (redwormcomposting.com)
- Homestead Chronicles (www.homesteadchronicles.com)

HOMESTEAD LADY SPEAKS:
So many times, I'm amazed at how much of our homesteading lifestyle finds its way into our home classroom and this is not something unique to home educators. What is the bonus of composting with worms? You get to study worms, of course!

What kid (apart from a few) doesn't love to watch a worm do its thing? How many times have you rescued worms from the pavement in a rain storm only to go on to discuss why they come to the surface when it pours? That leads to a discussion of what worms do in the soil, why we value them and why we keep a worm bin in our laundry room.

You're not just teaching about biology, you're also conveying a lot about what you value as a person. "See this small worm? It's miniature but mighty. Our planet would soon be at a loss without this one species. Never discount the power of one."

Nothing is ever too small to make a difference. Class dismissed.

Hickety pickety, my black hen
She lay's eggs for gentlemen
Sometimes nine, sometimes ten
Hickety pickety, my black hen.

- Classic Nursery Rhyme

For the Homesteadish: Make Your Own Chickens

I've chosen to write about chickens here because they seem to be the most common backyard farm bird. However, this information equally applies to any bird from quail (a fantastic choice for small space homesteaders) to turkeys (a fun way to ensure a healthy meal at your next holiday). I will say, though, that some varieties of bird naturally reproduce and make better mothers than others.

Guinea fowl, for example, can be just plain stupid about their mothering while I've seen some ducks willing to mother just about anything. Bottom line, as you breed your own poultry stock, get to know your animals and, if you can't have the mother bird raise her own babies, use the incubator and step in thereafter. Establishing your own breeding program will help you save money, ensure the quality of the genetics in your flocks and enable you to produce eggs, meat and manure on your homestead in ways that keep you close to your animals.

Chickens on Your Homestead

Every homesteader who has even a small bit of earth to call their own will most likely keep chickens at some point. Larger scale homesteaders often find them to be a good source of income or trade and have a large variety of poultry. The point of this section is NOT to teach you all about keeping chickens. There are many, many fine books on that topic nowadays and

I refer you to the excellent resources at the end of this segment. We are going to stay focused on learning to grow up your own chickens in various ways.

There are four places a new homesteader usually acquires chicks to begin their chicken keeping adventures. One is the feed store; a dubious place of health and breed identification. The second is a hatchery, either local or online, which may require shipping. Thirdly, you can find sweet deals on chickens of any age through your local classifieds after you've checked out the poultry's living conditions. For a fourth option, you can purchase chicks from your chicken breeding friend since you know and trust them. Each of these sources for chicks has merits and drawbacks.

COMMON SOURCES FOR CHICKS AND OTHER POULTRY

FEED STORES

Feed store owners love animals, of course! They typically have various systems set up to house baby poultry come spring and some are more effective than others. The babies are kept warm, fed and watered regularly and don't lack a single thing, except an owner to come along and claim them. However, because feed store employees deal with such large numbers of baby poultry, particularly in the spring, and because there are other things to do in the store, it can often happen that the breeds of the various chicks get mixed in their pens, especially with larger birds like ducks and turkeys. Also, more seriously, illness can occur in the animals waiting to be purchased; as well as stress from being in such a busy environment. I don't recommend you begin your poultry keeping days by purchasing your stock from a feed store. At least, not in large quantity.

HATCHERIES

Hatcheries are often the places supplying the feed store, so by going directly to the source, you cut out the middleman and some of the potential problems inherent with feed store birds. Hatcheries, whether local to you or online, can provide you with specific dates for hatching and shipment. They can also provide large numbers of birds shipped to your local post office with the utmost care. Mortality during shipping can occur and so there are usually one or two extra chicks provided to cover that possibility. The shipping charges can be steep given the delicate nature of the cargo, however. There are also minimum order requirements for nearly all hatcheries, because the birds act as insulation for each other and you must have enough present in the package to help them all stay warm. If you only need three birds, this can be an issue. There are a few places that will ship fewer birds, but they must spend more on insulation for the packaging and you will pay a pretty penny for that service.

LOCAL CLASSIFIEDS AND FRIENDS

Your local classifieds, especially those online, provide quick, easy to access groups of ads in your niche. Buying from a classified ad will mean that your purchase is basically local, even if you must travel a bit to get there. You will also be able to inspect the birds before you purchase—something you can't do with a hatchery unless it's local to you. The biggest

> drawback to purchasing from a classified is that you have no recourse if there's something wrong with the birds. If you're completely new to poultry, you may not be able to tell what looks healthy and what doesn't. Buying birds from friends or acquaintances may cause similar difficulties for the newbie so be sure to have an experienced chicken keeper come with you.

I've used the first three sources before and had mixed luck with all of them, the best birds coming from my hatchery purchases. The second time we ordered 30 meat birds (or was it the first time we ordered 30 layers?), I sat looking at the purchase price of each chick, plus the shipping, and realized that constantly purchasing chicks was not a financially sustainable practice for us.

Raising your own Organic meat birds or layers can be cheaper than purchasing Organic finished fryers and eggs from your local health food store. However, we were concerned about reliance on methods that required so much outside input for them to work. For us to raise our own meat and eggs, we necessarily had to rely on birds being constantly available from a hatchery and we also needed them to be shipped to us since we don't live near a large, local hatchery. Shipping is a notoriously shaky business—from natural disasters to oil prices, the availability of economical shipping was not something on which we felt comfortable relying. What, then, was our option?

We started looking into how to raise our own birds as it became apparent, as it so often does in homesteading, that it was really our only choice if we wanted to be self-sufficient. Perhaps you're thinking that growing up your own poultry stock isn't relevant to you, because you'll only ever have space for a few birds here and there. I have two words for you to consider: natural reproduction. It will make your breeding efforts worth it, even if you only need a few chicks. First, let's talk about the next best thing.

An Incubated Bird is Still a Bird

The next best thing to a chicken flock reproducing itself is a source of fertile eggs and an incubator. If you will only ever need a few chickens here and there, the smallest varieties of incubator are just fine. If you're wanting over 20 birds a year, I suggest one of the larger, home-use table top incubators. We use a Farm Innovators brand but there are several on the market and I encourage you to do your own research. Reading the customer reviews on Amazon can be helpful—I always read the five and one-star reviews for a nice mix of opinions.

There is a right way and a wrong way to use an incubator and you'll need to allow yourself a learning curve, but it's not rocket science and you'll figure it out. If you don't have a rooster and, therefore, do not have fertilized eggs from your own flock, you can usually find a local source. The online classifieds or your farmer friends come in handy for locating specialty items like these.

Chicken eggs are the easiest to find in most areas, but other poultry, like duck and turkey eggs, can also be found. Although they are usually a bit more expensive. If you want to incubate an egg other than chicken, make sure that your incubator can accommodate that—large or small eggs can require special parts or manual turning.

Once the chicks hatch in the incubator, you will need to provide them with everything a baby chick requires, just like you would if they'd arrived from a hatchery. Food, water, warmth, love and checking for pasty butt (a fun thing to have to do when your chick has no mother) are all in constant demand. Don't expect to hatch out 100% of your eggs in an incubator—even a chicken mother doesn't always hit 100%. Anything over 50% is considered successful enough, although you'd really like to have your hatch rate go up as you get better at using your incubator. Incidentally, in my casual observance of what others are doing and my very non-scientific experience, a mother hen will almost always out-hatch an incubator, if her nest remains safe.

It's hard to think about the eggs you've carefully tended not producing a baby chick after those long 21 days of waiting, but you need to know ahead of time that it will happen. Some eggs will simply never produce life, while others may begin the pecking and peeping process of hatching only to pass on before it's completed, not having enough strength to survive. Others will hatch out and die afterward for any number of reasons.

Living with things that can die on you is a big part of the homesteading lifestyle and you NEED to get accustomed to the idea of it. (I don't think you ever get used to it, but it can become a familiar pain.) I reminded my daughter of this when we lost one of our incubated chicks that she had worked with for days to save. You can only love something if you're willing to risk losing it; we'd rather have had our hearts broken than to have never loved and served that little chick at all. It didn't stop her from crying for a few days, but I know it was a lesson she internalized just by living with and loving that baby chicken and, indeed, all the animals on our homestead.

Where Have All the Mothers Gone?

> "You may want to breed just for the sheer pleasure of observing those tiny balls of fluff scampering about, shepherded by their adoring, watchful parents. Or you might find that there is a buoyant market for selling purebred chicks or ducklings. You may also find that you want to replace your good but old layers with their younger progeny."
>
> Alanna Moore, from *Backyard Poultry Naturally*

We use our incubator a lot but the best success we've had as far as hatch rate and overall health and well-being of our birds, has been when we've used an actual mother hen. Over the generations, and typically driven by commercial production needs, humans have bred out the instinct for lady layers to set their eggs and become mamas. The reason? A hen stops laying, for a time, once she's lain enough eggs for a clutch, or a group she plans to sit on and hatch out.

For a standard sized, fluffy hen that number can be around ten and for a bantam (half-sized breed) that number can be around six of their own (smaller eggs) or four standard size eggs.

It's all relative to the bird, of course—some of them are just natural over achievers when it comes to motherhood and will hatch out way more than these conservative estimations. Many chicken keepers, however, are not impressed with the number of chickens being hatched since they're too busy saying, "A hen that doesn't lay eggs?!!! Well, we can't have that!"

If you foresee a time when perhaps the poultry houses can no longer afford to ship chicks to you, or that your incubator's electrical cord can no longer suck energy from the plug in the wall, or you simply want to save some money and improve the health and well-being of your birds, you may want to investigate the why's and how's of finding a hen that can hatch out her own eggs. For this, you will need a willing rooster but, trust me, they're not hard to find. If you aren't allowed a rooster as part of your zoning, find a friendly farmer who is willing to share the services of one of his lads. You can also buy fertilized eggs for your broody hen to set. Where there's a will, there's a way; nothing is more obstinate than a broody hen.

Through natural selection of non-broody hens, we've reduce the breeds that tend to go broody down to just a handful—Silkies and other Bantams like Cochins and Phoenix, as well as Orpingtons, Ameracanas/Auracanas, Brahamas, Sussexes and a few others with spotty records. Not every one of the hens of these breeds will go broody but, if any chickens will, it's these ladies. It's worth looking at those few, remaining natural moms and deciding what you think about the tradeoff of baby chicks for eggs. Will we live to regret our demand for egg production over the capacity for our ladies to naturally produce their own posterity? I've concluded that I already regret it.

Incubators are a great option for hatching and the modern provision to produce more chickens. We will continue to use one for large quantity production—like with broilers—for as long as we can. However, I'm in awe of the process of how a mother hen produces chicks and, when they're willing, we let them do just that as often as possible on our homestead. Incubators and humans can't replace a mother hen. We can only step in and do our best. To illustrate my point and to, perhaps, inspire you to try developing your own breeding program, let me recount our first experience with a broody hen hatching out her own chicks.

AND THEN THERE WERE SILKIES...AND OTHER SETTERS

I'd raised Buff Orpingtons and Ameracana/Auracanas before with only a few Orpingtons ever going broody. It was my children who wanted to try Silkies, looking for a mother hen to begin our experiments in chicken reproduction. Silkies had several things to recommend them. For one thing, they're small—they are small, live small, eat small. They're about half the size of a standard breed chicken. They're also notorious for going broody, that instinct to just sit and sit until they hatch something. So, we purchased a few Silkie chicks along with our standard sized laying hens and waited around for at least one to decide to sit eggs.

I don't think of myself as an overly sentimental person; I'm not prone to crying too often at corny movies and I'm not exactly the most sensitive person on the planet. I usually won't even put a plant in my garden unless is serves some, typically pragmatic, purpose. As those Silkie chicks grew, though, I fell totally and completely in love with them. We're talking holding them all the time, cooing at them, bringing them treats—totally sappy stuff. And I wasn't the only one; every one of us fell under their spell.

These animals went from being chickens to beloved pets from their first little peeps. They had big personalities for such little things. We ended up with two roosters and two hens. One rooster, a darling lad we named Reepicheep, ended up being too crow-ey for our neighborhood and we gave him away to friends who had lots of space around their property. The other rooster, Samson, was gentle with his girls and appropriately fierce with encroachments into his territory—if you count snuggling up when he was held being fierce, that is.

One Silkie hen refused to do much in the way of chick production, but she ate bugs and laid little eggs and she was welcome to stay as long as she wanted. Then there was Snowy, a lovely, gray little bird that my daughter named, cuddled and adopted when Snowy was no more than a handful with a pom-pom of feathers on her tiny head. My daughter truly believes that chicken is her baby. Maybe that's why Snowy turned out to be such a good mother herself. As Snowy started to hatch out clutches of eggs, I began to see, for the first time, how much chicks learn from their mother.

WHAT A MOTHER HEN DOES

A chick brooder really is an odd environment from a chick's point of view. The padded floor is made of paper and shavings, no grass or bugs and certainly no fresh air since fresh air is too cold for a newly hatched chicken with no mother to hide under. There's food and water, but no other chicken to show you what to do with them—it's the huge, human hand that comes out of nowhere to dunk your beak into both troughs, hoping that you'll figure it out from there.

Watching Snowy take such time and care with those chicks was fulfilling. She snuggled them under her to keep them warm, allowing them to venture out but clucking them back in when they strayed too far, or were gone too long. She carefully showed them the water dish and reminded them to drink, instructing them not to fall in. The food on the ground they only picked at now and then until their appetites kicked in and Snowy explained that grain is food and wood shavings are...well, she didn't seem sure what they were, so they were just classified as "not food." Once they graduated to their outdoor pen, she was right with them, keeping up a constant, low chant of cluckings to let them know what to do and where to go. "This," she explained, "is a goat. Watch its feet, as it is not clever with them."

Even after several weeks, long after I moved on to other obligations, Snowy was still out there instructing and watching over her then adolescent babies. They were still small enough to escape through the chain link fence and go off under the apricot tree for some big adventure; this distressed Snowy to no end and she had a smack-down cluck she brought out for these occasions. As big as they thought they were, those baby Silkies would still get lost in the alfalfa and cry for their mother. She patiently picked her way over to the sound of distress, calmly assessing what was what and kicking a little feathered patooty if she needed to—figuratively speaking, of course.

I'll never forget that first experience and, just so you know, Snowy is still on the homestead and is, even as I write this, raising up a batch of five standard sized pullets. It doesn't seem to occur to any of them that Snowy isn't their biological mother—I guess love is blind even with chickens. In short, if you're curious about creating a sustainable way to replace and/or produce chickens on your homestead, I highly recommend you bring in a Silkie (or another broody-sensitive hen) and wait for them to be willing to hatch out a clutch for you. A Silkie can only set a handful of standard size eggs because of her smaller size, but what you give up in volume you'll make up for in quality. If Snowy is any indication, Silkies deliver in every respect!

TO ACCOMMODATE A BROODY HEN

You've never met anything so determined as a broody hen and they'll hatch their eggs just about anywhere they deem sufficient. However, there are several reasons why providing a broody hen and her eggs their own space is a good idea.

For starters, broody hens go practically comatose while they sit. Hardly willing to get up even to eat, a broody hen goes into sitting mode wherein instinct tells her she must keep those eggs warm and not rise for anything. To keep her strong, healthy and eating, you should pick up your hen, "fluff" her a bit and then set her down next to her water and food. "Fluffing" is the term I use to refer to the gentle up and down motion in the air I employ while holding my hen securely. Some hens will get up every few days of their own accord to eat and poop. Others will just slump back to the ground even if you do pick them up. This fluffing motion, two or three times, usually kicks the hen out of her stupor for long enough to drink and eat. She will certainly holler at you the whole time she's off her nest, but your hen will get her stuff done and return quickly.

Another reason to isolate a broody hen is to keep her safe from other chickens getting up in her business all day long; so, too, from predators who can easily sneak up on your hen while she's in her stupor. A large dog crate, a supersized tote box, even your bathtub are all fine places to keep a broody hen and her eggs. It's best to pick a spot where your hen will have enough space to walk around and tend her chicks for a week or so once they're hatched.

Finally, keeping your broody hen isolated means you control how many eggs end up in her nest. Once a broody hen has gone "into the zone" she may lay a few more eggs herself, or she may not. If it looks like she's done laying her own in her nest, you can add in your choice of fertilized eggs for her to hatch. If you leave her in the general population of your coop, other bossy hens may come up and lay in her nest at various times during the twenty-one-day hatching cycle. If the eggs get staggered in over too many days, you'll have eggs that never hatch because the broody brain starts to lift off the hen as soon as hatching begins. Your mother hen may abandon anyone not hatching after a few days.

> Don't be shy about adding fertilized eggs under your hen. Usually a standard-sized bird can handle about ten standard-sized eggs; more even, if you're adding Silkie eggs because they're much smaller. Just make sure she can keep them adequately covered with her breast and bum—some hens are better endowed than others. If you end up with brand new chicks from another source at hatching time for some reason, you can slowly introduce them to your mother hen for her to care for as soon as the hatching of her own eggs has begun. Place your already-hatched chick underneath the mother hen so she assumes that it hatched underneath her. She should care for it just fine—chickens are awesome that way.
>
> After a week or two, you can probably let your mother hen and her babies out to enjoy some fresh air and exercise. Most likely, your mama hen will bring her chicks out proudly and begin her instructions immediately. She may end up being something of a show-off and it's important to preen her and feed her treats and tell her what a marvelous job she did. She's just provided your homestead with a valuable and helpful service, after all.

ACTION ITEMS: After properly researching chicken keeping, begin locating local hatcheries and online hatcheries for large orders of birds and take notes on each in your homestead journal. Read the mission statements of each hatchery and breeder, which you can usually find on their "About Us" page. Read reviews from customers and price check—write everything down in your journal because you won't be able to keep all that information in your head.

If you are only looking for a few birds, still try to use a local hatchery, not only to support your local economy but also for their customer service. If you don't have a hatchery close by, plan to look in your local classified or at your local feed store—be advised, most feed stores only sell chicks at certain times of year. Hatcheries, too, will have specific months in which they ship to protect the chicks from the heat of summer and the cold of winter.

DIY: Investigate building your own chicken coop—take a deep breath and just investigate it. If you get overwhelmed, try converting an existing garden shed instead of building something from scratch. A coop will most likely be your biggest cost, so scrounge and upcycle as much as you can to mitigate it but be sure you give your coop a nice coat of paint, so it isn't shabby. If you must buy a prefabricated coop, you can at least research how to construct your own feeders, waterers and nest boxes. After the coop, a lot of the initial cost of chickens is the culmination of all these bits of equipment you need to have. The Internet is your friend as you search out other people's experiences making their own chicken-keeping items.

The least you can do is make your own chicken toy by shoving a large needle threaded with yarn through an overgrown zucchini and suspending it from a low hanging gate or post. At first, the chickens will be convinced that this flying zucchini has been sent to murder them in their beds (chickens are convinced that death lurks around every corner). But, eventually, they'll be pecking away at it and having a blast. You can do this with any veggie or fruit that you can get a needle through. I usually use one of those huge, hand-quilting needles that you can buy in most sewing notions stores.

BUILD COMMUNITY: If you can't have chickens where you live, try to find a friend or like-minded person with more land and fewer restrictions where you can build a flock together. Your friend can house the chickens on their property, you can contribute cash for whatever you decide is fair by way of feed and equipment, and you share the resulting eggs. Even if you have space and zoning for a small laying flock, this is still a great option for growing meat birds. Especially if you don't have enough space for a flock of broilers and/or can't slaughter on your property (some municipalities and HOAs have rules about harvesting animals on your property). You know what they say about birds of a feather and this is a great time to make some chicken friends.

BOOKS TO READ: For general chicken information try *Backyard Chickens Naturally*, by Alanna Moore, which will teach you how to raise up your backyard flock in more natural ways and has a whole section on allowing broody hens to hatch out chicks. *Free Range Chicken Gardens*, by Jessi Bloom, is a worthy title for those who would like to learn what to do when free ranging their hens, especially in urban settings. *Chickens From Scratch*, by Janet Garman, is good for the novice chicken keeper as it's very short and to the point.

For chicken health try *The Chicken Health Handbook*, (or anything Gail Damerow has written about poultry) for trouble shooting and advice. Most treatment suggestions in this book will be conventional, while still being reliable. If you want a more natural reference, I can recommend *The Accessible Pet, Equine and Livestock Herbal*, by Katherine Drovdahl, which is a reasonably good resource. You'll need to have a basic herb knowledge to use this book well, FYI. For the other stuff you worry about (and which gets controversial in the chicken community), I can suggest a little book called Chicken Hot Topics by Jessica Lane. If you're new to chickens and need help deciding big issues like what to put on the coop floor, whether you should clip wings and what to do about various egg issues, this book can help.

WEBSITES: There are a lot of chicken forums online and you'll find your favorites, but the popular My Pet Chicken site also has a blog (www.blog.mypetchicken.com/) and is an especially great resource for urban and beginner chicken keepers. Backyard Chicken Project (www.backyardchickenproject.com) is another quality source of information. Backyard Chickens (www.backyardchickens.com) has a large, online forum where you can ask questions from other chicken keepers and there are several threads that cover hatching your own chicks.

Really, there's so much online about chicken keeping these days, simply search for your question and you'll find ten different answers. In this case, as with everything online, be very aware that just because you read it on the Internet, doesn't mean it will be helpful to you. Some information is downright incorrect, but a lot of what you read is also just personal experience.

What works for someone may not work for others, so read a lot and carefully consider a course of action. More information exists than what's on the Internet, too—don't forget actual books and real-life chicken friends you know.

HOMESTEAD LADY SPEAKS:
Make sure you stay involved with your city council regarding backyard chicken laws, especially if you live in a big city. I know you're busy, everyone is, but being vigilant is the best way to guarantee that your viewpoint and the viewpoint of those like you is being represented. The needs of backyard dogs and backyard chickens really aren't so different, but mainstream people, including politicians, have more experience with dogs. It's up to you and me to educate our city-mouse friends and neighbors on why backyard chickens should be allowed anywhere and everywhere. We need to clearly and calmly explain the realities and benefits of a well-maintained, healthy flock.

Be prepared with facts and figures, especially if they bring in an "expert" from your local agricultural university or agricultural corporation. These experts have much to teach us about the biological needs of chickens, as well as being qualified to give advice to the broader (read: larger) poultry industry. However, despite experience in animal science studies, it often happens that not all of what they have to offer translates into real-life, common sense policies for backyard chicken keepers. Be respectful of their viewpoint, but don't take everything they say as gospel. Be prepared to politely challenge their recommendations if you feel your experience is different from their seemingly random recommendations for small farm livestock in municipalities, especially when it comes to fees, coop locations and space requirements. (Can you tell I've been in your shoes before?)

If we don't get involved and don't take the time to show up to city council meetings and voice our views, then we have only ourselves to blame. Honestly, the idea of backyard chickens is probably the easiest small farm livestock sell in our time—it's very chic to have backyard chickens, donchya know?

Our family members, including our children, have spent a great deal of our precious time down at city hall waiting through countless fence line debates and municipal squabbles until they finally got to the animal ordinances. A few of those proposals directly affected us and others would have had no bearing on our zoning but did affect the property and backyard chicken rights of others in our community. It doesn't have to directly tie back to you to be relevant to you. We homesteaders need to stick together, whatever the size of our homesteads or the size of our flocks and herds. Get involved, be proactive, and be polite. Even if you fail, it won't be because you didn't strive for it. And there's always the next city council meeting to revisit the issue.

Chicken people unite!

"I asked the waiter, 'Is this milk fresh?'
He said, 'Lady, three hours ago it was grass.'"

- Phyllis Diller, Comedienne and Actress

For the Homesteadaholic: The Dairy Animal Question

There are other dairy animals for the homestead besides cows and goats, of course—sheep, yak, water buffalo. (I wonder if the water buffalo knows it can be milked.) I haven't tried those other ones, but I have milked goats consistently and a cow for a bit. I certainly have my opinions on both animals, but I was curious what other homesteaders might say so I interviewed a few and would like to share some of their insights below.

> **IS IT SAFE TO DRINK RAW MILK?**
>
> I'm not looking to start a fight with the pasteurized milk people out there – do what you gotta do, I say. However, if you've ever asked the above question, here's some information provided directly from the Real Milk (www.realmilk.com) website, a sister site of The Weston A. Price Foundation.
>
> "Real milk that has been produced under sanitary and healthy conditions is a safe and healthy food. It is important that the cows are healthy (tested free of TB and undulant fever) and do not have any infections (such as mastitis). The cows should be eating food appropriate to cows, which is mostly grass, hay or silage, with only a small amount of grain, if any. The milk should be full-fat milk, as many important anti-microbial and health-supporting components are in the fat. The cows should be milked under sanitary conditions and the milk chilled down immediately."
>
> For more information and to read actual research into raw milk, please visit the site.

The Cow's Turn

My first question was: *What are the best thing about milking a cow?*

Across the board my dairy friends pointed out the quantity of milk. A cow has four teats, as opposed to two on a goat, and produces milk in proportion with its size. As Lee Ann from One Ash Farm and Dairy (www.oneashfarmanddairy.com) observes,

> "The best thing about milking a cow is the quantity of milk you can get! Our girls vary from giving 2 gallons a day to one that gives about 4 gallons. It sounds like a lot but when you start making dairy products for your family, and using the excess milk to feed farm animals, you can go through those amounts very quickly."

Jenna, of Flip Flop Barnyard (www.flipflopbarnyard.com), says,

> "It's hard to pick one thing as best. Aside from companionship, the amount of milk and cream per milking is great. Plenty for drinking, making cheese, making butter and much dairy yumminess."

Amy, formerly of Home and Farm Sense, agrees and observes,

> "You can make butter with cow's milk without special equipment, but you cannot do that with goats since their milk requires a cream separator due to natural homogenization. Cows eat a lot more than goats but are easier to fence. Finally, cow's milk stays fresher longer than goat's milk."

One final thing in a cow's favor is the amount of highly useful poop that each animal produces. This poop can be composted each year and applied to the gardens as amazingly nutritive plant food. Goat poop is great, too, but there's less of it. Never discount the value of properly managed dairy animal poop—the stuff is worth its weight in gold!

So, just to recap - the best things about cows include:
- High quantities of milk
- Companionship
- The cream separates from their milk without special equipment
- Versatility of milk to become other products like cheese and butter
- They're simpler to fence
- Their milk retains a fresh flavor longer

My next question was: *What are the worst things about milking a cow?*

Again, the vote was unanimous—cows don't care where they poop, including in the milk barn (or on the hay or on a calf) and including during a milking session. Their manure is much more voluminous than goat's manure, too. Lee Ann shares a tip for dealing with pooping during milking:

"We keep a 5-gallon poop bucket handy—when one of the cows raises her tail to get ready, one of us runs to her rear and catches the poop in the bucket. It's certainly a funny process but keeps the milk and the floor clean!"

It must be mentioned that, while cows rarely bust out of the fencing like goats, they do lean and push their way out sometimes. They can be hard on fences with the weight of their incessant hide scratching on posts and wire. Speaking of their weight, another complaint of cows is the size. Cows are just plain bigger than goats, even the smaller breeds. Because of their size, they kick harder and, yes, sometimes their foot ends up in the milk bucket. Goats are guilty of this, too, though. As Jenna observes,

"That is when it is OK to cry over spilled milk."

Absolutely! That spilled milk represents a lot of wasted hard work on behalf of you and the animal. Whatever you choose, goat or cow, be sure to purchase, breed and select for good temperament. Also, be calm while you milk and avoid loud noises; be patient as you train your milkers, too. Some days, with a cow or a goat, that is easier said than done. Cows are dopey, and goats are wicked.

So, to review – the worst things about cows include:
- Bigger animal equals bigger poop
- They're much larger animals which can be harder to handle, especially for kids
- They can tax a fence with their weight as they lean or scratch

The Goat's Turn

I then asked: *What are the best things about milking a goat?*

Each homesteader agreed that it must be the size—goats are much smaller and less intimidating than a cow. Amy illuminates a few other good traits, as well:

"Goats are easy for children to milk and be friends with—they are very social. Their manure breaks down easier as it is pelleted. Special cheeses and amazing soap can be made with goat's milk."

Because of the smaller amount of milk, they produce, processing the milk takes less time overall unless you're running it through a cream separator.

Another point in a goat's favor is her pure, white milk that, because it has a lot less lactose and a smaller sized fat globule than cow's milk, is much easier for most people to digest than cow's milk. Also, because their smaller size makes them easier to transport and handle, and because of their more successful breeding rates, goats are a lot easier to breed than cows. Goat dung is smaller, as Amy mentioned, as well as dry and ready to apply to the garden immediately. Finally, although smaller, goats have big personalities and are a little less dial tone than cows—of course, their personalities are kind of a double-edged sword.

So, to review – some of the best things about goats include:
- They're smaller and commensurately easier to handle
- Children tend to be able to work well with goats
- Goats are sociable animals
- Their manure is pelleted and classified as "cool," which means it can be directly applied to the garden
- Less milk equals less time each day to invest in the milking process
- Their milk is pure white and typically considered easier on the human digestive system
- Breeding is typically simpler

To finish up, I asked: *What are the worst things about milking goats?*

They're *goats*, was the basic answer from the ladies. Read that sentence with as much disgust in your voice as can. Ugh, they're *goats*. Amy says,

> "Goats are a pain in the neck to fence and when they do get out they can do a lot of damage to fruit trees, gardens, etc."

LeeAnn adds,

> "Goats can be very temperamental. If they don't like the feed that day, if there are bugs, or if they are in a bad mood, then be prepared for the milk bucket to be tipped over! They are adorable, fun animals, but you need to really know your milk goat to build a strong partnership."

Goats are a lot less placid, to be sure. I have a goat that will toss her feed dish off the milk stand if she doesn't like what I've given her. I had another who would sit down during milking if I took too long. There are ways to get around their opinionated behavior, much like we must assist our children in being obedient. My husband modified the feed dish so that it now locks onto the milking stand and when I have a goat that wants to sit, I put a small bucket under her rump or ribcage, preventing her from doing so. If she kicks, I restrain her with my favorite soft nylon hobbles. Because of their smaller size, you can sometimes reason with a goat. If a cow wants to do something badly enough, the cow gets to do it because of her size. Fortunately, most milk cows are bred to be docile.

So, one last recap – some of the worst things about goats include:
- They're goats
- They're difficult to keep fenced
- They're goats
- They can do a lot of damage in the garden, if they get out of their fencing
- They're goats

Milking Questions

You can milk a cow or goat as long as the animal stays in milk and a lot of that is determined by breed and the animal herself. You can milk one, two or even three times a day with most animals; I recommend once or twice a day. Here's how our homestead ladies do it:

HOW OFTEN DO I HAVE TO MILK?

From LeeAnn:

> "We milk both cows and goats twice a day, every day. Our belief is that this is closer to the natural way a kid or calf would suckle. The only time we have milked once a day is if the baby was still with the momma. Cows can be milked for up to two years, with declining production. However, if they are not re-bred within 3-6 months of calving, their fertility rate declines. You must dry up both cows and goats 2 months prior to having the baby.
>
> "Goats come into season [and are ready to be bred] every fall, and gestation is 5 months. So, you get milk from early spring until about August every year if you want to breed them again. Otherwise you are waiting another year. Cows have a 9-month gestation, so you get a good 7 months of milking once they are bred. (So, on average for a cow you have 7 months + 4 months or so until they are bred so very close to a year.)"

From Jenna:

> "The goats we milked twice a day. They lactated for around 4-6 months or so until drying off for kidding. My cow is on a once a day milking schedule. She has been in milk for 6 months and is not bred back yet. Once she is bred back we will milk her until 8 weeks before calving. I have heard of cows in milk for as long as 5 years before freshening again."

From Amy:

> "We only milk once a day. We milk our cows for ten months and then give them two months off to grow their calves and get ready for calving. Their gestational period is 9 months. Cows will re-breed at any time of the year. Goats are usually seasonal, going into heat in the fall and kidding in the spring. You can also milk goats for about ten months, giving them two months off but their gestational period is only 5 months."

We're currently in the market for a dairy cow and as I've been reading and talking to people, I've learned that milking cycles vary from cow to cow and can also be breed dependent. Goats are the same, for the most part. Our head dam came to us in milk and we milked her straight through for three years, milking every day, twice a day. At the end of that lactation cycle, she was down to about a quart a day from her peak (the time a little bit after kidding), of about a gallon a day. She's a solid mother, a smart herd leader, a wonderful milker and the biggest pain in the butt I've ever met. Besides me, of course.

A few of our other goats dried themselves up after about a year, despite our efforts to keep them in milk, because a long lactation cycle just wasn't in their genes or personal biology. The best way to know what kind of milker you're purchasing is to buy a registered dairy breed whose ancestry can be traced and documented. Otherwise, buy only from someone you can trust when they say you're purchasing the best milker there is. Even then, how long you'll be able to milk a dairy animal is still an individual discovery.

DO I MILK IN WINTER?

If you live where it gets particularly cold in winter, you may have concerns about milking during those months. I've lived in zone 5 the whole time I've milked (although I'm moving to zone 6) and I can testify that with adequate shelter and care, your animals will do just fine. You might feel like you're going to freeze to death, but the milkers will do well. Each of our milk-maids lives in a different area of the country, though, so here's what they have to say:

From Amy:

"We live in the Pacific Northwest and it can be very cold here. Our animals do fine in the winter except when it's rainy and cold (snow and cold is fine but the cold rain causes health issues). We have barns and shelters our animals can go into during inclement weather. Calving can be difficult in the winter if the cow decides to calve outside. (Our goats always have their babies inside). It's important to get the calf dried and in the barn on bedding for warmth. We do milk year-round."

From Jenna:

"From what I have seen with proper care, food, and shelter both species do fine through the seasons. I have milked through warmth and bitter cold. The temperature changes seem harder on humans than the animals."

From LeeAnn:

"We milk year-round. South Carolina is mild with temperature, but we tend to have dramatic changes. 70 one day, 35 the next. We also get ice storms. All our animals are outside, with the goats having a shed to hide in. They really like their 3-sided shelter. The cows can take cover under some pine trees. They all do well with minimal problems. The goats are more sensitive than the cows."

ADVICE TO WRAP UP

I asked our friends for final words of advice to help you make this important decision, so here it is:

From Amy:

"If you have iron clad fencing goats are probably easier until your hands build strength from milking. If you have less than ideal fencing or no fencing, get a cow (you can fence a cow with one string on electric wire, plus you get butter).

"We love both and cannot imagine not milking them. However, like to take time off from the milking, it's good to have a backup friend you can call. Or, keep nursing calves or kids that can be put back in with the milking animals when you need to not milk for a day."

From Jenna:

"Doing research, tasting milk, and weighing out the pros and cons for the individual situation is the key. I think that a home dairy animal is one of the biggest assets to a homesteading family. My advice to anyone looking to start with a dairy animal is to research a lot, get hands on experience first if possible, and be sure you are buying a quality, healthy animal.

"Also, ALWAYS make sure feed is securely locked up. A goat or a cow will find a way to get into feed bins or bags if they can. I know from experience; thankfully it turned out OK and we can laugh about it now. It can be very dangerous and even deadly for an animal to over eat."

From LeeAnn:

"For us, starting with a goat was perfect. We had been around animals for 30 years but had never milked. So, to get past the learning phase with goats was terrific. There are also only two teats on a goat versus four on a cow, so it's easier on the hands if you're hand milking.

"Craziest hint of all? With our first goat, we discovered that she calmed down when we would sing. The Happy Birthday song was her favorite but made up words worked too! To this day, when one of our dairy animals appears a little frustrated, we will sing to her and it helps to calm her down."

I hope sharing these insights from experienced milkers has helped you decide which dairy animal you'd like to start with. Maybe not, though—as you may have noticed, all three ladies milk both goats and cows. It's not so easy a decision, so just make the best one you can for your family and your land. If you are convinced you need both now, you can blame it on me as you're trying to convince your significant other. I've got your back.

> **IF YOU GIVE A HOMESTEADER A MILKER...**
>
> Milk animals can be a slippery slope. Just ask LeeAnn Perez, who thought she was just going to keep a few goats for her family and ended up running a dairy.
>
> "We started milking about three years ago. We were able to buy a Nubian goat in milk and thought that would be the perfect answer for our family of three. Well, I started making cheese and yogurt and before you know it we didn't have enough milk. So along came goat two. Once we were committed to the schedule of milking, my husband determined that we really like cow milk better, so why not get a cow? We found a perfect jersey girl in milk and brought her home. Then my daughter decided she would start buying day old cull calves and bottle feeding them to sell to homesteaders who wanted sweet cows. Then came cow two, cow three, and cow four.

> At that point, everyone we knew wanted to buy raw milk from us, so the natural reaction? Become a certified Grade A raw milk dairy. So here came our state partners, they approved our facility and we have been a micro-dairy ever since! We still raise the calves, still make great stuff for ourselves, but now also sell at the local farmers markets and directly off the farm."
>
> It just takes that first animal and it's all downhill from there. You've been warned.
>
> If you're in South Carolina, stop by and see for yourself:
> One Ash Farm & Dairy:
> 2886 Piper Road, Ridge Spring, South Carolina 29129

No-Yuck Milk

Have you ever taken a big swig of fresh-from-the-animal milk and had it taste...like that animal? Or, have you tasted goat milk from the store and wondered what on earth happened to it to make it taste so foul? Cow, goat, sheep – they're just not animal smells you want to *drink*. That strong flavor is not inevitable, though, especially with dairy goats. Our American taste buds are accustomed to cow milk and we're often more forgiving of off (ripe or cultured) flavors in their milk; with goat milk we can be merciless in our assessments. Taste is a completely personal interpretation and will vary from person to person, as milk flavor can vary from animal to animal. However, there are some simple steps you can take to ensure that your fresh milk is tasty every time. (Unless your animal got into the onion patch, or lost their head and ate a bushel of mustard weed. But that's an issue for another day.)

DAIRY LINES

The flavor of milk, any milk from any animal, is dependent on several factors, so let's cover just a few of those in brief. The first factor is genetics. The only real way to determine the quality of the dairy lines of a milking animal is with a pedigree; the easiest way to obtain reliable pedigrees is by purchasing a registered animal from a national association specific to that animal. There are several dairy goat associations in the United States alone, for example, and a simple internet search will provide you with their criteria and rules. When you're ready to purchase a dairy animal, if you want a pedigreed lady, be prepared to pay for her—at least double, sometimes triple what an un-pedigreed animal will cost.

SAFE MILK HANDLING

Once you've established quality dairy lines in your backyard herd, the next factor to concentrate on is clean dairy handling habits. Establish good routines and be strict with yourself by following through with them every time you milk. EACH time you milk, clean the teats well with a quality solution. EACH time you milk, use a strip cup to dispose of the strippings. Or you can dispose of them in a napkin or paper cup. I like to use the strip cup because I can check for chunks or off-color milk which might alert me to mastitis or other infections.

Also, milk ONLY with clean hands; your mom was right, nasty stuff lurks on seemingly clean-looking hands. When you're done milking, dip each teat in a bit of the cleaning solution for good measure. I also apply a mild, homemade, herbal antiseptic salve to my girls' bag and teats; it protects against harmful, microscopic critters but also moisturizes their skin. Yes, I spoil my ladies.

When you're done with your equipment, wash it in hot water with a soap you trust EVERY time. Seamless, stainless steel milking buckets are suggested so you don't have hidden gunk building up in the seams of your bucket. A lid is also desirable to cover your milk as you bring it in from the dairy barn to the house to prevent anything from getting into your milk. Whatever equipment you have, including reusable filters, clean it thoroughly EACH time you milk.

CHILL MILK QUICKLY

The last matter to tend to for sweet tasting milk is to cool it as quickly as possible once it's out of the animal. During my winters, this isn't a big deal for me since my mid-winter temps are sufficiently cold to rapidly chill the milk by simply putting it into a snowbank for about twenty minutes. However, during the rest of the year, and particularly in the summer, I make sure to take one of my children to the barn with me to immediately take the milk up to the house once I'm done milking. I finish up in the barn, while they filter and chill the milk as soon as possible. There are several acids in goat milk that give it that characteristic tang and, if left warm, they increase the caprine taste of the milk. Raw cow milk, too, will do nothing but culture rapidly while it's still warm. Some people enjoy these flavors and leave their milk warm on purpose.

Other people, hoping to cut that strong flavor, put the milk directly into the freezer after filtering. We used to do that, too, but I can't tell you how many glass canning jars I've busted in the freezer because I simply forgot my milk was in there. Cooling your milk in the refrigerator is not the best option either because it doesn't bring the core temperature of the milk down low enough, fast enough.

The best method we've found for cooling milk rapidly and, therefore, helping to ensure the best tasting milk, requires a glass canning jar or a similar stainless-steel container sufficient to hold your quantity of milk. Additionally, you will need a bucket with a lid that will fit into your freezer that is also the right size to hold your glass milk container. We use a medium size, food grade bucket onto which fits a Gamma Seal lid so that we can easily screw and unscrew the bucket lid. Finally, you'll need some salt and some water. What you'll be doing is creating a saturated saline solution that you'll keep in your freezer and only take out for chilling your milk. The salt lowers the freezing temperature of the water to produce a semi-frozen, chilled mixture that can cool your milk much quicker than the refrigerator, or even the freezer can.

> **MAKE A SALINE COOLING SOLUTION**
>
> Let me illustrate.
>
> If you're using a half gallon canning jar to contain your milk, then get a medium sized bucket (about a three-gallon, or eleven-liter, capacity). The most important feature of the bucket is that it fit both the jar and your freezer, so I hesitate to give you a specific size. Add about a gallon of water to a medium sized bucket.
>
> To that water, add around 1/3 cup of salt and mix until the salt dissolves. This will form a brine which acts as an anti-freeze; you may need more or less salt depending on the hardness of your water. If salt doesn't work for you, try an alcohol solution starting at about half water, half isopropyl alcohol.
>
> Keep this mixture in your freezer and every time you filter your milk, put it into a glass container like a canning jar, and submerge it to half way up in the semi-frozen brine solution. Make sure the solution doesn't come up over the top of your jar and into your milk. To be safe, you can cap the canning jar so that nothing falls into your clean milk.
>
> Since the entire process doesn't take long, I just leave my bucket in the kitchen sink while the milk chills. What takes forty-five minutes to cool to 60° F/16°C in a freezer will now take around ten minutes.
>
> When the milk is chilled, take it out and store it in the refrigerator.
>
> Put your awesome bucket back in the freezer until you need it again. Every now and then, switch out the solution to keep it strong.
>
> We've tried so many other methods and this is still our favorite for how quickly it cools the milk.

If you don't have the space for a whole bucket, you can keep plastic bottles of water in your freezer and, before you go out to milk, submerge them in a bucket of cold tap water. Once you filter your milk, follow the same procedure. It won't work as fast, but it will work and takes up less space than a small bucket in your freezer. It does require more water since you must use new water each time you want to chill your milk, but it will do in a pinch.

Perhaps more effective, if you already have the counter-top ice cream maker parts, is a suggestion made to me by Rebecca, an intrepid *Hobby Farms* online reader. She shared with me that, until her volume rose so high she could no longer fit her milk into the unit all at once, she used the freezing chamber from a counter-top ice cream maker to chill her milk rapidly. The liquid between the walls of these canisters freezes solid while they're being stored in your freezer, thereby producing your ice cream without salt or ice. Or, if you're as smart as Rebecca, rapidly chilling your fresh milk. As Rebecca points out, these bowls typically only hold 1.5 quarts, so you might outgrow them as your herd size increases but, to start out with, they might be useful to you. And, hey, you might just end up with fresh ice cream on a regular basis with one of these units around. Life is sweet with fresh milk.

Remember, no cooling technique can cover bad genes or bad milk handling practices (I should also add poor nutrition). If you've got those areas well in hand and would still like to improve the flavor of your milk, chilling it as quickly as possible will help. Using a reliable method, like the semi-frozen saline solution that rapidly cools and keeps your milk sweet and tasty, will save your taste buds...and the good name of fresh milk. Bottoms up!

> **ONE MAN'S POOP IS ANOTHER MAN'S TREASURE**
>
> A quick note for those in the city:
>
> "You might a homesteader if you've uttered the phrase, 'It's just poop, it won't kill you!'"
>
> - Lisa Murano of Murano Chicken Farm (www.muranochickenfarm.com)
>
> Right?! I have said that exact thing so many times! My younger children, who have grown up on an active homestead, have no idea what it's like to live in a place without animal poop. They'll give you an explanation of why and how we use the poop, too, if you stick around long enough. So, here's a quick note about manure for those homesteading in the city.
>
> When country mouse and city mouse are having a code argument, it's usually over poop. Sure, you may find yourself standing is a city council meeting debating how many small farm animals can "legally" fit onto your property and how those animals should "legally" be housed, fed and cared for depending on how involved your city is in wanting to tell you how to live your life on your land. But really—really, at the heart of the argument—it's the poop you're dialoging about. Country mouse sees animal waste as the great engine of the homestead; without that constant stream of high quality organic manure, the growing season comes to a halt. City mouse thinks of it as dirty and smelly. As we read in *Homesteading in the 21st Century*, by Nash and Waterman:
>
> "We don't think of manure as "excrement"—to a farmer, animal feces are a gift to be treasured, among the most valuable things on the farm, and a...good reason to keep animals, even if you don't intend to eat the meat. Manure makes it possible to grow food.... Animal manure is the bridge between the animal and vegetable cycles that power the farm.... Achieving the optimal carbon/nitrogen ration for good compost, without the additions of animal manure, requires a highly sophisticated awareness of compost science."
>
> We can debate all day about how short-sighted city mouse is and bout how sad he'll be if the grid goes down or there's a trucking/fuel strike and he's knocking on your door to beg a bagful of beans. However, the bottom line is, it's up to us country mice to educate and lead by example, especially if we live in a city. Many, many homesteaders reside in municipalities and that means we have neighbors with whom it's just easier and kinder to have good relationships. You and I can't expect our neighbors to have much patience with what we're doing on our land if we leave huge piles of festering dung near our property

lines so that they have no choice but to have that poop be a part of their outdoor experience.

Truthfully, it is just poop and it won't kill you, but it may just be the last straw when it comes to having a good relationship with your neighbors, friends and community members, including your code enforcer.

ACTION ITEMS: Locate people in your community or surrounding area who have a goat and a cow in milk and ask to have them teach you in exchange for a certain amount of time milking for them. Ask to taste the milk of every lactating goat or cow you meet. Check out the milking apparatus of the dairy people you know and figure out what you're interested in purchasing yourself. Begin hand muscle exercises to strengthen your hands against the time of milking that fast approacheth. You'll thank me later. If, after practicing milking, you can move your hands at all, write down everything you're learning in your homestead journal.

DIY: If you settle on goats, you're most likely going to want a milk stand for ease of milking. A milk stand is a simple, elevated platform with a brace attached that secures the goat's head, usually near a feed dish so that you can milk while your goat has breakfast. I've seen pictures in National Geographic of adorable South American kids blithely milking free standing goats, but I have absolutely no clue how they do that. My experience is that goats need a reason and some encouragement to stand still long enough to be milked. You can make your own milk stand if you're handy with a drill—there are lots of designs online, but you can visit the post from Fiasco Farms to get details and plans for a basic, sturdy milk stand design visit our links page under Livestock Wherever You Are: www.homesteadlady.com/homestead-links-page.

Additionally, you're going to need a place to store hay and a preferably weather-proof place to milk. A barn is ideal, of course, if you have the space and budget. If that's not possible, see if you can convert an existing shed to serve your needs for yourself and your dairy animal. If you choose a cow, you're going to need at least a 10' x 10' three-sided shed with about a 4' head-gate for milking. Though, if you're milking year-round, you may want a more enclosed structure. Your local online classifieds might be a good source for used or cheaper materials. We built both a hay shed, and a milking shed out of large, wooden shipping crates that a local man delivered for a very reasonable fee. My handy husband added hinges, handles and a simple roof structure over both sheds so we had a warm, dry place to work with the goats. Especially at first, milking is hard enough, so make your life easier in the long run and take the time to set up a good foundation of outbuildings and protected areas.

BUILD COMMUNITY: Organize a dairy field trip for your homestead or homeschool group—a small farm would be preferable to a large, commercial operation (good luck getting a tour of those anyway). Many small farms have field days or websites where you can find tour information. Do NOT just show up and expect them to accommodate your group. Farmers and homesteaders are busy, busy people and you'll need to schedule a time to visit with them. Be sure to ask what their rules are for visitors, especially regarding footwear. To prevent the possible spread of pathogens, quite often you'll be required to walk through a germ-killing foot bath before you're taken on a tour, especially if you live around livestock of your own. Have everyone prepare at least one or two questions ahead of time but let your guide get through the basics of their presentation, especially if they take you into a milking facility and show you equipment.

BOOKS TO READ: To naturally and holistically manage whichever animals you choose to bring home, I can recommend *The Accessible Pet, Equine and Livestock Herbal*, by Katherine Drovdahl, for those with some basic knowledge of herbal remedies.

For goat lovers, *Raising Goats Naturally*, by Deborah Niemann, is a simple, straight forward book that will empower you to raise those dairy goats right. Pat Coleby's *Natural Goat Care*, may also be of use to you. Even if you decide to keep goats only for meat, these books will have benefits for you because they cover a wide range of goat husbandry issues. Believe me, as you train a stubborn goat to be milked, it might cross your mind how tasty that goat would be with ketchup. Anything Gail Damerow or Sue Weaver have written will be of help to you, too.

To learn what goats are really like, though, you must read beloved children's author, Patricia Polacco's *G is for Goat*. This book should be required reading at every agricultural university on earth. Another must for all homesteaders, especially if they have kids, is Alice Provensons's *Our Animal Friends at Maple Hill Farm*.

If your interest lies in cows, hold onto your hats Mary Jane's Farm lovers because Mary Jane Butters has written a book for those of us longing for a dairy cow, *Milk Cow Kitchen*. Alongside quality dairy cow information for the backyard homesteader are a healthy helping of recipes, natural management tips and even crafts, always served up with her characteristic whimsy and farm-girl style. For a slightly more serious study, *Keeping a Family Cow*, by Joann Grohman, is a popular title for home-scale dairy cow management. This book will lead you through all you'll need to know about holistically managing your home dairy cow herd.

WEBSITES: For quality goat information from a clear-thinking, down to earth source, there is no place finer on the web than Fiasco Farm (www.fiascofarm.com). Because the American dairy industry is powered by cattle there is a lot of cow information online. Your local agricultural college is a good place to start finding information about common breeds in your area. Even if you'd like to employ more holistic management practices with your own dairy cows, there's much to be learned from people who live and breathe cows for a living. Raising grass-fed cattle and employing grazing practices like silvopasture are becoming more common every day, so ask lots of questions. (If you've never heard of silvopasture, look it up. It's really groovy.)

If you have questions or need products or advice, the Perez family (see the If You Give a Homesteader a Milker sidebar) is happy to help at One Ash Farm and Dairy.

HOMESTEAD LADY SPEAKS:

When we put our urban homestead on the market we decided to part with our goats. We gifted our small herd to some good friends who we knew would care for them and, most importantly, love them. It was a hard decision for me, but since I've never been too attached to animals, I figured I'd get over it soon even though I was going to have to purchase milk from then on. My concerns were mostly pragmatic.

Fast forward to a year later and I was still missing my goats. One goat in particular. Remember that goat I mentioned that tosses her feed dish when she doesn't like its contents? Her name is Maizie and she is MY goat. I know this because over the time we've lived together, we've formed a bond of a very real sort. The year of her second pregnancy was the year of my fifth and we were both feeling it. I'd waddle down to the barnyard quite often to visit and commiserate with her. When she threw those triplets (yes, triplets and all girls!), I was right there beside her, having been delivered of my own kid not long before. When we made the decision to relocate them and not take them with us to Missouri, I was sure I was doing what was best for all of us. Really, I was.

After a year of trying to sell the homestead, finally taking it off the market and putting it back on to show again this year, I realized I'd made a mistake and I needed my goats back. Fortunately, our friends had outgrown their need for the whole herd and I was able to retrieve Maizie and one of her daughters, an airhead named Lily. When we brought Maizie home, she came up to me, put her head down into my hip and gently pushed in. This was always her way of hugging me. After she was done and after I'd lavished her with rubs and scratches, she looked up at me for the longest time. I knew she was asking me, "Where have you been?" I explained to her, as I'm sure she speaks English, what was going on and that I'd been here all along, missing her. She forgave me, I think, but when we sold our homestead and our plans fell through for an immediate move to Missouri, we had to relocate them once again to a friend's barn. I keep visiting them and reassuring them that this is only temporary; that they're coming with us, I promised.

So much for not being much of an animal person. Turns out, I'm a complete milk sop.

"There are blessings in being close to the soil, in raising your own food even if it is only a garden in your yard and a fruit tree or two.
Those families will be fortunate who...have an adequate supply of food because of their foresight and ability to produce their own."

- Ezra T. Benson,
Religious Leader and U.S. Secretary of Agriculture (1953-1961)

For the Homesteaded: Points to Ponder on Meat

Taking the leap into meat production may not seem like such a big deal to some, but for others it can be a daunting prospect. When we grow an animal to become dinner we will be required to take the life of that animal at some point. This can be a deal breaker for some but bear in mind that there are professionals who can kill and butcher your animals for you, if you prefer not to do it yourself. There are a lot of other preparations and plans to consider, too — questions to ask and equipment to gather. I'm not going to argue the ethics of eating meat animals but, if you are going to eat them and you would like to start growing your own to ensure their quality of life and the quality of your food, here are some ideas to kick around before you get started.

WILL I SAVE MONEY RAISING ANIMALS FOR FOOD?

"You might be a homesteader if you raise a steer who drags you to within 2" of your death instead of buying meat at your local butcher!"

- Diane Coe of Pasture Deficit Disorder (www.pasturedeficitdisorder.com)

Yeah, you might be a homesteader if you find yourself working thirty times harder than you ever have for a good steak or a chicken dinner. People always ask, "*Will I save money if I raise animals myself?*" and the answer is not as straightforward as it may seem.

- Do you want to raise them Organically? You might save money, depending on how fastidious you are about it.
- Do you want to replace all your eggs, dairy and meat or just a portion? You might save money, depending on how much meat you're currently eating.
- Do you usually buy your meat from Walmart and that's your price comparison? NO, you will not save money no matter what you do.
- Do you currently buy your meat from an Organic or sustainable, grass-fed farm? You might save money, depending on how you factor in your time.

See. Not straightforward.

When we started raising chickens for eggs, we consumed way more eggs. Why? Because they taste wonderful, are an amazingly healthy whole food option when raised on backyard bugs and quality grains/sprouts and are abundant (at certain times of the year). When we started raising our own chickens for meat, we consumed way less chicken. Why? I don't know. More work? Taking a chick to a dressed-out fryer requires an investment in time, feed and equipment. It's a lot of work to deal with a steer that would out class you in a wrestling match, as Dianne says. It's a lot of work to haul the animal to the butcher or to harvest it yourself. Providing meat for your family from your own labor is just plain work. (FYI, dairy is equally strenuous—see *The Dairy Animal Question* section of this chapter.)

I think, too, we just consume a lot less meat overall. Our diets have changed dramatically since we started homesteading and, while we deliberately aren't vegetarians, we just don't eat as much meat during the year. My religion has a law of health that specifically recommends eating seasonally and we've started to take that to heart. That has meant that we eat most of our red meat when the seasons are cold (late fall, winter, early spring—about seven months of our year), also consuming poultry and wild caught seafood. Once the weather warms, we may still have a Sunday roasted chicken or a salmon fillet, but we're so busy eating from our garden and local farms that we've got fresh fruits, veggies and raw dairy products coming out our ears! Meat ends up taking a back seat in summer.

The goal for us has become to grow, harvest, hunt and store the meat ahead of time for each season and that takes some planning, no matter how many animals we decide is desirable. Regardless of your volume, you just might find that, if you're going to eat meat, you prefer it come from your own land, so that you can control what goes into the animal and how it's raised, regardless of cost. You may not save when comparing costs to your typical grocery bill, but you might find that you're saving on doctor's visits and trips to the pharmacy. You may discover a whole new way of looking at food; a new way to prepare and consume meats and other products like nourishing fats and bone broths. Your animals may contribute to your health, home, pocketbook and land in ways you can't even quantify yet.

An Exercise in Meat

If you do eat meat, and would like to raise or hunt as much of your family's year supply of meat as you can, then you're going to have to ask yourself two questions: how much and what kind of meat? How much is a question only you can answer for your family, so I'll use mine as an example. Please bear in mind, though, that each year is different both in what we feel we need and what we grow, buy or trade. This is just an example, not sacred text.

Zundel Family: Currently we have two adults (40-ish years old), four girls (ages 2 to 12) and one boy (age 11). We eat red meat, white meat and fish seven months of the year when it's cool. We eat white meat and fish sparingly five months of the year when it's warmer. As far as livestock typically raised for meat is concerned, red meat is generally considered to be meat from cows, goats, sheep and pigs (though some people argue that one) and white meat is generally considered meat from poultry and rabbits.

There are fifty-two weeks in the year with about 30 of those weeks being cold weather months and about 22 of those weeks being warm weather months where I live in growing zone 5, give or take some weeks. During the colder months, we might eat meat five times a week and during the warmer months we might eat meat three times a week. We usually eat one pound of ground meat at a time (meat loaf is two pounds) and we can clean the meat off a 6-pound chicken or roast. I usually prepare more than that when I'm cooking larger cuts of meat or a whole bird just so that I can have leftovers for meals during the week.

As we look at going off-grid, I may need to adjust how I cook up meat at one time to scale back if we decide not to use a refrigerator. If I do need to make that adjustment, I may decide to can up leftovers in small batches (meat requires a pressure canner which may seem like more work than a water-bath canner, but really isn't). We don't eat rabbit right now, because I'm still working on mentally separating it from being a bunny. Rabbits are most likely in our future, though, so when we start eating them, I'll include rabbit in my poultry totals since its considered white meat. Once we start hunting, I'll include those numbers, too. Since we're raising heritage pigs for the first time this year, our pork numbers will go up, as well. Pasture-raised pork has been expensive and hard to come by where we've lived before, so raising our own will give us easier access to it. Bottom line, this chart is very fluid so think of it only as an example.

Let's say I decide that this year our meat requirements will be as follows:

COLD WEATHER MEAT

Ground Meat	Whole Poultry	Poultry Parts	Roasts	Fish and Pork
30 lbs.	180 lbs.	60 lbs.	180 lbs.	60 lbs.
	Or 30 whole, 6 lbs. birds			Includes bacon

WARM WEATHER MEAT

Ground Meat	Whole Poultry	Poultry Parts	Roasts	Fish
5 lbs.	132 lbs.	44 lbs.	None	44 lbs.
If we grill a few summer burgers	Or 22 whole, 6 lbs. birds			

Ground meat could mean anything small like hamburger, stew meat, or sausage. Poultry parts are usually anything left over from a whole poultry meat—extra white meat, dark meat and organ meat. Roasts really mean anything without bones that's bigger than hamburger and larger cuts of bone-in meat. My "fish" totals include miscellaneous meats like pork cuts, ribs and bacon. If money is particularly tight at any time of year, we may eat meat only once a week.

I take a list like this and decide what animals each meat is going to come from, based on what I think I can grow on my homestead and what I'll need to trade for or buy. When I'm raising the animal, I also need to factor in how much I'll need to provide in the way of feed for each animal, which includes whether I have pasture to provide for their needs and/or if I plan to soak or sprout grain for their rations.

These factors often determine what kind of animal I'm going to raise with the simple, and very real, measuring stick of how much each animal costs to raise. For example, my white meat primarily comes from poultry now, but once we start raising rabbits for meat, how my white meat is produced will change a lot. Rabbits can produce a lot more muscle in a lot shorter time, and can be grown with a lot less feed, than a turkey. Deciding on numbers is the easy part; picking which animal to raise and how to effectively raise it without breaking your bank is the hard part.

I can't tell you the best or most cost-effective way to raise your animals on your land and in your area. However, here are some things to consider:

1. Are you going to feed grain to your meat animals? If so, can you grow it yourself or will you need to buy/trade for it? Can you do that locally?
2. If you're going to feed grain, have you thought about pre-soaking or fermenting it to increase its volume and nutrient density? Soaking partially breaks down grain

rations so they're easier on your animals' gut balance. (Pre-soaking grains for animals is about the same as pre-soaking them for you, just on a larger scale. See the *Soaking, Sprouting and Microgreens* section in The Homestead Kitchen chapter. Also, see the Website suggestions at the end of this segment.)

3. If you're thinking of pre-soaking your grain, have you thought about going the next step and sprouting it for the biggest nutritional bang for your buck and to save on the amount of grain needed overall? Sprouting not only increases the volume of your grain but it also increases the nutritional content and availability of nutrients for your animals' bodies to absorb. They eat less sprouted grain than they would dry grain but get more nutrition from it. I encourage you to research it and think about it.

4. Can you pasture your animals for some or part of the year? If you don't have enough land in pasture, can you create "pasture" in certain areas of your homestead? When we lived on an acre, we broke up sections of our yard to create grazing areas for our goats and poultry. I'd plant a parcel with pasture grass and, after it had come up enough, let the animals in to graze. At the same time, I'd start another piece of land growing up feed for the animals. We'd rotate them around the areas if we could over our growing season.

   We never had enough to graze them exclusively and ended up supplementing with hay and a little grain. However, it added variety and nutrients to their diet that translated into better health. Our gardens benefited. The fodder areas the animals had gone through the previous year and left their rich dung in became garden beds. We'd move everything back and forth and all around until we had the animals and the soil fed. Well, that's when we were on our game. Some years everything was a mess and there was no order and stuff just grew around. Strangely, those years worked well, too.

5. Do you have space to grow fodder crops for your animals—mangels and other root crops, extra squash, some clover or alfalfa? Do you know someone who has extra produce that you could relieve them of to feed to your animals? Any way you can offset your feed cost will be money in your pocket and happy protein in your tummy.

Hunt or Grow on the Homestead?

Growing our meat on our own homestead is a great way to ensure that meat is up to the quality we want it to be and that the animals are raised in a humane way. However, it's not the only way to provide meat for your family—there's always hunting! We have yet to embrace this option but it's something we see coming in the next few years. It's not that we're opposed to it or that we lack the experience required for weapons (well, I do but my husband is adept at firearms and has done some work with bows). Our main roadblock is experience. We're city kids who are moving at our own pace into the homestead lifestyle. We just haven't taken the opportunity to work with hunters as mentors to teach us this useful skill.

It's important to team up with ethical, honorable hunters who will teach you how to responsibly harvest these game animals so that they can be a blessing to your family. When you find these hunters, listen to them speak about the animals and their habitats—hear the appreciation, respect and reverence? That's typical of this group, so be sure to listen well and adopt their habits of constant weapons practice for accuracy and their commitment to only take down reasonable amounts of game.

Besides ethics, there's a lot we need to learn about hunting. For example:

- What time of year corresponds to which animal
- What our local, state and federal hunting laws are
- What kind of gear we'll need
- What kind of weapon is best for which animal
- How to stay safe and keep other hunters safe
- How to field dress and transport a felled animal and much more - see the sidebar below for more information

Some of you who already hunt may not need these few paragraphs, but I mention this very viable, worthwhile option for the benefit of those to whom it may not have occurred on their own. If it hadn't been for several homesteading friends who hunt on a regular basis to fill their larder each year, I wouldn't have thought much about it because it just hasn't been a part of my family culture.

HUNTING 101

As I asked my hunting friends for some advice, the most common responses were:

1. Get an experienced mentor and go hunting with them as often as they'll tolerate you.
2. Spend money on quality equipment like appropriate clothing for hunting season, effective weapons and sharp butchering knives.
3. They also reminded me to keep it legal and make sure I had a hunting license for my state and knew exactly when and where it was appropriate to hunt.
4. They also cautioned me several times on safety, suggesting that I take a hunter safety course and practice with my weapon consistently to prevent damage to people and animals.

Jenn Dana from Little House on the 100 Acre Wood (www.littlehouseonthe100.com) shared particularly pertinent advice:

> "Safety first! When you are carrying around a loaded gun, crossbow, or bow and arrow, the same potential to take an animal's life is there to take a human life. Always be cautious of your weapon!
>
> "Please don't trespass on other peoples' land. Get permission before you hunt.
>
> "You will wound an animal eventually. It's terrible and no one wants to do it, but you are shooting at a moving target. Try not to beat yourself up about it. Remember that if you fatally wound it and can't find it, it really does just go back into the cycle of life and other animals will benefit from it. If it's not fatal, it will recover and be fine. Practice, practice, practice to make the chances lower that this will happen."

I was curious to know how much meat you could bring home from the hunt for the year and everyone said that it varies by location and the conditions that year. Typically, you get a certain number of tags (permits to hunt a specific animal) per animal, per person each year. The goal is to fill the tags (harvest as many animals as you are legally allowed) and put up meat for your family by way of freezing, canning, smoking, dehydrating and otherwise preserving.

Of course, I asked my hunting-savvy friends which of the game was their favorite to eat. Jenn claims that there's nothing to rival venison (deer) tenderloins cooked up with mushrooms and onions. Randall Wilke, husband of Lesa of Better Hens and Gardens (www.betterhensandgardens.com), reminisces that the best game he's ever tasted was the squirrel that he brought home to his grandmother as a child. When I asked Randall, what made it taste so great, he said,

> "Probably the fact that I brought it home, and anything that grandma made I remember as being delicious."

Tammy of Trayer Wilderness (www.trayerwilderness.com) was rhapsodic about the virtues of moose and elk:

> "[My husband] shot a moose two years ago and it was the MOST amazing tasting meat we have ever had, hands down. It did not have to be cooked in anything special, it just was incredible meat. Now my second favorite is elk and I love placing a roast on my wood stove with our homemade chili sauce and just let it simmer ALL day long. The house smells wonderful and the meat is so incredibly tender and juicy in my cast iron Dutch oven. Our chili sauce* is amazing with any kind of game meat, including turkey."

*You can find the recipe for Tammy's chili sauce in the cookbook mentioned in the *Books to Read* of this segment, as well as in the Resources section in the back of this book. Eat up!

When I asked for parting tips or bits of practical advice, Heather of The Homesteading Hippy (www.homesteadinghippy.com) advised that you set your hunting blind or chair out on your hunting grounds in the summer so that the animals get used to seeing and having it be there. Randall cautioned that a common newbie mistake is,

"Buying too large of a weapon, both in terms of weight and power, and not practicing enough with it to be accurate."

Choosing and purchasing a weapon is where having a mentor comes in quite handy. Jenn directed her advice to the ladies saying,

"Women, join your husband in this sport. I guarantee you will love it. It is a hobby that my husband and I are both passionate about now. Your husband will love that you share his passion for hunting."

Tammy said,

"It is always important to watch out for other hunters and be aware of your surroundings. Dependent on where you are hunting, be extremely aware of your surroundings because such animals as wolves, mountain lions and even bears will come into the shot because it will offer them a free meal. So be very aware of other predators.

"Knowing your choice firearm is extremely important, such as how to load the shells, where the safety is and even how to clean your gun upon returning from your hunt. Remember, it is called hunting because you're doing just that, hunting for the animal, and you are not guaranteed a kill. I personally just enjoy being able to step away from the day to day grind and enjoy the quietness of God's country and all the beauties in it!"

I was surprised how many of my friends talked about the beauty of nature and the deep respect they have for the life that thrives in the forests as part of their hunting experiences. There's a strange stigma that's been attached to hunting and hunters in our modern culture—as if somehow the act of hunting is evil. There is certainly a difference between ethical and unethical hunters, but if you ever want to hear a bit of natural poetry, ask a dedicated hunter what one of their favorite parts of the hunt is and you might get something like this:

"I enjoy sitting in the woods and being still. It's the ability to see a part of nature that you cannot see while just strolling through. The wooded area becomes alive about thirty minutes after you have settled in your tree stand. Squirrels, deer, turkey, birds and many other animals carry on their business without fear, unaware that you are watching them. It's amazing. It's as close to God as you can get here on earth…"

-Jenn Dana

Harvest: The Polite Term for Kill

It's no longer a part of our modern, industrial culture to do much in the way of harvesting our own animals for meat so this is the most difficult activity for most of us. If it were still part of our yearly workload, there'd be a lot less animal cruelty and a lot healthier protein consumption! I freely admit that I'm a big sexist when it comes to division of labor on the homestead and, in our family, killing is my husband's job. As a woman, I'm biologically wired to beget life—not let it bleed out on my shoes. I have slaughtered before, or harvested (a more polite term), but so far only a chicken and only because I wanted to make sure I could do it. I felt like it was my duty to honor the chickens who fill our table by being a part of the whole process at least once. I will learn how to harvest other, larger animals eventually but, unless my husband dies before I do, it will never be my job. I simply don't believe it's part of my function and I do believe it's an important part of my husband's. (His tender heart does not enjoy the task, but he does it for his family because he's willing to serve us and provide for us.) You need to decide who will do the harvesting on your homestead or if you're willing to pay a processor to do it for you. Remember, there's a cost associated with that service.

If you choose to do it yourself, there are a few things to consider before taking the plunge into slaughtering:

1. Will it be one person's job to slaughter or will several have a hand? Will you need to hire a professional to come help you butcher? You may decide with larger stock to send them off or have an experienced group of people come slaughter on site for you. Or, at least, teach you how the first few times so that you can eventually do it yourself. All of these are viable options and it may depend on the animal. For example, a chicken is a lot simpler and quicker to harvest than a cow.
2. Do you have all the equipment you'll need for each animal you plan to harvest? Is all that equipment ready to go before harvest day? Sharpening your knives is NOT something you want to do while the animal is being brought in for harvest—that task should be completed well before you begin.
3. Do you have someone to help clean and dress the animals being harvested? In my family, that's me, friends and my older children. My husband is so busy with the harvesting that he doesn't have time to stop and dress each animal. Here's where my woman job kicks in and I set things in order. With a chicken, for example, he kills and beheads it and I pluck, eviscerate, butcher and otherwise "tidy" the product. This must be done as quickly as possible after harvest and is best done with a group of people. It's a much quicker task to dispatch an animal than it is to clean and prepare it for preservation, so having a group processing the animals is helpful.

As you begin butchering, let the harvester take a break when he needs it. Taking life can be a particularly exhausting process for some natures and personalities. In fact, you may want to factor that reality into your plans when you're deciding how much volume to harvest at once. There's a soulful difference between slaughtering thirty chickens and one hundred chickens at once. Not to discourage you from doing large numbers—it's easier overall, in my opinion—but if you do decide to do a lot at once, maybe plan to have a second-string slaughterer on hand.

There are a lot of things to consider as you begin to grow animals for meat and one important point is calculating what time of year you'll be harvesting them. We like to harvest in the fall, winter and early spring simply because the bug populations and temperature are both lower. My fingers ache, of course, but winter is a great time to be able to lay out meat to cool and freeze without using the freezer. (Outdoor cooling and freezing of newly butchered meat only works if you have actual freezing temperatures in winter.)

You may know you're busy at your day job a certain month of the year or that you have planned a vacation for your family this year. Do NOT plan your meat animal harvesting schedule so that it crosses over into these other events. You simply won't have time to get it all done, and you'll overly stress yourself. When you plan for meat animals, you're planning several months to years in advance and the grow times are different for each animal. Just sit down at the beginning of the year and calendar all your plans with your family.

STORE AND PRESERVE

The good news is, there are a lot of ways to store and preserve your meat for use throughout the year, whether you're harvesting big batches or small. If you're off-grid, or just prefer to use less energy by not utilizing a freezer, canning is a good option. You do need to use a pressure canner for meat, which is not difficult, but does take more time than a water bath canner. I've learned over the years that I just don't get big batches of meat canned fast enough to avoid spoilage, because of all the other demands on my time. So, I prefer to can smaller batches, more often. I'm planning to can more meat stews and casserole filling this year, too, which also require pressure canning but are a good way to use up random bits of dark meat and organ meat. Smoking is another good option, as is salt curing appropriate meats.

You'll need to build a smokehouse or smoke-unit of some kind, but it doesn't need to be elaborate. Dehydrating is a simple way to build up meats in your food storage without using as much space as canned meats. You can dehydrate in your solar oven, by the way, or build a solar dehydrator for preserving meat and other foodstuffs. Freezing is an option, of course, and convenient if you have a large amount of meat at one time, and don't mind using power to run it. Be aware that large amounts of frozen foods are subject to the dangers of ice storms, earthquakes, tornadoes and more shutting down electrical grids and causing freezers to stop working. Ask me how I know.

You can also "store" your meat in the homes of customers who purchase your extra amounts and in the homes of friends with whom you barter. Using your ability to grow your own meat as a source of income or incoming goods for the homestead can be just as profitable as if you ate all the meat yourself. The key is to find a rhythm that works for your family but allow yourself several attempts before you feel like you know what you're doing.

ANIMALS ARE MULTI-FUNCTIONAL

Traditionally, on any small farm of ancient times or even just a century ago, you would have found a mix of livestock. Not only did these animals provide food, they also provided an opportunity to manage the land in ways that created self-sufficiency. To grow food producing crops, the soil needs to be nourished with rich, organic matter, like manure. Animals produce just such manure. Animals were able to eat fodder and harvest leftovers and turn them into useful poop. Animals like the strong-snouted pigs who like to root around in the soil and scratchy-clawed poultry were also tasked with the job of lightly tilling and mixing the manure into the soil as they moved around the pastures and pens. Additionally, browsers and foragers like goats and sheep would play on and trample over large piles of brush and organic materials that would compost down and become more food for the soil. All the while making more and more useful manure.

Over time, our society decided it had enough of the noise and the poop of livestock and the idea of including animals in the business of common farming faded away into the specialized industries of pork, beef and chicken. That is a hugely oversimplified condensation of the history of how livestock was taken out of the process of crop production but there are other books you can read if you'd like to know more—it's a fascinating and frustrating study. A good place for the modern homesteader to start is Michael Pollan's, *Omnivore's Dilemma*. Suffice it to say, regardless of whether we live on a crop farm, for the majority of Americans, it's no longer an assumed part of our organic culture to have a feeder pig, or even a flock of chickens, or a dairy animal.

But times are changing.

ACTION ITEMS: Get out your homestead journal and start running some numbers.

- How many times a week do you eat meat during the year?
- What kind of meat?
- Do you want to be eating more or less?
- Are you interested in trying to grow a meat animal you've never eaten before?

I didn't think I could eat rabbit until an urban homesteading family we knew invited us over to help with their rabbit harvest and stay for dinner. It's taken me awhile to come to a place where I'm willing to raise rabbits for meat, but it's not because they don't taste delicious. That family dinner was enough to convince me to try meats that were new to me!

Figure out what animals you can legally grow or hunt where you live. If you want to grow or hunt something not present or legal in your area, find out if you have a friend with a piece of land you could use where it will be possible. What equipment will you need? Where will you find your stock, if you're growing the animals yourself? Do you know anyone near you already doing this that you can shadow for a while to learn their system? It often happens that, until you've spent time with the animals and their growers, you don't even know what questions to ask!

DIY: There are several places to start working here. First, start right now to save for a high-quality set of butcher knives. Go write it into the budget before you even finish reading this. Good knives will make or break your butchering efforts…and your hands. Learn how to sharpen your own knives because you will need to do this often.

Ladies, and even the gentlemen, you're going to want a butchering apron because blood and other "material" gets, literally, everywhere. If you're not big on sewing, but have an old jumper dress laying around, you can modify it to become a rockin' butchering apron by following the instructions from Fave Crafts found on our links page under Livestock Wherever You Are: www.homesteadlady.com/homestead-links-page.

If you're ready to build your own smoker for meats, there's a simple design from the University of Connecticut on our links page.

BUILD COMMUNITY: If you plan to take up hunting, make it your business to find a mentor within the next few months so you can start making plans for the year. If you don't know anyone of your acquaintance that hunts, you may have a local hunting club that can give you references. There are a lot of groups online at outlets like Facebook and Twitter. Grab a like-minded friend and take a hunter safety course as well as a basic firearm or bow hunting course. Take your spouse instead and make it a homestead date night.

BOOKS TO READ: For grass-fed meat growers, read *Good Meat: The Complete Guide to Sourcing and Cooking Sustainable Meat*, by Deborah Krasner. Grass-fed meats cook up differently and learning how to communicate about which cuts you want from your grower can be confusing. This book will help with all of that from recipes to diagrams. Also, *Nourishing Traditions*, by Sally Fallon Morell, has great recipes for offal and bone broths. Also good is *The Trayer Wilderness Cookbook ~ Homesteading The Traditional Way ~ Volume 1*, by Tammy Trayer, for traditional foods, game meats and other wholesome recipes from her off-grid homestead.

Butchering Poultry, Rabbit, Lamb, Goat and Pork: The Comprehensive Photographic Guide to Humane Slaughtering and Butchering, by Adam Danforth. He wrote a similar one for beef. Hands down these are the best books on this topic, in my opinion. The editing is clean, the photos are clear, as are the instructions. I didn't grow up doing this and have no intuitive knack for it, but Mr. Danforth always makes me feel like I can get it done.

Meat Smoking and Smokehouse Design, by Stanley Marianski, is a great book covering a wide range of methods, DIY projects and instructions for both domestic and game meat smoking.

WEBSITES: Anyone who's ever butchered a chicken since Herrick Kimball started blogging has most likely already visited his How to Butcher a Chicken site but here it is for those who may be new to it: https://howtobutcherachicken.blogspot.com. Be sure to check out his wonderful chicken plucker tool! Game and Garden (www.gameandgarden.com) is run by Stacy Lynn Harris and not only will you enjoy her game recipes, sustainable living and homesteading tips but you'll also enjoy her Southern charm. FYI, she also has print books available like Gourmet Venison.

Print magazines all have websites and most have editorial blogs which can often be of service when you're looking for answers to a specific question or if you need a tutorial. Try Hobby Farms, Backwoods Home, Mother Earth News, Grit, Urban Farms and many others specific to the animal like Backyard Poultry Magazine. These are all great in print, of course, but their online presence is highly beneficial for those who prefer to read on their laptops or devices.

HOMESTEAD LADY SPEAKS:
If you have children in your home, you'll want to decide ahead of time whether you want to involve them in harvesting your animals. For some homestead and farm kids, harvesting an animal is a rite of passage; a symbol of approaching adulthood. Some parents refuse to allow their children to participate for fear of their child's emotional well-being. If you're not sure what you'd like to do, know that there are many different tasks come harvest day and only one of them involves killing anything. Children make great runners back and forth to the house for baggies, sharp knives, towels, and drinks for tired workers and ice for the chest cooler. Older children can entertain younger children, so the adults can work uninterrupted. If you home educate or would simply appreciate a good biology lesson, have your kids come over and "dissect" an animal, finding its internal organs and naming the bones.

FYI, regardless of the animal, once the head and skin are off, they look more recognizable to us as meat. It might be easier to include your children from that point on if you're all new to the harvesting process. Although, I will say, kids make excellent chicken pluckers and mine have helped with everything so far except the actual killing. My son is getting older and his dad will probably invite him to participate in that sometime soon. My daughters will also be invited to learn how but, as I said before, I'll train them to other tasks for the most part. There have been some years where my children are in the middle of the blood and ice water and other years where they're more interested in watching the baby and putting supplies away. Since we have several years to teach them these skills, we usually let them pick their own pace and never force them to do anything that they really aren't comfortable doing.

As a parent, it can be hard to find the line between teaching them to work and requiring too much. Search it out for yourself and start asking yourself some specific questions now so that you have a plan for your children come meat harvesting time.

CHAPTER 5

Homestead Finances

Apart from land, animals and family, one of the main things any homesteader (or any adult, let's face it) needs to learn to manage well is money. Homesteaders have big dreams even if they're in tiny places. Turning those dreams into even partial realities is going to take a lot of sweat equity, but it's also going to take a lot of equity-equity. You need money and resources to homestead, especially when you're first getting established because startup costs for various projects typically represent a financial investment.

If money management isn't your strong suit and/or you feel the pinch of your wallet like most of us do, don't despair. Homestead finances are a subject that everyone can handle, even if you don't like math and don't feel your resources are enough to assist you in finding your dreams. Learning to manage money, just like learning to homestead, is more about the kind of man or woman you want to become than it is about what skills you already have. As homesteaders, not only are we striving to become good stewards as we manage organic resources, like the land and its inhabitants, but we're also attempting to become a certain kind of human being. A responsible, thrifty, grateful, useful sort of person. One of the best resources we have available to teach us to be wise resource managers on all levels is money. Or, even, the lack of it.

Here's what we'll be discussing in this chapter on finances for the modern homesteader:

HOMESTARTER level: This section is devoted to getting out of debt so that your homestead can run smoothly. Here we'll cover the two main debt repayment strategies, five measured steps to getting out of debt and the simple practice of learning to pace yourself.

HOMESTEADISH level: The next logical step in the financial journey of the homestead is to practice saving resources in every way we can. As a homesteadish type of person, you may be surprised at how much of this advice you're already following. For new ideas, this section covers three areas of conservation and savings: the kitchen, the home, and the homestead (land).

HOMESTEADAHOLIC level: What does learning to upcycle have to do with wise spending? We'll talk about that here, as well as four steps to keep a sense of balance in your financial homesteading life. Finally, we also talk about ROI (Return on Investment), which can't always be quantified in cash.

HOMESTEADED level: While homesteads aren't businesses like farms, creating income from the homestead may be something you already have at the back of your mind, or as an active part of your homestead vision. Perhaps you have children who'd like to build small businesses from your homestead. Here we'll cover some practical advice, initial topics to ponder and three areas of homestead business possibilities that you might like to consider: raw products, value added products and services.

As with every chapter in this book, be sure to follow along at the end of each section as you reach the Action Items, DIY Projects, ways to Build Community, resource recommendations and parting advice.

"The ultimate goal of farming is not the growing of crops,
but the cultivation and perfection of human beings."

— Masanobu Fukuoka, from *The One-Straw Revolution*

For the Homestarter: Homestead Debt

DRATTED MONEY

So how do we learn to manage the resources and liabilities we have now to provide for our homesteading future and improve our personal way of being? As my friend Rebecca Shirk of Letters from Sunnybrook (formerly a frugal-living blog), wrote to me,

> "Ideally it would be wonderful if you could save up for and pay cash for your homestead, animals, and expenses. I doubt that is a realistic picture for many of us. What we can do, is to use credit carefully and not take on more than we can handle."

Very sage advice. And by "more than we can handle" she means debt.

I am the last person in the world to be giving financial advice. Not that I'm irresponsible; I'm just not very precise. Managing finances is a lot like managing housework for me. I do it because it's important but sometimes my laundry takes days to finish and remains strewn around the house as the checkbook sits unbalanced while I nurse the baby and read to the older kids. Life is life and I'm easily distracted. Plus, finances are boring to me.

Being a grown up, though, means doing a lot of stuff that you'd rather not do or wrap your brain around. As I interviewed many, many homesteaders leading up to writing this book I discovered that, across the board, each one of them mentioned finances as an area on which they wished they had a better handle. Each one said how money and its management keep them hopping, trying to figure out what is best to do for the homestead, and family, on the often-limited funds available. Truth is, I hadn't planned a financial section for this book—it didn't even occur to me until I read over my interviews and realized I had to include this information.

Homesteaders are weird about money, too. We're not saving up for big vacations or jewelry. We're scrimping and saving to buy a pig. Many, many of us are working our day jobs and homesteading in our "off" hours. Anyone who homesteads will tell you right now that the homestead doesn't have "off" hours. We're working harder than we probably imagined achieving self-sufficiency, and if we're trying to manage the stress of a debt-filled, borrowing lifestyle, we're just not going to have any fun at all. Or success.

As Rebecca writes,

> "A lot of homesteaders have a dream of living off the land, a passion for animals, farming and sustainable living. However, that passion will quickly fade if they become overwhelmed with debt and cannot afford to provide for the animals in their care."

NEITHER A BORROWER...

Debt can be a crushing task-mistress as you try to accomplish your homesteading goals.

So, quips Shakespeare's Polonius in *Hamlet*:

"Neither a borrower nor a lender be,
For loan oft loses both itself and friend,
And borrowing dulls the edge of husbandry."

This advice is so particularly well suited to homesteaders so, let's break it down. Polonius believes that borrowing, particularly from family and friends, is never a good idea. Money makes personal relationships messy, as the loan that "oft loses both itself and friend" will tell you. He also claims that borrowing money will inhibit that most gentlemanly virtue of domestic thrift, husbandry. Husbandry is an old word but such an awesome one. It certainly has to do with the management and production of animals and crops—all farmers and homesteaders are husbandmen.

The word husbandry also denotes one who manages and conserves resources, like a steward. Stewardship is one of my most favorite concepts because it connects our personal responsibility to serve and protect to our very identity. We *are* stewards. Preservers and administrators of assets that belong, not just to us, but to our family and our fellow men, provided by a loving Creator. Managing these assets well is not just a temporal blessing in our lives, but also an integral part of making us better people. Finding a path through this process without taking on needless debt is a goal that will help preserve the very soul of what we're striving to build.

Be Free

That was a long, explanatory introduction to finally say, let's start with homestead financial issue number one—DEBT!

When you're looking at purchasing a homestead, or a smaller home that will become your homestead, you'll invariably have to talk to money lenders. It's important to keep your head on straight and think clearly during this process. Rebecca councils:

"When a lender tells you how much loan you can afford, they are giving you the maximum amount for your current situation. That means no wiggle room if other expenses increase, and no more potential to access credit should an emergency arise. You want to stay well below that maximum borrowing amount, so you aren't right at your financial limit.

"People talk about types of debt being good (mortgage) or bad (credit cards) or even evil. I just see it as a tool you can leverage to get resources. It comes at a price, and that is the associated interest rates and fees. The worst part of debt is the stress it causes when we agonize over how to pay it back. This usually happens because we are very close to or over that amount we can comfortably take on.

"When we feel that stress, it is time to look into reducing the debt. The less debt we have, the more self-sufficient we become as well. We won't have to put so much of our resources towards paying someone else"

ARE THERE TYPES OF DEBT?

"People talk about good debt and bad debt, but is any debt really "good?" Credit reports will even dock you points if you don't have a fixed loan, such as a car payment. Those loans are an opportunity to show how well you manage debt repayment.

"In view of managing your homestead, however, any debt means more of your precious resources will go towards repayment. It may also mean putting the home-stead itself at risk in the case of foreclosure.

"While creditors may keep dangling loans and credit card offers over your head, insisting you can afford more credit, only you can gauge what is manageable. The more variable your income, the less you want to take on debt.

"Paying down your debt should always be a high priority, so you have resources free to handle day to day needs and preparing for the future. Debt wastes valuable resources as you work harder to pay for items plus interest charges. The quicker you pay down debt, the less the cost of borrowing the money.

"So, view debt as both a resource you can leverage and a liability when making decisions."

-Rebecca Shirk

There are two main debt repayment strategies for paying it down quicker than just your minimum payment each month.
1. One is to pay extra on the debt with the lowest balance first, and then when it is paid off you can take that payment amount and use it towards the next lowest

balance loan. This is a very popular "debt snowball" method that people psychologically enjoy because you see individual debts being wiped off the list quickly.

2. The other method is to pay any extra money towards accounts charging the highest interest rates first. For those with very mathematical minds, this one makes more sense.

There are a myriad of debt consolidation programs and low interest balance transfer options out there. Some are good, and others are scams. Be very careful before taking on new debt to manage your current debt. Read the fine print and make sure you understand all the rules, fees, and payment structure before you commit. If you don't stay on top of it, you could end up in much greater debt than you started with. Says Rebecca,

> "You can't get out of debt overnight, as much as we'd all like to! It is a slow, steady process of planning, being disciplined in making extra payments, and being creative in freeing up money to pay it off."

Getting Out of Debt

The only "trick" I've found to getting out of debt, is to sacrifice to do it. Just be willing to give up all the stuff I want now to pay for all the stuff I got back then. That's it. In my family, we just did that until the debt was gone. Sadly, we got more debt because, apparently, the first time didn't hurt enough. We've never bought boats or taken fancy vacations, and the debt was smaller the next time, but it was still stupid. Rebecca has more measured advice for us:

"A reasonable action plan for getting out of debt would be as follows:

1. "Have a clear and accurate picture of your current financial situation. That means all income, expenses, debt, interest rates and loan amounts.

2. "Review expenses and see in which areas you can reduce costs.

3. "Look for opportunities to increase income.

4. "Put money aside in an emergency fund account to help cover unexpected expenses without incurring more debt. Many people suggest around $1000 initially, but I think it depends on what these types of expenses average for your situation. Mine tend to be higher when they hit. Also, it depends on how much credit you still have available to use in an emergency.

5. "Make a debt repayment plan and be disciplined about following it. Having family members in alignment with your goals is key to making the plan work."

Pace Yourself

Everyone knows we should spend less money than we bring in and put some aside for a rainy day. As a homesteader, my biggest struggle with debt has been to manage my enthusiasm for learning about and implementing a new project on my current homestead, bringing it in line with my real-life budget. I've been guilty of using a credit card to buy a new homesteading book or purchase an animal. Then there's feed and equipment. And animal housing. And the garden tools. Suddenly, I realize I'm an unhappy homesteader because I'm weighed down with unnecessary debt. Rebecca writes,

> "Pace yourself! It is so tempting to take on too many projects and animals right away. We want to feel like homesteaders and be self-sufficient. To be successful, we must also be patient. Map out a slow, steady growth plan. Take on new animals only after you are able to comfortably handle care of those you already manage."

Here's an example of that slow, steady growth plan on our homestead. We determined a few years ago that we needed to start consuming raw milk and raw, cultured dairy products on a regular basis. Raw dairy is extremely pricey, but we knew, after sampling various types of these products, that this was still something we really needed to do for our wellness. We started making a list of how much we spent each week on milk to drink and the milk required to make a conservative number of cultured dairy products like yogurt and farmer's cheese. We also added on the gas it took to go purchase the milk since it was quite a distance from us.

Then, we nosed around our local classifieds and started writing down how much a quality dairy goat would cost, as well as her milking equipment and feed costs. Now, no doubt, the startup cost of a dairy goat far exceeded the weekly cost of purchasing raw milk. However, over time (after the housing was built and the equipment was purchased) we saw those numbers could even out. We also factored in the manure that the goat would produce and added that in favor of the goat because our garden is so large, and we require so much quality compost each year to naturally manage our soil program (we don't use commercial fertilizers). Even the straw we purchased for goat bedding would eventually go into the compost pile as a very valuable carbon source so, although we had to purchase the straw initially, it ended up serving two very important roles on the homestead—clean bedding and carbon for the compost.

We decided to set aside a little each month until we could afford to purchase a quality dairy goat—we held out for one registered with the American Dairy Goat Association (so we could trace her dairy heritage) and already in milk. In our area, that year, that cost us about $400. The lady who sold us the goat also sold us her small, dairy barn at a deeply discounted price as well as her milking bucket, hoof trimmers and birthing kit. We got the whole caboodle for around $600. We quickly started enjoying the benefits of high quality, raw dairy coming out of our backyard in such abundance that we were able to provide for our raw milk and cultured dairy needs that year. I continued to milk that goat for a solid three years, every day, snow or heat. She's finally resting this year, after which she'll go again for another three years, I hope.

It was hard to make myself wait until I had the cash for the goat and some equipment. I would see quality goats come up for sale and feel an urgency to buy right then. The truth was, I was still able to find a great animal at a price I could manage when I was ready to purchase. I didn't need to rush, which is in my nature anyway and very hard for me to fight back sometimes.

If you're more of a planner that moves slowly toward any action, I'll go ahead and caution you not to wait too long. Once you have enough saved to avoid debt, take the plunge and make the best decision you can—the perfect goat does not exist no matter how long you wait for her to come. We ended up going through several different breeds and types before we found the kind we liked. Wait, this isn't the goat segment! (See *The Dairy Animal Question* section in the Livestock Wherever You Are chapter for a discussion on dairy goats.) I'll get back to the point now...

Says Rebecca,

> "Focus on incremental growth; benefit each year from what you have made, grown and learned from the year before, and build on that foundation. This is true whether it is adding a new crop, type of bird to your flock, or the resources to make your own soap."

Just remember that excessive debt will "dull" your husbandry efforts; the homestead will ultimately suffer if it's founded on the sandy foundation of debt.

> "I think it's important to note that changes don't happen overnight. We've grown up in a society that has adopted a lot of bad habits. Those are hard to break! But small steps matter. It may be too daunting to think of cutting out take-out meals entirely but switching over to a bagged lunch several times a week might seem more doable. Every small change a household embraces makes a difference, both in the moment and in the future, by removing those bad habits from our children's upbringing."
>
> - Kris Bordessa, Attainable Sustainable (www.attainable-sustainable.net)

 ACTION ITEMS: There are plenty of debt reduction ideas floating around the Internet and your grandparent's dinner table, so I suggest you write down all the ones that resonate with you and do them. Go, Team! However, many of us have, tucked away in our hearts, firm ideas about what we should be doing to recue debt. So, right now, let's make a list of things we already know we can do without, give up, sell used, or pay off. List at least your top twenty ideas.

Next to each item list the savings that will naturally go along with dealing with that point on your list. *However*, instead of using dollar signs, draw pictures of the number of chicks you will purchase with that vanquished debt or what fraction of a dairy goat that item represents. If you pay off your credit card, how many seeds can that previous monthly payment purchase for next year's garden? If poultry and potatoes, not dollars, are what motivate us then let's get visual with it. Use stickers, doodles or whatever it takes but make that debt reduction list relevant to your homestead goals with this simple, silly exercise.

DIY: Cut something off or up. Pick a debt creator that you already know is a weakness and chop it up. Is it a cable bill? 999 channels and there's still nothing on; pull the plug.

Do you need to take a credit card out of your wallet and shred it into teeny, tiny pieces? I'm not your mom so I can't really help you out on this one, but you probably already have something in mind. I know I do. I'll go get my big-kid pants and my scissors. How about you?

BUILD COMMUNITY: Here's another occasion where one of the most important communities you can build is the one that lives in your house. Get your family together and talk about your financial goals as they relate to the homestead. Get everyone's ideas, even if they're impractical—the conversation is the important point here. Make another list so that no one's input is lost. The goal is to work everyone around to be on the same page over time. Sometimes the language of adulthood suffers in translation when we're talking to kids. We often think we're being heard when what they really hear is the teacher's voice in the old Charlie Brown movies—wha wha wha whawha. Remember that?

One of my homestead friends just posted that her son was angry with her for not buying him the $45-dollar shoes he wanted and called her an embarrassment. She was so appalled, she was speechless. She wrote about how this child is loved and sacrificed for and fed and clothed in a way that he's never known true hunger or temporal want. Being the excellent homestead mom that she is, she told him that his current tennis shoes are good enough. If he wants those $45-dollar shoes, he can buy them himself with his own money, should he decide to earn some. Good parents begin with the end in mind, and homestead parents make prize-winning tomatoes out of rotted manure. Put the two together and our kids are going to be just fine if we hang tight to our principles.

Have some conversations about the changes coming as you get rid of debt and downsize your lives to make some better possibilities a reality. It's probable that not everyone will be on board with your homesteading plans, so you may not want to immediately tell your teenager that their I-device made the list of garage sale items so that your family can build raised garden beds. Just tell them that it's time for everyone to act like grown-ups and make some changes for the family's financial health. Give it time and patience.

BOOKS TO READ: *The Seven Habits of Highly Effective People*, by Stephen Covey—love him or hate him, he has a point. *The Total Money Make Over*, by Dave Ramsey—love him or hate him, he, too, has a point.

If you're really new to the whole idea of money management, *Live Well on Less Than you Think*, by Fred Brock, may have value for you. *Money 911*, by Jean Chatzky, can also help answer basic financial questions in regular human-speak. It's tabbed and formatted like a mini-financial encyclopedia.

WEBSITES: Turning to the Web for money saving ideas can sometimes overwhelm you, to be honest. However, Money Saving Mom (www.moneysavingmom.com) is a useful and comprehensive online resource for saving money and using resources effectively.

Frugal Mama (www.frugal-mama.com) may also speak to you and be helpful. Mr. and Mrs. Not Made of Money (www.notmadeofmoney.com) have good common sense for everyday people.

While none of these writers are self-described homesteaders, it's interesting to me how often their frugality sure looks a lot like homesteading with all that home-cooking, garden-growing, common sense living.

HOMESTEAD LADY SPEAKS

If you're interested in basically free curriculum on the topic of money management, I point you to the Self-Reliance Initiative created by the Church of Jesus Christ of Latter-Day Saints. (There's a nominal fee for each manual – around 60 cents, U.S. currency.) The religious culture of this church includes a huge emphasis on self-discipline and, as the initiative indicates, *self-reliance*. To better serve their members, the church created a completely free curriculum complete with education, goal setting, personal way-of-being challenges and local group networking meetings.

The Self-Reliance Initiative has four branches:
1. Finding a Better Job
2. Personal Finances
3. Education for Better Work
4. Starting and Growing My Business

As a sample, here's the description for the Personal Finances course:

> "For those who want better control over their finances. Group members will learn how to eliminate debt, protect against financial hardship, and invest for the future. They will create a financial plan and follow a budget. Spouses are encouraged to attend together."

If you're serious about getting control of your finances and your life, I encourage you to check this out! My family is participating in several of these branch topics and I can't tell you what a difference its made in how we think about and relate to money. Our gratitude has increased, but so has our ability to manage our resources. We've also seen the ways in which we were inhibiting our ability to be self-sufficient and solvent.

Full disclosure: this original program was written for Christians, and the LDS people specifically. So, if you have issues with scriptures, religion in general or "Mormons," this probably isn't the best curriculum for you, free or not. They quote scripture all over the place and talk about Jesus right and left – have you ever thought He might have something to do with your financial life? If you're cool with all that, though, it's worth a look. PLUS, at the time of this publication, the Church is putting together a non-denominational program. Look for that the summer of 2019.

To learn more and access the information, please visit:
www.lds.org/topics/pef-self-reliance/manuals-and-videos

"Money is a good soldier, and will on."

—William Shakespeare, from *The Merry Wives of Windsor*

For the Homesteadish: Just Save It

Ah, saving money. My first thought is to rant, "You have to have some money before you can save some money!!" Sometimes I laugh at my pathetic, first world idea of "how much" money I have. I could be as poor as a church mouse and there's always going to be someone getting by amazingly well on less than I have and doing it with way more style. I need to grow up and be better about saving money and resources each month. A homesteader is blessed because we are, even in small ways, creating product to use down the road, if we manage it well and store it against a time of need. Yes, we need to save cash, but we also need to be better about conserving and preserving the various harvests produced on our homestead each year.

Rebecca warns,

> "Prepare for the unexpected. If there is one thing you can be sure of, it is that you will be hit with frequent unexpected expenses. Being blindsided by them can derail your plans quickly and put you in debt."

Some in the bank, some in the barn and some under the bed has been our adage lately. That means, we have some money saved in a conventional place like a savings account in a bank. Some money we have tied up in assets for the homestead—tools, animals, and infra-structure—all of which is vital to our lifestyle and is useful whether the grid is up or down. We also have a bit of cash saved "under the mattress," so to speak; like small bills in our 72-hour kits and places like that.

A PENNY SAVED IS A TOMATO PRESERVED

How many times have I squandered a big basket of tomatoes because I just didn't get them sliced and into the dehydrator in time? How many wonderful, wild herbs have I wasted because I just didn't forage for them soon enough in the season when they were ready? How many more herbs have I allowed to sit uncared for even after I managed to harvest them? I'm not lazy or ungrateful, but I am disorganized a lot of the time. I could claim that homeschooling five children is about all a brain can be expected to organize but, regardless of the *why* of my disorganization, the result is the same—waste. I do NOT hold with waste.

So, yes, we need to save money, but we also need to preserve what we produce or glean. As Rebecca writes,

> "One of the biggest areas you can reduce expenses as a homesteader is growing, raising, preparing and storing your own food. Each year, you can fine tune the amount you need to produce to feed your family and add new categories for variety. You keep food costs very low when you are diligent in using up what you have. Meal plan, keep an inventory of what you have on hand, and learn to substitute ingredients. Trade your surplus with others for ingredients you don't produce yourself."

SIMPLE TIPS FOR SAVINGS

The following are some simple homestead tips to saving money, energy and other resources. Do NOT fall into the trap of thinking these things are too small to make much of a difference—they matter! Aside from the potential these ideas have to save you cash, these lifestyle choices will slowly begin to change how you think about consumption. You'll begin to see how you can become more of a producer and less of a consumer.

How do I know? Because they've worked for me. I grew up in a frugal home but, in many ways, I was still a wasteful person out of plain ignorance of how I might choose to live differently. Take or leave these ideas but let them spark your creativity to find what works for you. If you've read other sections of this book, some of these ideas may sound familiar to you but they absolutely bear repeating.

Saving in the Kitchen

Here are some simple energy and money saving ideas for the kitchen. Are you already doing some of these, or were there a few that were new to you?

1. Use a Crockpot instead of your electric stove to prepare a real foods meal. The Crockpot requires a lot less energy to run, and a Crockpot meal frees you up from manual labor in the kitchen on days when you have other boats to row.
2. Barter with friends and neighbors using your garden produce or fresh baked breads. Identify what they have that you need and vice versa. This will help you think outside the box when it comes to grocery acquisition. I traded goat's milk for

haircuts for a long time from a friend who was a professional hairdresser—we looked awesome and she drank wholesome milk. A win for all!

3. Learn how to forage wild plants, berries, mushrooms, nuts and herbs. This is FREE food—who can say no to that? This kind of activity will help us be more food-conscious in all areas of our lives. Learning to forage well requires some study and a willingness to try new recipes and nutritious types of food. Understanding which foods are nutrient dense, and which are less filling will help you and I to spend our food budgets more wisely on foods that nourish. We begin to see that all these principles are connected. Once we begin one energy or money saving initiative, we find ourselves involved in four more. That's homesteading in a nutshell. Self-sufficiency requires work, but the rewards are nourishing in more ways than one. (For more information on that see the *Foraging* section in the Green the Homestead chapter.)

4. Keep a flock of chickens AND learn to grow fodder and/or sprouts for them (see the *Soaking, Sprouting and Microgreens* section in The Homestead Garden chapter) instead of simply feeding commercial feed. With the cost of managing your flock you may not be able to beat standard, grocery store egg prices, especially if you decide to manage them Organically. However, if you compare buying free range or even certified Organic eggs at real foods grocery stores to what you can produce at home, you'll most likely save money. Especially if you learn to feed your flock in the most nutrient dense way possible, saving you money on feed (get a link for the online course "Feeding Your Hens" Right on our links page under Homestead Finances for more information: www.homesteadlady.com/homestead-links-page). You can certainly break even and maybe even have some surplus eggs to sell! Plus, consider the chicken manure you'll have; a valuable product that will create health in your soil, so you can grow a larger, better garden

5. Cook from scratch. Start by choosing one store bought item that you can commit to no longer purchase. Learn to replace that store-bought item by making it yourself. For example, stop buying granola. Stop it. It's so simple to make and by making it yourself you can buy the ingredients in bulk to save cash. Make big batches and dehydrate them to store. Use your homemade granola for breakfast and power snacks. Eventually, you can stop buying the oatmeal that goes into your granola and start rolling your own oats. No, I'm not crazy—you can do this and it's also a simple process. You just need a grain roller to attach to your manual wheat grinder. Oh, did I mention you'll probably start grinding your own grains at some point? Your mother-in-law may think you've lost your mind, but your bank account will applaud your efforts.

6. Use coupons for Organic foods, when you can find them. Sometimes we just can't keep up with making everything from scratch and that's when I look out for

coupons for Organic foods. I have the same problem with Organic coupons that I do with regular coupons and that's that they're mostly for what I consider junk food, Organic or not. Sometimes, though, the kids get their Organic cereal on Saturday morning because I'm just NOT making something from scratch. Period. That's when I'm glad I have my little pile of coupons for fair trade chocolate granola bars and Annie's® fruit snacks. Judge me if you wish, but I'm just keepin' it real here.

WHY BOTHER?

There are a lot of reasons why a person might seek out a more self-sufficient lifestyle. To be honest, unless I had to, I probably wouldn't ever figure out how to make my own toothpaste and use a solar oven because those things take effort to learn. Each time we've lost a job or had to cut back for whatever reason, I've had to learn how to make this or that myself, just to save the few pennies that made the difference between eating fresh eggs and powdered eggs—or powdered eggs and nothing. I love central heat and take-out and store-bought pasta. It's just that, so often, cutting wood from our forest is so much cheaper than setting a thermostat. I can't tell you how much I miss twenty bucks worth of Chow Mein delivered to my door on days when I'm so exhausted from working all day that the idea of making dinner from scratch makes me want to cry. Don't even get me started on rolling out pasta dough.

When you don't have options, you can't take them. So, you roll up your sleeves and figure out how to make it work. The best part about these forced frugal times in our family is that we pick up better habits and learn that we really can make our dollars stretch further. Left without the options of convenient, store-bought ease, we do it ourselves and discover, in the end, that we survived. We haven't lost anyone yet to the lean times and what we've gained has been impossible to assign a monetary value. It's important to remember that our strength-building trials are crafted with as much love as our blessings.

Savings in the Home

Saving at home isn't just for penny-pinching moms; everyone needs to learn this skill!

1. Look for free family events to attend in your community. Even if you're single or your family is small, these enjoyable, local events can save you a pretty penny, while still providing for your entertainment. Often local farms and libraries are hubs of free, neighborhood activities and classes. Zoos, community gardens, and children's museums often have half price or free days. Churches, park systems, and your university extension, can all be quality sources of free events. These opportunities are also helpful in that they create close ties with your immediate area and your neighbors. By establishing local relationships, you are opening up the possibility of future enrichment in countless ways. Building a good relationship with your community is its own reward.

2. Pick up change off the street. No joke, just do it. We have a jar we keep in our kitchen that's just for rummaged change. We use it once a year to go out to dinner as a family. It doesn't always cover our entire bill but its money we didn't have before and didn't let go to waste by leaving it a drawer or on a street corner. If you save your pennies, your dollars will take care of themselves, as the saying goes.

3. Analyze the media coming into your home and find places to pull the plug. What about the number of cell phones in your family? Is your thing HDTV that you never use? Are you ready to ditch the TV altogether? I'm not here to tell what to get rid of but I am strongly suggesting it's time to get rid of something and not just to save money. Homesteading is a way of life and it takes time to live that life. This time is spent in the three-dimensional world and the work will not do itself. Don't stop reading and researching and relaxing at appropriate times but do ferret out the weasel in the digital portion of your home life. Trap the weasel and chop its head off. You won't be sorry.

4. Use LED lights and buy them slowly, as you can afford them. Also, use smaller screened electronics in your home—bigger screens suck more energy and, therefore, more dollars.

5. Make sure your windows are insulated well. If your windows are old but you can't afford to replace them, manually seal them in the winter months. I taped my windows in the winter when I lived in Russia.

6. Hold a clothing swap with friends, neighbors, church ladies or your school group/PTA. I LOVE these events! I keep boxes of clothes in storage and change out mine and my children's wardrobe as the seasons change. My kids don't know any different and it's a great treat when we go to the thrift store to buy some "new to us" article. Clothing swaps, and outright asking for people's cast offs, have saved us hundreds of dollars every year. An extra tip: when you do purchase used clothing, get as close as you can to the wealthier neighborhoods. Wealthy people often don't wear their clothes for very long and typically purchase more expensive brands. Expensive clothing is often made with higher quality materials and will last longer. If you can buy wealthy people's secondhand clothes, you're golden!

7. Hand-make gifts. Yes, this will take longer and will require more thought, but it can save you money and change how you look at gifts and their purpose. (The cash savings in this exercise is very dependent on what kind of gifts you're accustomed to giving, and how many people you have on your list.) Everyone has something from the heart that they can share, and you are no exception. Try this on your next gift-giving holiday but plan for it now because it will take more time. Word to the wise, don't plant to make EVERY gift by hand – you're only human. Be realistic about your handmade plans.

> **MAKE HANDMADE HAPPEN**
>
> The following is a handmade example from our family. Every year my children and I hand make gifts for most of the people in our family Christmas drawing. I'm not going to lie; this tradition exhausts me since several of my children are still young and all of them require some amount of help from me. Why do I do it? Why do moms do anything? Because it's a valuable lesson for all of us, and because it really is fun. To prepare for the adventure of handmade gifts with five small children, we hold an annual Leon Day celebration. "Leon" is "Noel" spelled backwards if you were wondering.
>
> Every June 25th, exactly six months until Christmas, we get out some holiday decorations and music. With Bing Crosby crooning in the background, we draw family members' names out of a hat. We keep it simple and include the family members we're closest to and with whom we're most likely to spend Christmas. (I'd love to make gifts for all our cousins, aunts and friends, but I'm only human and we only have six months to get this accomplished.) After we draw names and know for whom we'll be crafting, we jump on Pinterest and start looking for handmade gifts kids can do—I even created a board just for that so I can pin stuff I see throughout the year.
>
> The children get excited making plans for their assigned people and I love to watch them really think about what the recipient would enjoy having. It's fun to keep it a surprise for the next six months as we gather supplies and work away at our projects. Regardless of how the gifts are received, the gifting is far more meaningful and frugal for our family.

Savings on the Homestead

The whole point of our homesteading lifestyle is to learn to live sustainably – or, in other words, frugally and intelligently.

1. Produce your own feed, as much as you're able. Grow extra veggies and greens for the livestock. Harvest your weeds to give them, too, but check to make sure they're not toxic and never spray them with herbicides. Save your kitchen scraps and beg from friends and restaurants you trust. Learn to sprout or grow fodder indoors. Ferment your grain feed to increase its nutritional content and fill your animals' bellies more efficiently. Never compost something if it can reasonably be fed to an animal.
2. If someone is getting rid of usable wood, wire or fencing, take it if you have the space to store it in a tidy way. You WILL need these items for chicken coops, garden stakes and creating animal pens. Do not buy something if you can repurpose instead. This is a learned skill for some and, for others, it means getting over a feeling of embarrassment or pride. Let me share with you what my grandmother always said to me when she'd invariably do something too loud or too silly to go unnoticed by complete strangers in a public place. Here it is. "Relax, Tessa, we'll never see any of these people again anyway." As a teenager, I wasn't so sure this was enough

compensation for the embarrassment, even if I was laughing in good humor. Now I understand the infinite wisdom in her counsel. The world is full of people and people are all going to make their own judgments, some generous and some downright mean. Regardless, don't sweat it. Do what you gotta do to get where you want to go and let the negative just roll of your back.

3. Look for free or reduced-price canning jars in your local want ads or on online garage sale sites. I never pay for new canning jars anymore unless they're the half gallon size because nobody in their right mind gets rid of those. You can look in thrift stores for these, as well, but quite often I find them for free when an older lady decides she needs to pass on her stash to the next generation. If you acquire jars from such a lady, be sure to drop by next canning season with a freshly put up treat to let her know you put the jars to good use. Trust me, she will be grateful to know the jars that kept her family in food and health for all those years have gone into hands who appreciate them.

4. Over time, make your own of everything. I once analyzed my recycling bin to see what it was I was putting in there. Without exception it was plastic wrapping from something or other I wasn't producing myself—local meats, olive oil, grass fed butter and more. Here's a quick breakdown:
   - Take item number one, local meat: I'm not raising beef yet, but we can do enough chicken every year that I don't need to purchase any. We can also eat less meat.
   - Item number two: I'm not making my own olive oil because I'm not nearly that cool and olives don't grow in my climate. However, I've investigated pressing my own sunflower seed oil. Maybe it's time I tried that out this year.
   - Item number three: I can totally make my own butter by getting local cream and making my kids shake their arms off. (I don't try to get cream from my goat milk, too much work.) I CAN do all of these things so…I probably should, huh?

If you're new to all this homesteady stuff, pick one simple thing you can make yourself. How about making your own deodorant or toothpaste? Does that seem wild? There are a lot of recipes online so go find one that doesn't look too weird. When you don't like it at first, try again in two months and see if you've changed your mind a bit. Or, try a different recipe. As you work to change your lifestyle, your lifestyle will change. Have confidence in your ability to adjust and adapt and work hard. You can do this.

WHAT'S THE CASH-VALUE OF CHARACTER?

"Exercise resourcefulness. Barter, trade, share knowledge of what works well locally. Be active in community. Be kind, humble and grateful. People will want to help you succeed. Prioritize and master new skills."

- Rebecca Shirk

There's very little in capital acquisition that can compensate for practicing resourcefulness and generosity. Saving, repurposing, trading, and learning to think outside the box and to make something even from nothing, all contribute in their own way to an abundant lifestyle. That abundance naturally moves itself outward, as you touch others' lives for good by gifting your neighbors part of the harvest, helping a child prepare an animal for presentation at a county fair, or getting together to can with your friends.

For example, there's something of value for the modern homesteader in building a report through service and genuine affection with the elderly within your circle of influence. They came up in a time when self-sufficiency was still a desirable and expected virtue and they have a lot to teach you. Sharing your knowledge, questions, experiences and flops provides a way for you to continually pay it forward. Building community is vital to your personal success on the homestead, as well as the broader modern homesteading movement. All that reaching out comes back to you tenfold and you're enriched in ways that you can't achieve by staying miserly and closed in a corner.

And what is it about people who are naturally grateful?! Sometimes they're really irritating. How come they can be so chipper when everything may be falling apart? Why don't they wallow a bit? Why not complain when life is unfair? How do they keep such a wholesome perspective?

I've studied history and scripture as carefully as I know how, and I think I've discovered a truth about these grateful people. Gratitude is empowering. It means you're never a victim of anything. Now I know that's a bold statement given the atrocities that occur, particularly those perpetrated against the innocent. But the capacity to look around your life and find things for which to be grateful, no matter the circumstances, is a gift that means you will always win. The bad guy or the dark time lose their purpose if you take the power out of their hands and the fear out of your heart. The ability to even be grateful for the trials because of what they teach you takes it one step further. Gratitude is like a sword, cutting through those who would tear you down and the situations that threaten to overwhelm.

In its simplest form, being grateful can take a "bad hair day" and turn it into an intense gratitude for your unruly locks as you watch a friend go through radiation treatment for her cancer. Sometimes the greatest gift gratitude gives us is perspective.

Maybe our success boils down to our ability to resourcefully, generously and gratefully move through our journey toward self-sufficiency without settling for the mediocrity of those who only seek and worship the dollar. There is so, so much more to life.

ACTION ITEMS: Keep track of what works and what doesn't as you save money and resources in your homestead journal. Write down prices, irritations, successes and recipes. In a few months, stop where you are in your journal and go back to the first entry and start reading your notes with more experienced eyes. For activities like this, your homestead journal becomes a narrative, descriptive, expository and persuasive essay all rolled into one great education. Your homestead journal will keep you motivated and moving forward. You WILL flop a lot of the time, but you will learn from those times—sometimes more than you do from your immediate successes, if you're anything like me. The only way you can learn from your experience, though, is if you WRITE IT DOWN!

DIY: If the lists in this section exhausted you just by reading them, go back and pick one thing only in each segment to focus on for the next three months. With your homestead journal in hand, write down anything you can think of having to do with your three selections. For example, let's say you pick:

- Kitchen - use a crock pot for dinner two nights a week
- Home - analyze and ditch some media
- Homestead - find canning jars for the coming harvest season

Your journal entry might look like this:

Purchase a crock pot and find two crock pot dinner ideas online. Are there cookbooks for crock pots?

Make a list of all the devices for each member of the family; plus, all computers, TVs and players of any kind. How much does each one cost? What are one or two we could do without or reduce? Who will complain the most? What can we put in place of that device—what can we do instead of use it?

Figure out where to find free canning jars—look on Craigslist? What do I even want to can this year? Should I buy it from the farmer's market or grow it myself?

I need to go to the library and see what books they have there that can help—maybe cookbooks or a canning book? It would save me time if the library sold chocolate, too. Oh well.

I'll get on the Crockpot this Monday and start making the media list this week. We'll go over the media list next week, though, after the older kids get back from camp. I'm not going to look for canning jars until I know what I'm going to can so I'll decide that first—by the end of the week. I'll go to the library this Wednesday after I take the girls to church group. I'll find jars by the end of the month and buy new lids and rings to be ready to use them.

BUILD COMMUNITY: Email your neighbors, friends, school group, homestead group and/or congregation, sending out feelers about getting together to organize a clothing swap. Your local church may be able to offer a venue large enough and moms are great at organizing stuff like this. You'll need tables to display items, racks to hang more delicate clothing, hangers, and recycled grocery bags for people to use who forget to bring their own and name tags for all the volunteers helping.

It helps if you have a venue with separate rooms for different types of clothing and accessories, just to keep things organized and crowds under control. You'll need a few people who are gifted with media and marketing to create a flier, send out announcement emails asking for donations and inviting people to come swap when it's time. Typically, those who've donated items to the swap get to come first to take what they need from the stash—usually a few hours to a day before the main event. After that, anyone from the community is invited to come and take what they need. Amazingly, you still may end up with items at the end of the swap, which you can donate to your local shelter or thrift store.

To read a short article on how to organize a clothing swap please visit our links page under Homestead Finances: www.homesteadlady.com/homestead-links-page.

BOOKS TO READ: *Leadership and Self Deception*, by The Arbinger Institute, is technically a business book but there are some amazing personal insights to be gleaned here if you're honest with yourself while you read it. We are the leaders, the CEO's, of our families, after all. *The One-Page Financial Plan*, by Carl Richards, is a quick read that connects the idea of financial planning back to the human factors of emotion and value. Read it if you need to get your financial priorities straight.

Dave Ramsey's Complete Guide to Money, by Dave Ramsey, may be of help to you. I've read a few financial books and they all annoyed the tar out of me. Mr. Dave doesn't. He does have courses to sell (this book is the handbook for his Financial Peace University), but his books and radio show have been enough to inspire me. Or convict me on several occasions.

The e-*book Hope: Thriving While Unemployed*, by Angi Schneider, is great if you're struggling with under or unemployment. It's a small publication that has brought my family a great deal of comfort and inspiration during similar times. Filled with frugal tips, practical advice and a good dose of common sense, it's a great book for hard times. You can find it on Amazon or on our links page under the Homestead Finances chapter.

WEBSITES: For all around frugal living and from scratch living, please visit Little House Living (www.littlehouseliving.com). Visit Penniless Parenting (www.pennilessparenting.com) for some serious savings tips—we're talking extreme frugality here. So much to learn!

HOMESTEAD LADY SPEAKS:
Like I mentioned, I believe that learning to save is more about becoming the kind of person we want to be than it is about dollars. Making gifts by hand has taught me to be less wasteful in general and to be far more grateful for what I've been given. Saving energy has made me careful to appropriately carry out my stewardship on the earth by being more aware of what

I'm consuming and what I'm planting in return. Anytime I connect with my community in a service-oriented event, I'm enriched beyond measure—especially fun stuff like swaps! The best part about this kind of event, in my experience, is that it's such a blessing to people. I've been provided for by it to be sure, but it's an uplifting thing to see that others are enriched, too.

My sister was once volunteering at a swap we held at our church building to which we invited the public. One mother had to ask several times how much certain items cost before she finally understood that everything was free for her to take, if she needed it. Her face lit up as she took the large, plastic bag my sister offered. With a big smile, she burst out, "Ah, honey!" and went off to collect for her family's needs, laughing in happiness. That experience was a gift in which my sister and I still find joy.

"Between stimulus and response there is a space. In that space is our power to choose our response.
In our response lies our growth and our freedom."

— Viktor Frankl, Physician, Author and Holocaust survivor

For the Homesteadaholic: Wisely Spend

Choosing what to spend your resources on for the homestead is a perpetual issue—it's always coming up in every season and with every crop or animal. I loved this advice from Rebecca,

> "Before making a purchase, look at what you already have. Can you use something else, make it instead, borrow, or trade? Often just by delaying a purchase we come up with an alternative and find we really didn't need it."

This is such beautiful advice, I may just paint it on my pantry wall.

My husband works full time off-site and I work full time on-site. The bulk of the daily homestead works falls to me and the kids. Our main job during the day is education (homeschool) and it takes up most of our time. So, when we go to do the work of the homestead, we need systems that work well and easily. We also need equipment and animal housing to be easy to use, maintain and clean. I am far too prone to put that need above my need to spend my money wisely. I'll sometimes say, "Let's just buy that completely assembled and new item to save us time and effort!" A good example was our chick brooder.

I wanted to purchase a small chicken house, place it on cement, put a light inside and just call it a chick brooder. When I needed to clean the brooder area (chicks poop a lot), I thought I'd just slide the little house to a new place and clean the old area. No big deal. Well, even used, prefabricated coops in our area were several hundred dollars. My engineering husband came to the rescue.

He salvaged a weird, wooden shelving unit made from pallets—one man's junk is another man's chick brooder. My husband brought it home and, using materials we already had around the homestead, slowly turned it into the most user-friendly, two story chick brooder the world has ever seen. The unit was tall, and we were able to create two separate areas for brooding, complete with hinged doors to make clean up a cinch. We started many a happy chick/poult and put many a broody hen with her eggs into that practically free brooder. If we'd listened to me, we would have gone into debt to buy something that was quicker to obtain but no easier to use. My husband wisely spent less money, while using skills and product he already had to create something far superior to what we could have purchased ready-made.

> "A farmer depends on himself, and the land and the weather. If you're a farmer, you raise what you eat, you raise what you wear, and you keep warm with wood out of your own timber. You work hard, but you work as you please, and no man can tell you to go or come. You'll be free and independent, son, on a farm."

— Laura Ingalls Wilder, from *Farmer Boy*

It's All About Balance

Rebecca poses a very powerful question:

> "How can you balance the sense of freedom and joy in the natural world, with the pragmatic and necessary daily business operations?"

According to her, there are four things we can do to accomplish this balance and two of them we've already talked about a bit.

REALISTIC PLANNING. Rebecca says,

> "One of the first things every homesteader needs to consider is: what is a practical and achievable plan? The picture in their minds of what the ideal homestead experience is, may have to be adjusted because of their individual circumstances. Age, health, location, responsibility to care for aging parents, and desire to have children all should be carefully considered."

HAVE A CLEAR PICTURE OF YOUR FINANCES. Rebecca reminds us,

> "Don't let it be a surprise when you are sitting down with a lender. Before you decide to invest in or grow your homestead, make sure you have a good understanding of what you can reasonably afford to take on. Have an accurate working budget. Check your credit reports. Give yourself some wiggle room in the budget by financing less than you are approved to borrow."

PACE. To be successful we must also be patient and use our plan to stick to measured growth. If we will resist the temptation to take on too much too fast, we will be able to manage our homesteads successfully.

PREPARE FOR THE UNEXPECTED. Rebecca reminded us in our wise spending section that there's one thing you can always count on and that's that stuff happens. When stuff happens, it always requires resources to fix. If we will make planning ahead a part of our homesteading, we won't get into unnecessary debt to stave off crisis. We don't want to let a lack of planning put our homestead at risk; that would be an unsound business practice. And the sooner we look at our homestead like a business, even if we're not producing an income, the more organized and thorough our efforts will be.

Spending Wisely and ROI

I'd heard the term ROI before but didn't really understand what it had to do with my homestead until Rebecca reminded me that,

> "All major purchases and investments should be looked at in terms of what the Return on Investment, or ROI, will be. Here is where having a growth plan in place is especially important so you don't buy impulsively. Even if you can get a great deal on a piece of equipment, if it isn't good timing, you could use up resources you need for something else along the way. Always keep the plan and long-term goals in mind so you aren't sidetracked.

> "When looking at ROI, make sure you take all costs into account before calculating potential return. Careful research can save you a headache, and money. Talk with others about their experience with a product or crop in your area. Borrow and test out a tool before buying one to make sure it is the right fit for the job."

The idea of spending only when it's necessary, even if you can get a good deal on something now, to preserve funds to keep them in-line with my overall homestead plan was powerful to me. This is why the plan is so important. You can use the plan as a measuring stick to provide guidance in every day matters, but it can also give you a consistent feeling of well-being and control. You can be sure you're not spending too much or spending unwisely, which includes in an untimely way, because your ultimate goals are always in front of you. I don't know about you, but my peace of mind down the road, in my busy lifestyle, is worth a little planning now. Sometimes the ROI is just in how you feel.

> **SHABBY CHIC**
>
> "You might be a homesteader if you give up changing coats before going somewhere public because they ALL are covered in hair, fur, straw, feathers." - Rebecca Shirk
>
> You may not be at this place yet but, trust me, you will be eventually. I used to have good shoes, I really did. I used to be kind of a fancy lady. However, five kids, several gardens and a whole parade of farm animals later I've just learned to be careful with my limited time and means. Does that mean all homesteaders are cozy flannel advocates? A good number of us probably are but well-worn and much-loved aren't the same thing as dirty

and shabby. I think of Caroline Ingalls and other pioneer women, endlessly sweeping the dirt floors of their dugouts and prairie cabins in their starched aprons with their work-worn hands. Beauty, like love, has many different definitions and most of them are completely lost on the modern world's sensibilities.

If you're concerned that your family is too much in the world, I submit that a slight shift in thinking is all you may need to start the journey back into a more grounded lifestyle. So many modern parents find themselves and their children swamped in hectic obligations and over stimulated by media nearly everywhere they turn. Life these days is so loud, you hardly hear yourself think. The worst part is, no matter how hard we work we're just spinning our wheels. We're constantly moving but never really getting anywhere; always learning but not really knowing anything. It seems to me there's something just plain unnatural about the kind of lifestyle our culture is trying to achieve. We keep moving farther and farther away from the earth and are finding more and more automations to replace the work of our hands. Ha! Listen to me, I sound like an anti-progress, anti-industrialist!

I'm not really talking about progress in the sense of computers and foaming soap. It's more a mental shift than a physical shift that's at the core of the problem of our "modern" lifestyle. The difficulty is the notion that conveniences are the point of life, the reason we're here and all that we're really meant to experience and create. Across religious lines, I would hope that the family of man could agree that there IS a purpose to our being here and there ARE meaningful things to experience and create. Maybe even something more meaningful than computers and foaming soap, as groovy as those things are.

I've found that service and work are two of the greatest tools we have in learning who we are and what is truly important to us. The homestead lifestyle can provide an opportunity to become really accomplished at both. I don't mind my comfy shoes. I hardly notice them anymore because I know they've been broken in running to and fro as I work endlessly at the tasks set before me. Each time I put on my work shirt, I square my shoulders to another day of schooling my children or bringing in the harvest or...insert whatever task is meaningful to you. There's very little that's convenient about work and service but there is everything you could ever desire in the way of purpose and self-definition.

My coat is, indeed, covered in hair, fur, straw and feathers and I think it looks more like the raiment of a queen every day.

ACTION ITEMS: Use the short but serviceable Homestead Planning sheet in *The Do It Yourself Homestead Journal* or draft your own homestead planning sheet. Include sections for goal planning for the next year, two years and five years. Plan further, if you like (my attention span is short). Be sure to make a list of assets and liabilities on your homestead, including family debt as a liability and each member as an asset. You may also want to have a space for each part of your homestead—animals, family, gardens, green living, community building, kitchen, finances and emergency preparations. Get a working list of the current cost of needed items, plus a wish list (what you want is important, too, even if it's only to dream) and a list of activities/ventures you'd rather NOT do. Let's get real, let's be honest and let's get 'er done.

If you'd like a more thorough set of homestead planning sheets, I can heartily recommend the ones I use from Quinn at Reformation Acres; to find a link for those, please visit tour links page under Homestead Finances: www.homesteadlady.com/homestead-links-page.

 DIY: If you didn't do the DIY homework from the *Savings* section, go do it now.

BUILD COMMUNITY: As I've said before, the most important community you'll ever build will be inside the walls of your own home. As you create your homestead plan include every member of the family. This planning must be done strategically, as any mom who's ever had to coax a toddler or a teenager to do a chore will tell you. For one thing, your plan won't be written overnight—this is going to take some time. Inspiring vision in the people over whom you have stewardship is a skill to learn. Simply demanding that everyone gather for two hours each week until this plan is hammered out won't encourage the free flow of ideas if you have family members who aren't on board with the whole homesteading thing or don't take it seriously. This can include your spouse.

If you need to create an inspiring atmosphere, I highly suggest you start with food. Make something as delicious and wholesome as you can. Share it with your family and talk about how you enjoy cooking from scratch and how you'd like to do more of it. Move on from there to talk about the idea of homesteading, wherever you are and how you really need their input and talents to make the venture work. Do this slowly, over time, if you must. You don't want to do this without their vision joined to yours if you can help it. For some people, homesteading will never be their thing, but family can always come first—even an annoyed teenager can tell you what he *doesn't* want to eat from the garden. If you're really struggling, especially with a spouse, I highly advise you learn how to use prayer and/or positive meditation to keep calm and focused on your goals. Stick to your plan and do your best; not all your rewards will be monetary, but you WILL see results.

BOOKS TO READ: *Smart Money, Smart Kids*, by Dave Ramsey and Rachel Cruze, may be of help. Yes, another one by Dave Ramsey. I'm not a financial expert and I don't pretend to be, but I've read several books on home economy and money management. Mr. Ramsey suits my personality and how I learn best. If you don't like him, I won't take offense and just ask that you let me know if you find a title you love.

For a very actionable, practical and fun book on teaching your kids about the real world of money, try *Earn It and Learn It*, by Alisa Weinstein. This book will particularly resonate with home educators or those who enjoy hands-on learning with their kids. The book is full of projects full of learning, accountability and reward.

The American Frugal Housewife, by Lydia Marie Francis Child, dates to the 1800's and is a particularly interesting read for the homesteader. Much of our equipment and expectations have changed but there is a lot of value for homesteaders to read books like these and glean all we can in the way of ingenuity and frugality. Incidentally, this authoress was a noted abolitionist and activist, as well as writer and may be best known in our day for her Thanksgiving poem, "Over the River and Through the Woods." This poem is like a little piece of American history and common sense all rolled into one.

WEBSITES: One Hundred Dollars a Month (www.onehundreddollarsamonth.com) is a great read for anyone wanting to learn about frugal yet wholesome shopping, gardening and general frugality from the fun personality of Mavis Butterfield. Heather at The Homesteading Hippy (www.thehomesteadinghippy.com) has a lot of information you will find valuable on thrift and other homestead virtues.

REBECCA SPEAKS:

"Your time and energy matter, too! Sometimes the investment far exceeds the return when you factor in your labor. In many cases this is fine because the labor is enjoyable. For instance, it takes me a long time to knit a pair of socks. It would be cheaper to buy them, but I enjoy knitting as an activity. You may find, however, other activities you don't enjoy where a lot more time and effort go into making something than the payout. If that is the case, let yourself off the hook. Homesteading is a labor of love, but if you don't love what you are doing, then it isn't worth it for you. Find alternative ways to handle or purchase those things."

- Rebecca Shirk

"The greatest fine art of the future will be the making of a comfortable living from a small piece of land."

-President Abraham Lincoln

For the Homesteaded: The Homestead Side Hustle

As I associate with homesteaders, I've listened as they've spoken of their trials and triumphs while they move through the various phases and learning curves this lifestyle provides. It's been my experience that, eventually, we all come to this place where we'd like to make a bit of money, or tangible profit, from the homestead and all it produces. Some of us are merely interested in earning some pocket change from various activities we already participate in on the homestead, without too much bother or extra effort. Others of us are looking to bring home our overly-stressed, corporate-climbing spouses so that the homestead can be a true family affair and, therefore, require enough income to make that goal a reality.

Whatever your motivation, there are several very sell-able and trade-able items you and your homestead can create. Before we get into that, let's get our focus adjusted.

THE BUSINESS OF HOMESTEADING

It never occurred to me to think of my homestead as a regular business until Rebecca Shirk sent me this high-quality advice:

> "Viewing the homestead as a business is key to ensuring its long-term success. Unless you are planning on it being a hobby, supported by full time work elsewhere, you will need to treat the homestead as a small business. This means keeping accounts, paying taxes, and controlling operating expenses.
>
> "According to the Small Business Administration, seven out of ten new employer establishments survive at least two years and just 51 percent survive at least five years.

This is not because people lack vision, but because often their ideas and enthusiasm surpass their business savvy."

I repeat Rebecca's insight from an earlier section,

"A lot of homesteaders have a dream of living off the land, a passion for animals, farming and sustainable living. However, that passion will quickly fade if they become overwhelmed with debt and cannot afford to provide for the animals in their care."

This was eye-opening advice for me and something I'm still working to do, especially as we're relocating to a new homestead. Having a plan and treating our homestead as a producing entity has made a positive difference in how we work and view our efforts. For us, it ends up being less about producing an income and more about managing our assets, including time and effort, to the best of our ability. We are careful with waste, especially in how we spend our capital as it relates to the homestead.

For those who haven't really thought much about it before, here is Rebecca's parting encouragement:

"Homesteading provides so many income opportunities. Be creative!

1. "Sell what you raise and grow, but make sure you preserve enough for your own family first.

2. "Make products from your own resources to use, trade and sell.

3. "Write about what you do for blogs, magazines and books.

4. "Host tours and open your homestead for others looking into the lifestyle.

5. "Teach others homesteading skills at your home or community workshops.

6. "Lease space in your homestead to others that need space for a garden, crop or pasture."

Homestead Business Foundations

Remember that any concept you want to develop can be started small and grown as you have time, ability and desire. When I suggest you sell transplants, for example, I mean start by selling your extras, to the tune of 5-50 plants. Don't go out and buy a commercial greenhouse and fill it for sale unless you're ready for that. You don't have to start off elaborate as small things, over time, can really add up. If you want to go hog wild, jump in, but have a plan in place before you do.

A few more tips before we begin—these are just things for you to mull over as you consider points like legalities and marketing. Not all of what we chat about here will make much sense to you, if you're completely new to all of this, but it will eventually. Just keep these ideas and points in your head:

This section contains numerous links to online content for further study and tutorials; please visit the links page under Homestead Finances where each one will be clearly labeled: www.homestead-lady.com/homestead-links-page.

1. **PACKAGING:** As you package or advertise anything, label it clearly for price and contents. Develop a logo and use that brand all over your website, social media, packaging and business cards. It will be worth whatever you pay for it in the long run. Many times, your state will have specific information that they require you to include on a product label, too. Also, if your state has a "Made In _____" label, be sure to use it. For example, in Utah, the label reads "Utah's Own." I love that! Label your goods with the one from your state and look for it when making purchases.

2. **ONLINE:** Absolutely start a website for anything you decide to sell—if it's not online these days, it doesn't exist. There are free platforms and those that require payment for hosting. I pay for my hosting through Wordpress.org to ensure that I own all the content and so that, barring technical problems, my site isn't shut down for some random reason. To learn more about self-hosted Wordpress.org vs. free Wordpress.com, please visit our links page. Even if you list your products with a presence like Etsy (www.etsy.com), an online artists and crafters market, or Ebay (www.ebay.com), I still advise you have your own site and connect back to your Etsy shop with a widget (a piece of computer code) on your site. Make sure you have a branded logo for your Etsy store, which can be the same as your homestead logo and may require a professional artist or designer to develop. For more information on how to start a website and/or Etsy shop, see our book recommendations below.

3. **SOCIAL MEDIA:** Set up accounts all over social media, particularly Facebook, Twitter, Pinterest and Instagram. Again, if you're not online, you don't exist when trying to sell something. Engage there as often as you can to build a following that knows and trusts you. To learn more about using social media, and about a particularly helpful site called Social Media Examiner (www.socialmediaexaminer.com), please visit our links page.

4. **FREEBIE OFFER:** Whenever you go to market a new product, remember the power of the freebie. As a customer, I love getting a few free things tossed in with a purchase because it makes me feel like I'm really getting my money's worth. It also assures me that the seller is confident enough in his products to allow me to sample a few gratis, assured that I'll come back because of quality. Especially online, the power of the freebie with the launch of a new product is certainly something worth considering. A freebie could be something as simple as a coupon to an online store. It could also be a free publication of some kind. When people sign up for the

newsletter at my blog, Homestead Lady, they receive a short e-book; the one current at the time of this publication is *Herbs in the Bathtub: Growing Herbs Wherever You Are*. Try to make your freebie of use to your readers and in keeping with your own personality and voice.

5. **LEGALITIES:** Always, always check legalities before you begin any project and don't assume that just because other people are doing something, it's legal. Check federal, state and county laws. If you have an HOA or CCRs, be sure to check that anything you're doing on your land won't be in violation of those rules. Don't forget to double check zoning for your property and insurance, especially if you invite people onto your homestead and/or sell from your homestead. Know your state's cottage laws by heart and be prepared to comply with state inspections if you need to set up something like a certified kitchen to sell baked goods at a venue like a farmer's market.

6. **PROFESSIONAL ADVICE:** You may want your lawyer and tax specialist to look over your business plan before you begin as well. Have those experts advise you about the pros and cons of incorporating or becoming an LLC (Limited Liability Corporation). Don't invest in a project before you've covered every possible legal angle and make sure you speak to actual human beings to double check anything you read online. You don't want to risk losing everything you've worked for on your homestead because you missed a technicality. It's no good fuming that this stuff is required in our litigious, regulated society—it is what it is. Just get it done.

FARM TO CONSUMER LEGAL DEFENSE

The best resource I know of to keep yourself informed and protected as an educated consumer, homesteader or small farmer, is the Farm to Consumer Legal Defense Fund (FTCLDF). According to their website, FTCLDF:

"...protects the rights of farmers and consumers to engage in direct commerce; it protects the rights of farmers to sell the products of the farm and the rights of consumers to access the foods of their choice from the source of their choice. FTCLDF is a true grassroots organization and receives no government funding and little or no corporate funding."

For consumers, artisans, co-op organizers, homesteaders and small farmers the FTCLDF exists to defend your rights and broaden the freedoms of family farms. We joined at the Homesteader level this year and have enjoyed our association with them. To learn more or to join, visit our link page.

Homestead Income Three Ways

So, let's begin from those ideas of Rebecca's previously mentioned, and the others you've been mulling over, and expand on them to make a list of things you and I might do to make money from the homestead. Please bear in mind, that for me, creating an item that can be traded for another product or service is just as useful as having cash, to a certain degree. The mortgage company only takes cash and so as long as I have a home loan, I'll need a certain amount of capital for that. However, there are so many other areas where having a product (like food) would be just as helpful to my family as having cash.

The following are three areas of homestead cash or trade generation that you might want to consider.

RAW PRODUCTS

These are items you can sell or trade that an animal produces, the ground grows or is essentially created by something other than you.

1. **HEIRLOOM SEEDS:** Save seeds from your crops every year and market them locally or online. This will be especially useful if you start your own seed saving group. (For more information on starting a group see the *Seed Swap* section in The Homestead Community chapter.) Find a few garden bloggers whose writing and style appeal to you and offer to give them seed in exchange for an honest review. Celeste from Simple Gifts Farm (www.simplegiftsfarm.net) sells seed garlic from her homestead and she made me just such an offer this year. Her cloves are tucked neatly into my garden beds awaiting next year's harvest and blog post to review them.

2. **TRANSPLANTS:** This will be simple to do if you're already starting your food crops and herbs from seed. Grow up extras of plants you observe people putting in their gardens in your neighborhood, or heirloom varieties you enjoy growing yourself. You can also do this with bedding plants, fancy annuals, and pretty perennials for ornamental gardens.

3. **BABY LIVESTOCK:** You can simply sell off your extra chicks, piglets, and bees or you can begin your own breeding program. Often surplus males of various species are sold to become dinner later and several species are inclined to produce multiple offspring in one go. Registered dairy and breeding animals like goats listed with the American Dairy Goat Association (ADGA) will usually fetch more money than unregistered animals because people can be sure of their pedigrees. Small animals, like rabbits and chickens, usually sell in higher volume.

4. **FERTILE EGGS:** If you have a rooster in your flock, sell your fertilized eggs on your local online classifieds. Be sure to test for fertilization rates and decide if you want to ship, which will lower the hatch rate. Even healthy, static clutches rarely hatch out every egg, so just be ready to give a rough estimate of hatch rate/fertility to your customers. Try raising a breed that's rare for your area and determine if you can create a niche for it.

5. **BERRIES AND OTHER FRUITS:** These items always sell, berries especially. Fruit is easy to market because it's sweet. Instead of planting three blueberry bushes, plant six and try to sell your surplus. If that's successful, plant more.
6. **VEGETABLES AND SEEDS:** You know how delicious homegrown veggies are because you've purchased them from farm-stands yourself and grown them on your homestead. Try to grow a variety that is a favorite in your area, as well as something that's new. To determine what's locally popular, visit the farmer's market and local nurseries. Maximize your space and energy by using wholesome, nutritive growing cycles full of biodiversity and lots of good compost. Think square foot gardening for commercial production, on whatever scale you are able. When you go to market the plants, offer an heirloom tomato and provide an information sheet with each purchase on how to save the seeds so that your customer can plant them next year. It might seem like a counter-intuitive move to teach them how to grow their own, but they'll remember you as being helpful and invested in their success. Remember that we do what we do because we want to create abundance everywhere, not just in our wallet. To get inspiration about saving and selling seeds, see the *Seeds For Generations* sidebar in the Homestead Community chapter.
7. **EGGS AND MILK:** These items are always popular and in demand.
8. **GARLIC AND OTHER HERBS:** Herbs are a great niche product. Braided garlic is visually appealing and potted or fresh cut herbs have both culinary and wellness value. Include a small information card with each purchase that details the value of the herb in the kitchen and in the medicine chest.
9. **HAY:** I put this here because if you're already raising hay/straw, the odds are you have a bit extra you can sell. It typically requires baling, which can be laborious if the amount you're marketing is large. However, in rural areas you can often find people with large baling equipment who will bale your hay for you for a share of the hay.
10. **DAIRY SHARES:** If it's legal in your state and you already have the livestock, this can be a relatively simple idea to bring to fruition. Cow or goat-share programs exist to allow consumers in non-raw-milk-friendly states to still acquire raw milk by paying a set fee for a "share" of a dairy animal that is then tended and milked by a farmer. Technically the animal belongs to the customer and so she is drinking the milk from her own animal, while merely paying the farmer to maintain it. This protects the farmer from liability issues where raw dairy laws are concerned and provides a way for raw milk drinkers to legally acquire the milk they want. To learn more about the helps that are available with dairy share programs, please visit our links page under the Homestead Finances chapter: www.homestead-lady.com/homestead-links-page. There are always complications when you're

dealing with laws, codes and people, BUT if you have the animal and the animal is already producing milk anyway…

11. **FISH:** If you have an aquaponics set up that is really thriving, and you're tired of eating fish, then selling your surplus can be a blessing for both you and your customers. Aquaponics is a hydroponic grow system designed to produce fish whose waste, in turn, becomes food for crops through a specialized watering system. Be sure to grow a type of fish that is generally pleasing to consume, if you plan to sell them. Also, check the livestock/pet ordinances for your town to be sure your system is legal. To learn more about setting up your own aquaponics system visit our links page under My Aqua Farm.

12. **MUSHROOMS:** If you're growing these already for your own use, or you can harvest them ethically with a bit to spare from your land, spread the joy and sell the 'shrooms. To learn more about growing your own mushrooms read the very reliable *Organic Mushroom Farming and Mycoremediation*, by Tradd Cotter. For foods, especially raw milk and mushrooms, it's perfectly logical to require your customers to sign a liability waiver before they purchase product. Keep these on file in a safe place.

13. **HONEY:** While honey does require quite a bit of input from you to get it from the hive to the customer, still, it's the bees that make it. You extract, filter and bottle, but you don't technically have to create this very, very popular item to sell from your homestead. For quality information for keeping bees for honey, read *Backyard Beekeeper*, by Kim Flottum.

14. **STUD SERVICES:** If you have the space to keep males, renting out their services as a stud is a simple way to provide a service and make a little side money. You'll need to keep tabs on your buck's fertility in your own herd and decide if you want to offer a money back guarantee. Be sure to talk to experienced stud managers to develop a program for mitigating pathogens and diseases being introduced to your herd from visiting females. Only you can decide what's best to do for your boys but separate pens for each male and a quarantine time for the ladies may be helpful to you. A registered sire will probably bring in more money than an unregistered one. As a goat owner, I prefer the local market for this, so that date night for the goats is in person (or, more properly, "in goat"). However, you can educate yourself on selling the semen of your male. You'll most likely be working with a semen collector, whom you'll need to pay, and you may need some certification/registration to be credible enough for people to work with you. Again, that's why I prefer local romancing.

15. **FIBER:** Honestly, I wasn't sure where to put this one! Fiber animals require shearing and their fleece requires skirting and cleaning, so you do a good deal of work to get this product marketed, but the animal does the growing. You can sell raw

fiber, too, for a discounted amount—some spinners prefer unwashed fleece to wash as they see fit. You can sell clean fiber by the pound and you can also have it turned into roving. If you're handy with a wheel, you can spin it into yarn of different weights and thicknesses. As one experienced with textiles, Sarah Dalziel (www.wearingwoad.com) suggests that novelty yarns are more cost-effective/time effective as compared to spinning generic worsted weight yarns. She also suggests trying to find a unique marketing angle like naturally dyed yarns, unusual fibers and anything to make it more unique. These special touches will justify charging a sufficient price to compensate you for your time and investment. To see a photographic encyclopedia of more than 200 animal fleeces and the fibers they can produce, read *The Fleece and Fiber Sourcebook*, by Ekarius and Robson. This thing is so gorgeous it will make you want to own every fiber animal on planet earth.

16. **WORMS:** If you're already vermicomposting then you know that your red wrigglers reproduce well when they're healthy. Turn that excess into some income. For more information on vermicomposting, please see that section in the Livestock Wherever You Are chapter.

17. **COMPOST:** Sell your vermicompost and/or your garden compost. If you have animals on your homestead, be sure to include their dung in your compost to make it even more robust. They don't call compost black gold for nothin'! Consider whether you want to bag it or sell it loose. Will the customer need to come pick it up or will you deliver?

VALUE ADDED PRODUCTS

These are items for sale or trade that you create from the raw products that come off your homestead. These require some manipulation on your part, but they are often items you already enjoy making and of which you can produce just a little bit extra without too much hassle. Remember what Rebecca said about ROI?

> "Sometimes the investment far exceeds the return, when you factor in your labor. In many cases this is fine because the labor is enjoyable. For instance, it takes me a long time to knit a pair of socks. It would be cheaper to buy them, but I enjoy knitting as an activity."

About what are you passionate?

1. **BEAUTY PRODUCTS:** Women will always be interested in products like soaps, lotions and creams. Was that sexist? Men need them, too, of course but ladies like to make an experience of purchasing these items. Find a way to make them as wellness-focused as possible and you'll probably increase your market, even if

you must increase your cost; so many of us are looking for a way to avoid chemically laden products in our wellness and beauty regimens. Visit The Nerdy Farm Wife (www.thenerdyfarmwife.com) or Curious Soapmaker (www.curious-soapmaker.com) to learn more about homemade soap and other natural care products. Always stay abreast of state and federal laws regarding the sale of any health and beauty products.

2. **HAND-CRAFTED FIBER ITEMS:** Hand-knitted, crocheted, felted and woven articles are so lovely and rich. Many people are unwilling to pay what these items are worth, but you can find niche markets, especially if your pieces are unique in some way. A great venue to connect with other fiber artists, network and even sell patterns and yarns is Ravelry (www.ravelry.com).

3. **OTHER HAND-CRAFTED ITEMS:** Wood working, metal crafting, sewing, candle making are all skills that are vital to your personal enrichment on the homestead, but they are also useful in generating a bit of income. If you are fortunate enough to be proficient in one or more of these skills, feel grateful for the abundance in your hands.

4. **COOPS, BROODERS AND HUTCHES:** If you're clever with a hammer, build a few prototypes and see how they sell. Sometimes when we're gifted in a particular area we have a hard time realizing that not everyone is like us. This causes us to downplay the potential market for something because we assume that no one will want what we have to sell since they can just build their own. Not everyone can "just build a chicken coop," and those who can't will very much need what you can provide. There have been times when I've purchased something I could make myself just because I didn't have the time to do it. Your local, online classifieds can be a good place to start marketing these types of items.

5. **KITCHEN DELIGHTS:** If you're already making up a batch of your grandmother's recipe for blueberry lemon jelly, why not make a bigger batch and sell the extra? Learn all you can about perceived value when it comes to packaging and presentation; it's human nature that a big factor in what we're drawn to in a purchase is how the article looks. That's not to say you can create shoddy product expecting to hide it in pretty paper and grow a steady and loyal business. However, your hard work deserves to be show-cased as effectively as possible, so do some research. An easy way to conduct a beginner's study is to go on Pinterest and observe what's visually appealing (that's what Pinterest is all about). Remember, that apart from state laws, your farmer's market will also have bylaws that need to be adhered to. Make sure you know what they are ahead of time.

6. **LIVESTOCK GUARD DOGS AND TRAINING:** Now here's something I'm in the market for right now. I don't know anything about using an LGD yet, but I'm set to learn this year and I'd love to have a market for quality stock, as well as a mentor

to help me learn how to properly work my dogs. Herding dogs require a lot more training and, if this is a skill you have, people will be willing to pay you to learn! Training other animals can be useful as well, especially horses and any draft animal.

7. **FEATHERS AND PELTS:** If you have guineas or other fowl with pretty plumage, you may take some time to collect the feathers, clean them and sell them for crafts and jewelry. Remember, too, that many of these things can be done by children; what a fun side business for your kids. If you already harvest rabbits for meat, it may be a simple thing for you to sell the pelts. There are people out there who use them to make clothing, as well as Boy Scout troops that need to learn to tan a hide. Why throw something away (or neglect it in your to-do pile) if you can sell it?

8. **FIREWOOD:** If you have the trees, and a strong back, especially if you're already harvesting it for your own use, firewood can be a simple thing to market. Some of us use wood year-round but, for most people, it's a seasonal purchase. Firewood, to burn properly, also needs to season for a time. Plan harvesting and marketing schedules with those things in mind. I would consider something this physically demanding to be a short time gig until you have the financial footing to do something less taxing, but that's just my opinion.

> **HOW TO FIND YOUR COTTAGE LAWS**
>
> Pick Your Own has a great synopsis of what Cottage Laws are; the laws that determine how and what you can sell from your homestead. Most states require that you have a dedicated kitchen that passes an inspection. They also limit where you can sell and what type of items. Be sure you know these laws; to be certain, always talk to your lawyer. To get you started, visit our links pages.

HOMESTEAD SERVICES

These suggestions aren't necessarily for tangible things to market, although there is some of that. These ideas will most likely require an investment of time and energy from you but may serve to build up your brand and authority in a particular area. Just like all these ideas, they also provide a legitimate and needed service to your community for which you deserve to be compensated.

1. **ANIMAL BOARDING:** If you have the barn space, this can be a great project for your older kids. Boarding requires feeding, watering and, depending on the animal, a certain amount of maintenance like brushing and hoof trimming. Be sure to plan for separate housing for visiting animals to protect your own from disease and bacteria. This is particularly relevant with rabbits, as they will catch anything—even human colds and flu.

2. **RENT PASTURE OR GARDEN SPACE:** Peterson Family Farm in central Utah is a farm local to us that rents out garden sized plots in their own fields every year. The plots are marked off by Farmer Luke, the owner, and he flood irrigates every week. The only real requirement is that the plot-user keep the space weeded. This is a genius idea for several reasons. There are neighborhoods in our area where the homes really don't have much space to grow food and others who have HOA's that prevent much of that, so having this local to us is a boon. It also provides a sense of community and gets a lot of local people on site and frequenting his farm store which has been put together with care and taste. There are many community events with local growers and craftsmen on the farm every year and since many people are there tending their plots, they learn about all the fun stuff going on and invite family and neighbors. I've watched over the six years we've lived in the area as more and more land is eaten up by developers, but Peterson's Farm is working hard to stay relevant and vital to their community. A lot of that success begins with these simple garden plots.
3. **GOAT BRUSH-EATERS**: If you have a herd of dairy goats, consider advertising their services as weed-eaters. For specifics on how you might manage your herd to control brush and weeds, I suggest you visit sites like Goats R Us (www.goatsrus.com) to study their business model. In brief, you need to be prepared with a way to move your goats to and from the site to be cleared. You're also going to need movable fencing because goats are...well, they're goats and they wander. Goats can quickly overgraze a site, too, so you'll need to plan to be close enough to steadily check on their progress through the weeds.
4. **TREE FARM OR NURSERY:** I remember going to a Christmas tree farm every year to cut down our own tree; it was such a huge part of Christmas! It still is for me today and I've toyed with the idea of starting one just for the pure joy. It's A LOT of work, just ask any farmer you know who runs one, and it takes a good deal of time but, oh, what a thing to grow! By providing a tree farm local to your customers, you help prevent the devastating sight of already harvested trees languishing in tree lots, unwanted on Christmas day. To encourage your customers to return, you might offer a tree chipping service on site where they can bring their finished trees to have processed for free. This would be a bonus in that you would then have wood mulch, after a bit of time, which you could use or sell. While they're there having their trees recycled, perhaps you could offer them a few indoor plants for sale from your greenhouse and a coupon to return in a few months for bedding plants and vegetable starts.

   Focusing on a one-time sale is fine, but it's short-sighted; part of why we do what we do is to build community and life-long partnerships with the people around us. Find a way to get your visitors to return repeatedly and, before you know it, they'll be loyal customers. Lewis Hill's Christmas Trees is a quick read for the novice grower to get their feet wet with this venture.

5. **U-PICK FARM, CORN MAZE OR PUMPKIN PATCH:** There's always risk when you invite someone onto your land but if you've researched that risk and covered yourself with insurance, these enterprises can be fun. U-picks are season-dependent operations; you advertise based on what seasonal fruit or veggie is available. You need to decide each year the cost of product per pound and/or the price of admission. Be sure to set firm boundaries between what land is appropriate for customers to populate and what part of your property is off limits. Carefully guard your personal time as well and make it clear that you're working when you have customers present—you do NOT have the time to chat for three hours with each person.

    As a customer of U-Pick places, my biggest advice is to be very specific with price when you market. Be sure your customers know why your product is unique and why they need it. It can be helpful to provide a small sample of what's on for harvest at any given time. Mississippi State has some helpful u-pick publications in the form of PDF files that cover several different topics, including what you might grow. To find these publications, visit our links page.

6. **TEACH LOCAL CLASSES:** For some homesteaders, this is their most steady form of side income. Again, don't downplay the importance of what you're an expert in; so many of the homesteading skills have been lost over time and people are scrambling to revive them, truly yearning to know how to do what you already do. Community colleges, agricultural extensions and church congregations are some venues you might explore. I caution you against repeatedly working for free, but there may be times when you feel strongly about donating your time. I've done it before and have been happy to serve, especially in my congregation, but there are absolutely times when it's appropriate to expect to be compensated in some way or other. Again, never discount the beauty of trade items! Sometimes I'll host the class for free but charge a fee for any printed material from the class, if people would like to buy it.

7. **FARM TOURS:** This can be a big pain in the butt that provides dividends over time. Utah Natural Meat is a grass-fed meat provider in our area and part of their farming facility is local to us. Farmer Shayne will give tours to any interested party, with a bit of notice. This is a family farm—it's him, his wife and their four little kids doing most of the work, so taking time out for a tour is an investment. However, they are committed to helping their community understand the value of humanely raise meat and they strongly feel that a personal education lasts a lifetime. So, they invite the community onto their farm, where they can show off their greenhouse, their heritage pigs, plus cattle, goats, sheep and poultry, their

sprouting system that grows all the rations for their livestock and their awesome draft horses that work the farm. You can ask questions and get real life experience from an intelligent gentleman farmer and his family. We first learned about sprouting grain for livestock (instead of feeding dry rations) from Farmer Shayne and were able to pick his brain about re-creating the system on a homestead-sized scale.

On my own land, I've done homestead tours for Weston Price chapters, homesteading groups, seed groups, homeschool groups and just interested friends and neighbors. It can be a lot of work but it's good to know that you're having a positive impact on your town. Sometimes you charge for the tours, sometimes the tours are free, and you ask for a donation, or provide a homestead product for them to purchase afterward. Farmer Shayne has his farm store on site which he invites you to visit. We have found that food, and water bottles, are always a big seller on our homestead tours. Be sure to check with your insurance company to determine what liability coverage you might have. You may want to have your lawyer draw up a waiver form for every attendant to sign.

8. **WEDDINGS AND PARTIES:** If you have the tailored space, a lovely barn (or other indoor facility) and some pretty views, this could be something you really enjoy doing. I'd rather walk on my lips than deal with brides (several years in the floral industry killed the romance of weddings for me), but some personalities really, really enjoy this kind of thing. If you have the housing, you could also host family reunions, corporate getaways and more. One of our favorite family vacation spots is Briar Rose Farm in Hot Springs, North Carolina—a working farm, with working farmers, where we stayed in their 1800's cabin and helped out with the goats and cows. It was sheer bliss! Again, time for a call to your insurance agent and your lawyer, if this is something you'd like to explore.

9. **START A BLOG/WRITE A BOOK:** All of these require a huge investment in time so don't jump into them thinking you'll get rich quick. Remember that as you're engaging in any of the ideas in these lists, you're also living your regular life and maintaining a producing homestead—you're only human! I'm a firm believer in the power of the written word, as much for the benefit of the writer as for the reader. If you do decide to write, commit now to be true to yourself and your own voice. When people buy into your writing, they're coming into your head and getting to know you better; be genuine and don't let anyone tell you what's right to write. Write from the heart.

If you decide to write online, learn all you can now about SEO and keyword writing to make what you're creating visible in the digital world. I also suggest you join a network of bloggers in your niche and learn all you can from those precious contacts who become friends and mentors. Create digital products to provide your readers that genuinely help them achieve their homesteading goals. Once you write a book or create an online course, these products can be sold to and profit your readers without further maintenance from you. For the book writer, there are several networking forums (as on Facebook) for self-publishers that can help you on your way. To make the formatting of your print and e-book even simpler, I heartily recommend using Joel Friedlander's (www.bookdesigntemplates.com) Word templates. Someday a group of self-published authors are going to write songs about Joel, his templates are that helpful

FROM FIVERR TO FABULOUS

What is Fiverr (www.fiverr.com)? If you've got any interest in working online, listen up. In the words of one Fiverr vendor, Sarah Dalziel (www.sarahdalzielmedia.com),

"Fiverr is an online platform that connects sellers of micro-services to buyers of the same. For a buyer, every service is offered for $5, a "fiverr." From the seller's perspective, they offer services for $4, or multiples of the same, the other dollar goes to Fiverr. Fiverr takes a 20% instant commission from every sale. Fiverr is designed to help the buyer find good services, with keyword or category searches, buyer feedback, and seller ratings."

There are several online marketing services for various goods; Etsy is one we've already mentioned, and Fiverr is similar but with a slightly different angle. As Sarah says,

"If you have a skill to offer that can be translated to a digital service, you are ready to explore Fiverr as a potential income generator. Other gigs on Fiverr include web design services, transcription services, video or written testimonials, reviews of books or products, photo editing, book cover creation, personalized video recordings, translations services, video voice overs, handmade jewelry, original mini-art, promotional products, and more."

This kind of platform works itself very well into the business end of the homestead. Analyze what skills you have, especially those in your off-site, professional life. Are any of those skills marketable in a forum like Fiverr?

ACTION ITEMS: Read through the lists above again, this time with your homestead journal in hand. Mark down the ideas you hate first—you're just not going to do them, period. Now go back through and scribble notes on the ones that seem feasible or, at least, interesting. Put each idea in the center of its own piece of paper and start brainstorming absolutely everything associated with that idea, including known legalities or questions about it. You'll start to see connections between the ideas coming out of your head—this is called a mind map.

After a few days, come back and start organizing your thoughts. You need those few days for your brain to mull over what you've just come up with. Take that time to talk to your spouse or partner or friend or stranger on the bus who'll listen. Spend some personal time meditating or praying over your ideas and slowly, they'll start to shape up into something that could be awesome. Keep that list of ideas you decided you aren't going to do and refer to it every now and then to determine if your sentiments have changed. Sometimes the things we think of as our weaknesses end up being our greatest strengths in the long run.

DIY: Unless you're an artist or a graphics designer, you may need to get some outside help to complete this project. If you've reached the stage of wanting to sell homestead items, you're going to need a logo. After you come up with a name for your venture, it's time to brand yourself and your products so that people will recognize you immediately by your logo. Marketing is so huge when it comes to retail success that its importance can't be over-emphasized—you must have a picture to go with your name so that people know who you are instantly!

Sketch out a few ideas in your homestead journal—what comes to mind when you think of your homestead? Are there landmarks or organic symbols built into the landscape of your homestead that your community may already recognize? If art just isn't your thing, start with a tag line to go with your logo—some clever sentence that sums up what you're all about and what your homestead represents. Even a few keywords and phrases that come to your mind will help an artist and/or a copy writer (ad writer) help you craft a logo and marketing information for your homestead business. Just get some of your ideas out of your head and onto paper so that you'll have an easier time articulating your vision.

BUILD COMMUNITY: If the local farmer's market is on the horizon for you, start cultivating good relationships now with whatever governing body exists for that market. Each market, in each city, is different. There will be rules about doing business at the market, rules about what and where you can sell while there. There will also be municipal, state and federal laws of which you'll need to be aware when doing business at the farmer's market, but there will be people there to help you. You're not the first newbie grower to show up at a market, and there are systems in place to help you out.

Farmer's market or not, if you haven't already set up a website and social media pages for your homestead, it's time to start building online communities around those entities. There are free platforms, as well as paid ones, for websites and blogs. Signing up for Facebook, Twitter and Pinterest, as well as a slew of other social media, is free. Start posting and gathering interested followers. Just be true to yourself and write what you know, sharing other people's good information and tips. Provide a way for people to contact you for ordering or with questions. Make it easy for customers to buy from you. You don't have to sell online to find value in an online presence. If you're not on the Internet, most people won't be able to find you easily. That makes customers, and potential customers, very grumpy.

BOOKS TO READ: Several books were mentioned above where pertinent, but the following are a few more that might be helpful. *Sustainable Market Farming*, by Pam Dawling, can help you get a sense of what growing a market garden is like, with an emphasis on varieties to grow, tools you'll need and how to keep your garden growing through various seasons. *Grit to Great*, by Thaler and Koval, is a self-help book that applies to business minded homesteaders because of it's no nonsense approach.

The Market Gardener, by Fortier and Bilodeau, *Homemade For Sale*, by Lisa Kivirist, and even *Mini Farming*, by Brett Markham may all have value for you as you learn about growing produce to sell.

To learn more about blogging, including how to make a profit from it, read *How to Blog for Profit Without Selling Your Soul*, by Ruth Soukup. To get started with Etsy, try *Starting an Etsy Business for Dummies*, by Gatski and Shoup. Etsy itself has a large and useful help section which can be found on our links page.

Ox Cart Man, by Barbara Cooney; learn it, live it, love it.

Gaining Ground, by Forrest Pritchard, is both inspirational and instructive. *Gaining Ground* is really the story of one man's journey to restore his family farm through common sense, and sustainable methods. It's also extremely eye-opening about the realities of farming, and, by association, homesteading. Though Forrest is a farmer, every homesteader can relate to his struggles and triumphs. I simply love this book. (Please note, there's a bit of salty language every now and then.)

WEBSITES: Again, several sites were mentioned above but the following are a few more that can help as you learn how to speak computer-eaze. Get an education on marketing with social media, visit Social Media Examiner (www.socialmediaexaminer.com). To learn more about the two most reliable free blog platforms (though there are others), please visit their main sites at WordPress (www.wordpress.com) and Blogger (www.blogspot.com). Remember, I suggest you bite the bullet and pay for a self-hosted site at Wordpress if you are concerned about owning your content and/or if you ever intend to make money from your site.

To help you get an idea of what you can do with marketing online:

Our site, Homestead Lady, has a shop button at the top of the home page that will show you all our publications, both print and digital, as well as online courses as they become available. It's simply to navigate and clearly shows viewers what their buying options are. Teri of Homestead Honey (www.homestead-honey.com) has a direct link on her homepage to her books and courses, as well as a link to her husband's woodcrafts shop on Etsy. Kelly of Simple Life Mom (www.simplelifemom.com) does the same.

You can also check out Cobble Hill Farm Apothecary (www.cobblehillfarmapothecary.com) which features a market page, with a link to her blog, which is opposite of Homestead Honey and Simple Life Mom. Both styles are effective.

Janet of Timber Creek Farm (www.timbercreekfarm.com/shop-2) has a homemade shop on her site that takes people to homespun yarns and even farm-inspired post cards.

Both Reformation Acres (www.reformationacres.com/shop) and Tenth Acre Farm (www.tenthacrefarm.com/shop/) have shops on their blog that include products from Amazon with which they are affiliates. Establishing affiliate marketing relationships with vendors like Amazon allows readers to find great products without changing the purchase price for them while allowing you to make a small commission. Amy, of Tenth Acre Farm, provides a personal endorsement for each item in her shop so that the reader knows how she uses it on her homestead. Quinn, of Reformation Acres, includes her own homemade items right alongside her affiliate items, integrating both into her shop beautifully. This takes work and a great deal of computer coding, but both are a good example of why a website is important should you decide to sell items from your homestead. Incidentally, there are professionals who can help with the technical end for very reasonable fees.

If you're interested in running a Bed and Breakfast Inn from your homestead, you can go visit Woodland Farmhouse Inn's website, at www.woodlandfarmhouseinn.com, to get an idea of where to begin. Notice the clear listing of the rooms offered, the About Us page that features some history of the homestead, an easy way for visitors to check room availability and their state's B&B seal of approval. I've stayed at this homestead B&B and appreciated the atmosphere, gardens, history and charm. Not to mention the goats begging treats and the pigs escaping. To advertise your B&B, you may want to check out Bed and Breakfast (www.bedandbreakfast.com).

HOMESTEAD LADY SPEAKS:
Please note that you aren't obligated to ever sell or trade a single thing from your homestead. It's not a level of achievement that you must reach to have "made it." Many, many homesteaders are quite busy and content with simply providing for their own needs and the needs of their family. Some homesteaders prefer not to mess with the headache of figuring out cottage laws and city codes and would rather just use their handmade creations as Christmas gifts and birthday treats. That is a lovely, quiet way to live and you shouldn't feel compelled to do anything differently, if you don't want to.

Your homestead, your choice.

CHAPTER 6

Family Times on the Homestead

There are some certain rules of health;
Take them, they're better far than wealth:
Don't overeat, don't overdrink,
Don't overwork, don't overthink,
Be not afraid of honest sweat;
Run like a deer from shame and debt.

> Beware of bigness of the head.
> Get bigness of the soul instead.
> - From *The Rural Cookbook* published in 1907

What is home but a unique place to learn these truths of health, work and "bigness of soul?" Compared to all other classrooms one could find, there truly is no place like home. Family is a complicated and sometimes messy thing and, although our times are unique in many ways, we are connected to our ancestors and all of history by the struggle to build a happy, nurturing family. Though families are all over the map, and define themselves in different ways, I hope we can agree at least that creating a community of love and support within the walls of our own home is the ideal to which we strive, even if it seems elusive sometimes.

For a homesteader, the heart of our endeavors is the home. We loved the idea so much we even put it into our identifying label, *home*steader. While acknowledging the heartache of those who struggle with broken hearts and homes, the focus of the following discussions is to build up the family and put it in its rightful place as the focus of our homesteading lifestyle. After all, for what do we labor if not our family, wherever and whomever they are?

I was raised by a single mother, but I enjoyed the love and support of many grandparents, aunts and every imaginable assorted relative and friend throughout my childhood. Our church community also formed a kind of family. I was loved and supported and given good examples of hard work, resourcefulness and love. I remember that every home of my childhood—and there were many as we moved around—had some sort of garden. My mother was always exhausted, but there was invariably zucchini growing somewhere in the yard and homemade bread cooling in the kitchen. I get weary of Mother's Day exultations that paint mom as some sort of saint; my mother was not a saint. She was a good example of endurance and industry, and she always communicated her love for me. For what more could a kid ask?

You may not be the perfect parent—or sibling or child—but if we stop thinking of perfection in family life as somehow living it without making mistakes, we'll be far more satisfied. The aim of family life is to become whole and complete people, both individually and as a family unit; perfection is a long way off for all of us. We must stop wasting our precious energy pursuing idyllic pictures, and spend more time creating lasting masterpieces. Your homestead will most likely never function entirely like you hope and your family will probably never quite look like it just stepped out of the pages of *Parenting* magazine. However, if you choose wisely how you spend your time and with whom you spend it, you may just discover that you live in a house full of your best friends. Even if this is not your family experience at the moment, let's figure out how to build on and improve what we have at home.

Here's what we'll be discussing in this chapter on family times and the modern homesteader:

HOMESTARTER level: Dinner together can seem like an impossible ideal if you're working outside the home and have a teenager or two in your house. It's tempting to think that the dinner hour is somehow a negotiable point in family life; like it's a small thing that won't make much of a difference if you skip it. Yet, oh, how powerful meals together can be, especially when we've labored to produce our food! Here we'll share a few things to consider before you eat dinner tonight.

HOMESTEADISH level: To everything there is a season and a tradition! Don't let the magic of each month of the year pass you by without considering some wholesome and edifying traditions that can be adopted from your own family history or even from other cultures. Make it a habit to formally observe at least a few special holidays each year. Doing so will provide numerous ways to bring your family closer, making the work you do on the homestead throughout the rest of the days that much more meaningful. Knit your hearts together with happy traditions that make the cycles of home-life stable, predictable, and joyful.

HOMESTEADAHOLIC level: There are so many jobs to be done, especially for the homesteading family. Our many projects outside and on the land require our time and attention, and sometimes the inside of our home can turn into chaos—especially during spring planting and fall harvests. Here we will share some simple ideas to make chores go a little smoother—and, no, it has nothing to do with charts. In fact, I encourage you to ditch the chore charts!

HOMESTEADED level: Because of the maintenance needs of a homesteaded lifestyle, it's necessary to involve everyone in the work, from kiddo to grandparent, or whoever else may be kicking around the homestead. We'll cover five specific areas of homestead labor, and meaningful ways to engage all age groups in the tasks at hand. There is something for everyone to do, and helping all ages contribute makes them an asset, instead of a liability.

As with every chapter in this book, be sure to follow along at the end of each section as you reach the Action Items, DIY Projects, ways to Build Community, resource recommendations and parting advice.

"The sun looks down on nothing half so good
as a household laughing together over a meal."

- C.S. Lewis, Author and Professor

For the Homestarter: Dinner Together

When we talk about eating dinner together, please know that it doesn't have to be dinner that we're eating, and we don't have to all be together. Some families have grown-ups who work night shifts and are only available for lunch or breakfast. Some families are missing members because of long-term military deployment or other work that consistently takes them away from home. Regardless of the whys and who's, let's not enter this discussion on the importance of preparing and eating meals together by assuming that the dialog isn't relevant because we *can't* eat "dinner as a family." The word "can't" is not one that a homesteader really has the luxury of using. We can do just about anything; we are that awesome.

A Time to Teach

Dinner is the meal we spend the most time with at our house. All our meals are homemade—from scratch and all that jazz—but dinner is the one we really think about. I am blessed with several children who enjoy cooking and, as I write this, cooking is a part of our home school curriculum this year. I spend a good chunk of most mornings giving instructions and helping with technique for breakfast preparation, but dinner is where we get together and really teach the principles of nutrition and healthy food preparation.

Ever heard of "Kid-of-the-Day?" To learn more, skip ahead a bit and read the section in this chapter on Chore Systems. Dinner preparation is an area where having a Kid-of-the-Day rotation really comes in handy. When the dinner preparation is part of our homeschool for the day, then everyone is present and learning. If I'm making dinner on my own while the kids are busy with their own thing, I can still count on having my Kid-of-the-Day kitchen helper. I love the Kid-of-the-Day designation because it comes with privileges and responsibilities; in the case of cooking with mom, the Kid-of-the-Day enjoys both. Sure, they're the one chopping and stirring, but they're also taste-testing and giving me their opinion on what to make. So, whether everyone is with me in the kitchen, or it's just one kiddo, I know I'm passing on some of what I know, as well as getting some help preparing the meal.

While your children are all small this "help" is often more of a hindrance, so expect dinner preparations to take longer. Because of that, only do it as often as your energy allows; at the same time, don't be lazy about it. Cooking with very small children is an absolute chore, but it's also an absolutely necessary one if you want them to be comfortable in the kitchen later. It's hard, but you can do hard things. If the other parent can be present to deal with babies and broken glass (yes, it will happen), that would be enormously helpful.

Speaking of broken glass, one evening, in a span of three minutes I started a dinner omelet with my six-year-old, helped the ten-year-old sweep up the mason jar he'd just knocked over, fixed the handle on the chicken bucket for the twelve-year-old, cut up avocado for the starving baby and tended to the eight-year-old who'd just slammed her finger in the door. Just three minutes. Three. I know it was three minutes exactly, because I was literally counting down until my husband got home from work.

Sometimes I turn off the stove, give everyone a banana and read books until Daddy gets home. Because, of all of my many flaws, the biggest one appears to be that I'm only one person. I can't do it all and neither can you, but we can do something. And we should. Dinner is as good a place as any to begin. (If you're a single parent, you're a rock star for just cooking dinner at all, so keep the meal preparations very simple while your children are young. And use your older children to help mentor the younger ones. Think, "team effort" when you think about dinner.)

To make instruction easiest on everyone, especially if you have a large family and/or several younger children—apart from using the other parent—I like to give every child their own job. They're all still present and hearing me explain and teach, but they can concentrate on their own task. For example, the other night was a meatball lesson. My twelve-year-old chopped the onions, the ten-year-old ground the sourdough bread cubes into crumbs, the eight-year-old added the spices to the meat, while the six-year-old practiced her egg cracking skills and then stirred everything together. The older two oversaw reading the recipe out loud and making sure everything was in before we started forming the meatballs, which everyone did because it was squishy and fun.

All I did was get down the blender (for making bread crumbs) and get out the glass casserole dish (for baking the meatballs). We all forgot to preheat the oven (what can I say, they're my kids and I always forget), so we stood around talking about why we used grass-fed ground beef and what the difference in flavor is between yellow onions and purple. By the time the preheated oven beeped at us and we put the meatballs in, we'd also discussed which varieties of onion and carrot we'll try planting this year. As the meatballs baked, we made the salad and heated up our home-canned marinara sauce, which lead to a discussion about why we preserve food.

See what I mean? It's not just cooking. Preparing dinner together is about sharing yourself with your family. You have ideals about health and nutrition; you've studied and pondered (or you will soon) and you've discovered principles about which you're passionate. You could sit everyone down and deliver a lecture (or a sermon, depending on how you want to look at it) about why everyone should eat healthy or do whatever it is you're currently contemplating. However, you might get farther if you just show them, with you in the kitchen. What is that Ms. Frizzle from the Magic Schoolbus® always says?

"Take chances, make mistakes, get messy!"

> "Once upon a time, food was a family affair. Not just fathers and mothers and children, but aunts, uncles, grandmothers, and great-grandmothers. Knowledge and skill were priceless commodities. They could not be bought or traded easily, only taught and learned over decades and lifetimes.
>
> "In those days, recreational eating was a rare event, and food served an important and solitary purpose: nutrition. There was no time or energy to be wasted preparing food that did not provide maximum nutrition with minimum side effects...
>
> "Children were raised with the same principle in mind. Every member of the family had to be 'worth their salt,' or worth the resources required to keep them alive. If you were a little girl, by the time you were ten years old you were making the family bread. Not only were you capable of this work, but you already had all the training and experience needed to handle the responsibility. These young girls grew up at the knees of their mothers, grandmothers, and aunts, benefiting from decades of living experience and centuries of wisdom preserved through generations.... They needed no research, no scientists, and no expensive technology to give them permission to proceed. They had the most reliable, fundamental form of research available then and today: they had centuries of unbiased, 'peer reviewed,' conclusive results."
>
> - Melissa Richardson, from *Beyond Basics with Natural Yeast*

Getting to the Table

In our modern times, eating dinner together has become the focus of study groups that analyze the positive effects it can have on a child's relationship with food, and even their vocabulary. Eating as a family also improves table manners and conversation. Go ahead and do an Internet search of the phrase "Family Dinner Together" and look at everything that comes up. Outside of a study, as families, we have our own experience that tells us that eating dinner together is rewarding and brings us closer together.

Or, maybe we don't have that experience. If the joyous dinner hour has NOT been part of your family life, here are a few ideas for getting to the table together:

1. Commit to eat dinner together as many days as you reasonably can. My suggestion is all seven days of the week, but just do your best. If there are activities that can reasonably be assigned to another time of day than the dinner hour, then reassign them. If you need to move dinner time around in the evening to accommodate a necessary activity, be willing to be flexible. Make it work as consistently as you can, and family members will see that it's important, eventually making it a habit to be home for dinner.
2. Turn off the TV, radio (although quiet music in the background can be nice), and all mobile devices during dinner. Even the grown-ups need to leave their phones in their rooms and be present at the table. Whatever it is, whoever is trying to reach you, you'll get back to it in an hour. We can't expect our children to do what we're not willing to do ourselves. If the dinner hour is important, prove it. Unplug.
3. Say a prayer over the food, expressing thanks and asking for the meal to be cleansed of unworthy elements and to provide nourishment to your bodies. If you're not a prayerful family, have everyone holler out at least one thing they're thankful for before you eat. Beginning the meal with positive energy, in whatever way is appropriate for your family, will set the tone and make everything about this time spent together more beneficial. Sitting down to a full meal is not a blessing of which every family in the world gets to partake. Perhaps there will be less waste and fewer complaints if we wire our brains for gratitude the moment we sit down to fill our bellies.
4. Converse. We've all seen the pictures of a group of friends sitting around a table and, instead of talking to each other, they're glued to their devices—they may even be texting each other! As a nation, we're slowly losing the ability to converse with each other face to face. If you've unplugged and then sit staring stupidly at each other the whole meal because you can't remember how to have a normal conversation, don't give up! Family conversations can be fun, and to help you as you begin there are several conversation-starter games for sale. The children's toy company Melissa and Doug has a game called the "Family Dinner Box of Questions," but you might want to try several to determine which your family enjoys most. You can also just fill a canning jar full of your own ideas scribbled out on slips of paper and place them inside for each family member to draw out and read. Let everyone participate and share ideas, right down to the toddler. The point is to pose questions, introduce topics of general interest and simply engage each other in conversation. You live together and share the genetic makeup of thousands of years of family history; you have way more in common than you think.

5. Tackle hard topics. Not all our dinner hour conversations will be light but growing closer as a family is what this time is for. Don't shy away from hard questions or problems. Sometimes you'll be struggling through a particularly dark or difficult time as a family, and your conversation will be limited to brief comments in between tears. Other times, a tricky moral or ethical question will be posed by a child struggling at school, or an issue between siblings will rear its head at the table. Stay seated, when it's appropriate, and work through the issue together. Some conversations should take place in private, between parent and a child, but other discussions are beneficial for all to hear. Family means that, even when we're flailing or failing, we're doing it together. If you're a praying family, pray together for the child who is struggling; there's nothing quite like hearing someone you love lift up your name in a prayer of supplication.
6. Have theme nights and celebrate every holiday with food. Make Tuesday taco night and provide sombreros for everyone. Have a dessert-for-dinner night every now and then. Make one meal centered on a specific vegetable, like tomatoes. Have a corn cob eating contest or a costume dinner night. Make every holiday a reason to celebrate with a meal, even if the holiday is completely obscure and one you've never celebrated before. You'll all learn something new and maybe even start a family tradition that will last forever. Make dinner an event as often as you can; don't wear yourself or your family out but keep the dinner hour fun.
7. If you need more advice, ideas or convincing on the importance of the dinner hour, just visit The Family Dinner Project online at www.thefamilydinnerproject.org.

IS IT DINNER OR AN EXPERIENCE?

Let's face it, some of us need a bit of retraining when it comes to how we think of meal preparation. There are some simple motivations for preparing our food at home and eating together. If we're used to eating out, then it will save us money and wellness if we bring it back home. If we use a good deal of packaged products when we do cook, then let's get tough and donate them to the food pantry in favor of fresh ingredients. (Nutritionally, it's sad to give the pantry packaged food, but they aren't allowed to take homemade products, and something is better than nothing.) Once we achieve one goal, we move on to the next, improving our skills and becoming more adept at creating nourishing recipes together.

If you need to make the transition to cooking and eating dinner at home, I suggest you begin with just four designated nights. If you have a Sabbath, then that might be an easy one to start with since nearly everyone should be at home anyway. Our family also has a Family Night, once a week. Do you have one, too? If not, put it on the calendar, because that would be another good one for a DINNER TOGETHER NIGHT! It may mean that someone must reschedule a meeting or leave soccer practice early, but those activities are not the point of your homesteading life. Home is at the heart of the word "homesteading" and what you're building involves much more than cattle stalls and garden beds. Sacrifice to have this time with your family. You won't regret it.

Besides, you'll probably live longer and enjoy a higher quality of life. Ever thought about the ingredients in most restaurant food? It's like a Foodie's Most Wanted list: GMO's (genetically modified organisms, including plants), table salt, corn syrup, CAFO meats (confined animal feeding operations) ...All of a sudden, I have indigestion just writing this. Packaged foods made at home are not much better than restaurant food. Be wary of terms like "natural" and "healthy"; advertisers are prohibited from printing outright lies on their packaging, but they aren't prevented from being very...creative in their descriptions. That's not to say that there aren't high quality, packaged food providers out there. I know because I still avail myself of them quite often. I don't make gluten-free pasta yet and am always grateful that I can buy a good brand. I do make sourdough products and tomato paste—two things I eventually learned to create through my efforts to improve the dinner hour.

Forget about comparing yourself to other people, including bloggers and your grandma. Begin from where you are and let the idea of cooking dinner at home with your family inspire you to make changes. You've probably been wanting to make them for a long time but just haven't gotten around to it. Even if only a minority of family members are willing to come together to prepare and eat a meal, that's a simply wonderful place to start.

ACTION ITEMS: Get out your homestead journal and pick at least four days a week that you can all commit to eating dinner at home. Remember, not everyone has to agree with the idea at first; over time, those days will become part of everyone's "normal," and, if you cook it, they will come. (As I said, I suggest a goal of dinner together every night but that might feel impossible for some of us right now.)

After you pick your days, choose at least four meals to prepare together. If you've never cooked much with your children before, the first thing I'd suggest is to get your patience on, along with your apron! Allow an extra hour of time and think ahead to assign everyone a job that you think they can accomplish reasonably well. The children (or spouse) may not do everything exactly the way you would do it but—and this may come as a shock to some of us more perfecting types—there is more than one way to slice a tomato. For your first forays, pick recipes that are simple to assemble: salads, pasta dishes, casseroles, etc. Don't forget the dessert!

DIY: After you've chosen four meals to make together, brainstorm the rest of your menus for the month. There are a lot of online resources for meal planning, including recipes and whole menu plans. Using your homestead journal, make notes on what kind of food your family already enjoys. If it's not as healthy as you'd like, go online and find similar recipes that have been "healthified" - search keywords and terms like real food, gluten free, organic, GAPS or Paleo. For example, if you'd like to find a healthier macaroni and cheese recipe, try searching real food mac and cheese or Paleo mac and cheese. You'll get variations on that recipe that might inspire you to try new ways of preparing old classics.

Study cookbooks as you would textbooks for this project, planning meals for an entire month. I must admit that I have a slight addiction problem when it comes to buying cookbooks—I simply love them. If you have more self-control than I do, use your local library and your mom's collection until you have a decent pile of books on hand. Go through and write down every recipe that looks interesting to you, with ingredients you know your family will enjoy. You'll have way more than you need for the month, most likely. I suggest you fill up your month with about one-half favorite standby recipes that have been "healthified up" a bit and the other half new recipes.

By adding in recipes that are new to all of you and introducing them slowly, coupled with family favorites, you'll be more likely to win over the taste buds of your family. The method you choose to present new, and especially healthier recipes, will have a lot to do with how your family reacts to them. It's best to feed your children healthy food when they're young, because changing over a teenager's palette is harder. If all you have left at home are older children, be sure to set a good example yourself. If you don't like green beans and broccoli, too bad; get over it yourself and eat your vegetables.

You've already scheduled four days a week that you'll be eating at home and you've chosen two meals. Now it's time to decide who will oversee making these meals. If you have younger children, you'll need to help guide eager young assistants to jobs they can actually accomplish. If you have older children or no children at home, then everyone can just pick their passion. Passion should be a big part of food preparation; food is emotional and sustaining and vital and comforting. After you've assigned tasks, make a pantry inventory and a grocery list, and try to keep nearly everything on that list as minimally processed as possible.

Here's a meatball example: If you're making meatballs already, try making a red sauce to go with them from garden or farm stand tomatoes and some herbs. If you need to purchase sauce, try to find an Organic, non-corn syrup-y source. You could also make your own bread crumbs from homemade bread; this will take planning and time. But you're planning right now, so you're half done. If you raise your own meat, make sure you have ground meat on hand. If you're purchasing meat, try to find a local, grass-fed source. If you can't, Organic will do. If doing all those things isn't possible for whatever reason, pick only one of those ideas. Just begin from where you are, and that includes the limit of your pocketbook and work day. Schedule your time, make a plan, but don't plan beyond your current abilities.

BUILD COMMUNITY: Host a dinner for some extended family or a group of friends—no, not just grown up friends. There should be kids in attendance to keep things real.

Sometimes, the best way to get me to clean my house and make a high-quality meal, complete with dessert, is to invite someone over, because the accountability goes way up. I should have the passion with every meal I create for my family, but I'm a from-scratch-cooking realist and the reality is, I'm just too pooped sometimes. So, let's make it an event!

Have everyone in your family oversee some part of the evening and absolutely everyone should help in the kitchen. Sit down and plan your meal, and maybe an activity. You can use inspirational books like *Handmade Gatherings: Recipes and Crafts for Seasonal Celebrations and Potluck Parties*, by Ashley English. Particularly pay attention to Ms. English's ideas about potluck events, because good guests always ask what they should bring, and, with a plan, you can be ready with an answer. Also pay attention to slowing down and appreciating the process of what you're doing. This special dinner does NOT have to be elaborate, just hands-on and together.

BOOKS TO READ: Check out three books from the library: one on cooking from scratch, one on whole foods and one on food as medicine. (For example: *The Elliot Homestead: From Scratch*, by Shaye Elliot; *Nourishing Traditions*, by Sally Fallon; *The Heal Your Gut Cookbook*, by Hilary Boynton.) Why these topics? Because, in thinking back to when I started on my foodie journey, I realized that these are the books I wish I'd read first. The topics covered in these books are foundational to a whole host of others that you will find as you move forward. Don't worry, I promise that you won't miss anything worthwhile because each principle is connected to another. Keep your whole foods journey slow and steady, especially with any dietary changes you make. Everyone's bodies need to acclimate to what you're trying to do.

For simple fun in the kitchen with a literary twist* try:

- *The Tasha Tudor Cookbook*, by Tasha Tudor – a must have at our house!
- *The Winnie the Pooh Cookbook*, by Virginia Ellison (illustrated by E.H. Shepard)
- *The Little House Cookbook*, by Barbara Walker but with excerpts and historical information from *The Little House on the Prairie* books by Laura Ingalls Wilder.
- For you Regency fans, how about *The Jane Austen Cookbook*, by Maggie Black?
- Ever read Jan Karon's beloved Mitford series (some of my favorite books on earth)? There's a cookbook for you admirers out there called Jan Karon's *Mitford Cookbook and Kitchen Reader*.

*You may need to adjust or substitute some ingredients in these books if they don't suit your dietary philosophy. No worries, easy to do and they're still a lot of fun to use in the kitchen.

By the way, you're going to start reading cookbooks. I felt like such an odd duck when I discovered I was collecting cookbooks from the library and reading them like I would a compelling novel. Then I met other foodies and realized, I wasn't the only weirdo.

WEBSITES: For just enjoying family time in the kitchen and on the homestead, I suggest that you try The Elliot Homestead (www.theelliothomestead.com) as well as Deep Roots at Home (www.deeprootsathome.com).

For practical advice from the garden and kitchen and a traditional approach to food as medicine:

- Brown Thumb Mama (www.brownthumbmama.com)
- Wellness Mama (www.wellnessmama.com)
- Learning and Yearning (www.learningandyearning.com)
- Butter For All (www.butterforall.com)

HOMESTEAD LADY SPEAKS:
I'm going to be honest here. I think the phrase "quality time, not quantity of time" does NOT apply to family life. Family is all about time. In my experience, you spend a whole lot of time with your family just getting through the day. There is so much work to do and only so many hours. They're happy hours full of purpose and meaning, but they're often just packed full and hurried. Even if it's only you and your cat, or whatever your family looks like, I'm willing to wager that your days are bursting.

Yet, like little miracles, slipped in here and there—and sometimes so quickly played out that, if you blink, you miss them—there are also sweet moments of bliss. I spend hours and hours reading to children, and finally, the baby repeats what I'm saying. I spend huge amounts of time planning, digging and cultivating the garden. Finally, someone picks a tomato and runs with pure glee up to the house to show me. A lot of work, and only precious seconds of simple payoff. The work itself is rewarding, but those flashes of easy joy are hard-earned.

My foremost advice is, be present and don't miss it. Be there. Be there even if other people want you elsewhere. Be there even if you could be doing something else. Learn to say no to some of the *good* so you can say yes to the *best*. Your time is valuable, but your family's is priceless. You may not think you have time to cook dinner at home but, please believe me when I tell you, you don't have time *not* to. Children and grandchildren do nothing but grow older—so very fast.

"While raising our children, we establish traditions within our home and we build patterns of communication and behavior within our family relationships. ...Lessons taught through the traditions we establish in our homes, though small and simple, are increasingly important in today's world."

— Steven R. Bangerter, Religious Leader and Dad

For the Homesteadish: Traditions

Something that comes with living close to the land is a natural tendency to fall into the rhythms of the seasons. It's a gift the land gives back to you in return for your love and labor. Once you start to notice these rhythms, creating family cycles around them becomes a simple matter. For example, by September the fall harvest has begun in earnest—in some warmer regions it's been coming in like gangbusters, and you're already drowning in fruits and veggies. All this abundance means you have the best ingredients for nourishing recipes, as well as food waiting to be preserved to serve future needs.

No doubt you're busy, each family member with some task or another to perform—usefulness is in the air in autumn. Even if your homestead is one tomato plant and a rabbit, as the evenings cool and wood smoke starts to scent the air, your body begins to gather into itself, preparing for the hibernation of winter. It's natural to eat and seek warmth this time of year. It's normal to come in and roost together. It's comforting to be grateful for every blessing surrounding you. Our very souls are crying out for Harvest Home.

Celebrate a Harvest Home

Have you ever heard of Harvest Home? You certainly have if you're British, as this is a decidedly English holiday. But all cultures, across the globe, have harvest celebrations. Americans and Canadians call theirs Thanksgiving. The idea is the same and it's completely tied to this special season of the year. Sara Josepha Hale, often called the Godmother of Thanksgiving, described a New England day of gratitude as follows:

> "The table, covered with a damask cloth...was now intended for the whole household, every child having a seat on this occasion; and the more the better, it being considered an honor for a man to sit down to his Thanksgiving dinner surrounded by a large family. The provision is always sufficient for a multitude, every farmer in the country being, at this season of the year, plentifully supplied, and everyone proud of displaying his abundance and prosperity. ...[It] is considered as an appropriate tribute of gratitude to God to set apart one day of Thanksgiving in each year; and autumn is the time when the overflowing garners of America call for this expression of joyful gratitude"

At Harvest Home celebrations, it was traditional for the lord of the manor to host a great feast for all his tenants. Long tables were set up at the village green or in a great hall, and enormous amounts of foods were prepared. Fall is the traditional time to harvest animals for meat and lard, so roasts and other joints of meat were in abundance. So, too, were the grains of the season—breads and meads and sweets were prepared. Fruits and vegetables were also presented, in all their glory. This feast was not just for the wealthy, though; Harvest Home was a time of community. Of gathering. Just like the season itself.

WHY CELEBRATE A HARVEST HOME?

Why was this an important event in the year? Because at the heart of any community is the people. People of all stations would come together at Harvest Home to sympathize and to help; to gossip and to boast; to share and to learn. Winter rapidly approaches after the harvest and is such an insular time, especially if you live where the weather can be harsh. This fall season of thanksgiving is an important time of strengthening and nourishing needed even, and especially, in our modern times.

The fall harvest season of thanksgiving is an important time of strengthening and nourishing needed even, and especially, in our modern times. If you're looking to extend your Thanksgiving celebrations beyond one day in your calendar year, schedule a Harvest Home with your family.

It doesn't have to be an elaborate affair. Here are some simple steps for pulling off a successful Harvest Home that you can tailor to your needs and situation:

1. Invite friends and gardening neighbors for the evening, asking them to bring their extra produce to donate them to the swap table. The swap table is a place where everyone's extra produce can be traded or simply given away to all the Harvest Home attendees. If you have anything leftover at the end of the evening, donate it to your local food pantry by the next day.
2. Have everyone sign a party register with their family name and at least one thing they're grateful for this year. These are so fun to read afterwards! If your gathering

is of close friends and family, compile these into a list and email them to your guests by next week.
3. Using the recipe for Cattern Cakes at the end of this section, prepare a simple treat.
4. Gather the children together and crown a Queen of the Harvest - be sure to crown her in a wreath of flowers. Have the children take turns being the royalty every year. You can do this with boys, as well, if you think you can get them to hold still. Try a wooden toy sword instead of a flower crown for those daring lads who would prefer one.
5. Do a service project for the wild birds and make a peanut butter bird feeder from natural materials.
6. Provide the makings of a simple craft like Mason jar lanterns (see below). Have a parade around the yard afterwards.

TO MAKE MASON JAR LANTERNS

MATERIALS NEEDED:

Colored tissue paper

Mason jar in any size – we prefer half or full pints since small children have a hard time with the quart jars

Battery operated tea lights or small candles (with supervision)

Glue sticks

Scissors

Mason jar rings – you don't need the seals, just the ring to fit over the top

Yarn or sturdy string.

HOW TO ASSEMBLE:

Wash the mason jars and match each one with a ring ahead of time.

Put all craft materials on a table for the children to access freely.

Children can cut out shapes from the tissue paper and/or simply layer different colors of tissue paper around the outside of their jar.

It's easiest to run the glue stick over the jar and then attach the tissue paper. Putting the glue on the tissue paper usually leads to ripping.

The main thing is for the kids to have fun, so you can let them design freely. However, we have discovered that lighter colors work best – and only in one or two layers. If dark colors are used, it's best to have them included over the top as cutouts. For example, a light blue

> background around the Mason jar with a dark blue moon cut out and placed over one section. The quiet light of a tea candle has a difficult time making it through layers and layers of dark colored tissue paper. However, it's most important that the children enjoy this project.
>
> It is not necessary to add tissue paper to either the bottom or the top of the Mason jar lantern.
>
> **MAKING THE HANDLES:**
>
> We tried many handles over the years, but the quickest and sturdiest seem to be these yarn handles secured to the Mason jar lantern with the canning ring.
>
> Cut a length of yarn about 18 inches long and fold it in half.
>
> Tie a square knot at each end – nothing fancy, just make a knot at each end of the yarn. This knot prevents the yarn handle from slipping through the Mason jar ring.
>
> String the handle through the center of the canning ring so that each end of the yarn is sticking out from under the ring. The loop of yarn forming the handle should be through the center of the canning ring. Place a finger over each knot to secure it in place with one hand and twist the canning ring in place with the other hand. Tighten the ring well and pull up to use the yarn as a handle. The knots of yarn will be protruding slightly from just under the canning ring and that's how it's supposed to be to keep the handle secure. You did it correctly.

What a powerful time the harvest can be! But then, so is the spell of renewal in the spring. Easter, Passover and Equinox are rich with the symbols of spring and new life. You can observe the evidence of that renewal all over the homestead, and in your own body each year as winter gives way to the vernal awakening.

There are so many times throughout the year whose passing our cultures have chosen to mark with special feasts and festivals. Traditions abound.

Make the Time

Don't neglect creating righteous traditions in your family and on your homestead. They're pleasant things to do and lovely ways to connect with your family, but they are also one of the tools your family has to prepare against hard times. When sorrows come, jobs are lost, disaster strikes, or other maladies afflict your home, the rhythm of your traditions will provide a peaceful constant. I've had this lesson taught to me again and again. Recently, our family struggled with an emotionally and physically draining experience. As the mom, I'm in charge of making sure that things get done, whether I feel like it or not. But I just didn't have energy to come up with fun things to do to distract us from the stress of the situation. The children reminded me that, because of our family rituals and traditions, I didn't have to come up with new ideas. We already had traditions in place, established over the years of our life together. Sunday, we have a family night, Monday night is game night, and Friday night is homemade pizza and movie night. All the nights in between, we sit and read together as a family. Those traditions may sound simple, perhaps too simple to even bother with, but I can assure you that they have saved my sanity many times over the years.

On another note, I have heard farmers and homesteaders lament that, despite their best efforts, their children aren't interested in the self-sufficient lifestyle. Some children may never feel the magic of digging in the dirt or helping a goat to kid—and that's alright—they have a different vision. BUT, each child should be taught the rhythms of the earth on which they live and from whence they sprang. Those cycles must become relevant to them, and teaching that is on us, their parents and mentors.

The modern child is more connected to plastic than to dirt and, let's face it, the plastic isn't going anywhere any time soon. Instead of delivering a lecture about how their iPhone is poisoning their brain, why not tempt them with a pumpkin moonshine party? (FYI, "pumpkin moonshine" is another term for jack-o-lantern, not an invitation for underage drinking.) How about a Russian kulich bake-off or Three Kings Day caroling? Ever left gifts for the fairies at Midsummer or a popcorn garland for the wild birds at Christmas? How about a watermelon eating contest in the heat of summer? Food, music, stories, dancing, costumes, games—there are so many tools for building traditions at our disposal.

Don't waste a single one of them by being too tired or too busy to create traditions with your homestead family. I'm not going to lie, it can be exhausting for already overworked parents. But you can do hard things. Hard things, tiring things, things that make you wonder if it's worth it. That's parenting in a nutshell. You're growing a different breed of kid: the homestead kid. They're not like their non-homesteading peers - they live parallel lives. The key is not to teach them to fit in, but instead to stand out. The people outside your family circle will want the self-assuredness and competence your family members possess and will ask about your lifestyle. Homestead kids are multi-talented and multi-faceted. They know how to DO so many things, and that knowledge makes them more secure in an uncertain world.

Homesteading is the doctrine, but traditions are the "marketing." You wrap each principle in traditions and secure them with a bow of family sharing. In this way you teach and bind your family to the seasons, the cycles and the rhythm of the land. Sell it, Baby!

THE IMPORTANCE OF TRADITIONS

All throughout my childhood, my family was adamant about maintaining our traditions. The first fire of the season was lit on Halloween, even if that meant we turned on the air conditioner, too. (I'm from California so the air conditioner was on many a Halloween evening!) We read scriptures and a "fun" book every night, with Mom lying in the hall between our bedrooms as she read out loud. We camped every year at the same KOA Kampground in Watsonville, California and visited the freezing, cold Pacific Ocean rain or shine. Of course, like every other family who keeps them just for special occasions, we never ended up using the "good dishes" under any circumstances.

> I have a veritable army of memories that sweeps down my consciousness like Fiddler on the Roof's Tevyah belting out his "Tradition!" song at the top of his lungs. Childhood was a tricky thing to maneuver sometimes and knowing what came with the next month or the next season was always a source of comfort to me. Part of my religious culture includes a directive to seek after the best traditions and leave less illuminating ones behind. Even as a child I knew that, one day, I would want to give the gift of a steady stream of righteous traditions to my own children. I also knew that I'd need to include traditions from my husband's family and new ones from other cultures and peoples that were also uplifting. The question became, how do I sort through all the possibilities to find the best ones that would help me build my family?

ACTION ITEMS: Make a list of all the traditions from your childhood—all of them. Include camping trips, holiday activities, decorations, foods and each family reunion. For some of us, this list will go on and on. For others of us it will be short, if you can think of anything at all. No worries, this is more about what you're building in your current family than it is about just copying what your parents did. There may be a lot of what your parents did that you want to keep doing, though. Be sure to list your spouse's family traditions if that's applicable. If there are any traditions on this list that cause you too much stress or that are simply not useful to your family, I give you permission to cross them off without remorse.

Visit the library and get a few of the books mentioned in the resources section at the end of this segment and write down some celebrations worth exploring. Holidays are easy to begin with because they usually come with their own built-in menu and activities. You could also choose a life event that you really want to honor this year, like a birthday or the first day of school. Even the daily rituals of bathing and reading aloud together are worth celebrating, by designating a certain snuggly towel or a special cushion and blanket. Celebrating the everyday is a beautiful tradition to add to your family culture.

DIY: As you begin to formally observe various holidays, you create mementos and crafts to remember and celebrate them. Over time you will build up quite an inventory of decorations. A fun way to rotate those hand-made treasures, and to make sure that everyone's is seen and admired, is to make a holiday tree. Specifics for this useful fixture in your home can be found in detail in *Mrs.Sharp's Traditions: Reviving Victorian Family Celebrations of Comfort and Joy*, by Sarah Ban Breathnach.

In brief, you create an inverted conical frame, like a small, wire tomato cage. Your frame can be made of any sturdy wood or wire, even the aforementioned tomato cage. Around this, wrap vines, burlap garland or any other material you think will look pleasing to form your tree. Whatever wrapping you choose should allow for ornament hooks to hang so you can display some of the handmade decorations. Once the tree is fleshed out with your "vines," stand it in some prominent place.

We change out the decorations on our holiday tree pretty much every month, as new festivals and family times progress through the year. I have a lot of small children so, at any given time, our holiday tree features several salt dough ornaments, paper dolls, pipe cleaner shapes and watercolor masterpieces. If you have only older kids or adults in your home, your tree might be a little more chic. The point is that the holiday tree sits in its own special place from season to season. It reminds you that, no matter what else is going on in your life, there's going to be something to look forward to this month.

For more ideas on a variety of hands-on projects and learning activities, please see *The Do It Yourself Homestead Unit Study* addendum (www.homesteadlady.com/shop). This is a year-long unit study pack created specifically to go along with the book. If you're a homeschooler or just looking for some fun projects to work on with your children or grandchildren, this is the resource for you. See the back of the book for more details or visit www.homesteadlady.com.

BUILD COMMUNITY: You already know what I'm going to suggest you do, right? Pick a holiday that's new to you and celebrate it. The library and Internet are great resources, but so are the people you already know. Send out an email to everybody in your address book and list the cultural celebrations, festivals and religious holidays that you're interested in studying this year. Ask people for their help and advice and ask if they're willing to come over and talk to your family about what they know. You will discover some amazingly fun facts about the people around you, including their cultural and religious backgrounds. I'll wager that, not only will you find a mentor for your holiday/tradition project, but you'll also find a willing celebrant to join your festivities. Many a new friend is made by asking the simple question, "Can you help me with this?"

BOOKS TO READ: *Mrs. Sharp's Traditions: Reviving Victorian Family Celebrations of Comfort and Joy*, by Sarah Ban Breathnach, is so lovely! I received it as a wedding gift from my step-mother, Deborah, and it was so like her to give it. The book goes from month to month through the year and offers stories, crafts, party ideas, family time fun and so much more. It may not be suitable for everyone's personality, but there is a good deal here to choose from in the way of making your family life deliberately joyful and intentionally comforting.

Festivals, Family and Food, by Diana Carey is one I use all year round for planning feast days and fun activities. Similarly good are *All Year Round*, by Ann Druitt and *The Children's Year*, by Stephanie Cooper. All three books are appreciated by nature-loving, hands-on, crafty, whimsical Waldorf-type parents but they have appeal for everyone—especially those who aspire to be hands-on and whimsical. Let's face it, for some parents, we have to learn how to have fun; either "fun" was not how we were raised, or it just doesn't come naturally. I recommend you begin with *Festivals, Family and Food*.

Kids Around the World Celebrate: The Best Feasts and Festivals From Many Lands, by Lynda Jones, is a nicely formatted, easy to use book that provides simple crafts and lots of education.

We've adopted several customs from lands far outside our native one using this book. The best part about celebrating feast days from other countries is that you immediately realize that the human family is just that, family. We all love our children, enjoy a good dance, make memories with tasty food and can't wait to dress up and have fun. There's more that binds us than separates us.

We also humbly recommend our own book, *Homestead Holidays*, which is full of history, crafts, recipes and practical tips for seasonal celebrations on your homestead and with your family all around the year. Look for this book in 2019 – we can't wait to share it with you!

To connect your family to the seasons, try the booklet called *The Homespun Seasonal Living Workbook* (see our links page under The Homestead Family chapter). This will take you through twelve weeks of seasonal living with crafts, journaling and foods. Kathie (www.homespunseasonalliving.com) created this course and others like it to help modern people reconnect themselves with the earth, the seasons and the simple joys of home. Her blog is a true delight as well, and she's always creating new courses and books on similar topics.

Also, enormously helpful to my homeschooling is Rooted Childhood's monthly seasonal packets of stories, poems, crafts and recipes. These learning resources are beautifully created by Meghann with lovely graphics and high-quality materials. The information is technically geared to stewards of young children (think pre-school to elementary age), but my teens love the information and projects as much as my littles do. Honestly, I'd probably buy these packets just for me because I geek out over this seasonal celebration of cozy simple living stuff! (See our links page for how to get involved with her subscription packages.)

WEBSITES: For inspiring the simple, handmade life and rhythms of the family you might enjoy visiting:

- Soule Mama (www.soulemama.com)
- Imagine Childhood (www.blog.imaginechildhood.com)
- Strangers and Pilgrims (www.strangersandpilgrims.blogspot.com)
- Deep Roots at Home (www.deeprootsathome.com)
- Rooted Childhood (www.rootedchildhood.com), formerly Practically Hippie

HOMESTEAD LADY SPEAKS:

If you read the above books and put them down saying, "Yeah, that's not happening," don't dismiss the idea altogether. You and I probably just have different styles and personality types. You may not be inspired by what inspires me. Fortunately, crafty people are loquacious and there are, literally, thousands of craft and holiday titles floating around the library and book store. Also, there are religious and cultural holidays from countries around the world in every month of the year. You're bound to find something you'll enjoy.

Don't give up on the idea of starting a few traditions because glitter glue and ribbons aren't your thing. If being outdoors in a handmade shelter is more your style, try celebrating Sukkot with your Jewish friends because outdoor, handmade shelters are a big part of that festival. If it's in your nature to love a big, loud, colorful mess of fun, how about celebrating Holi with your Hindu friends and get, literally, covered in a rainbow of chalk and paint? What about honoring your ancestors with the Japanese celebration of Obon? If you're a garden nut, you'll probably appreciate the Yam Festival traditions of Ghana and Nigeria because some fun activities and yummy food accompany this festival!

I've celebrated all these holidays and many more with my family and friends. The experiences always reward my preparatory efforts with good memories, full tummies and a happier circle of loved ones. On the subject of full tummies and happier loved ones, here's the promised recipe for Cattern Cakes.

CATTERN CAKES

Try these simple pastries brought to us by our British Isles friends. Cattern Cakes were traditionally made on the lace-maker's holiday, St. Catherine's Day. This feast day happens to fall on my birthday, so I'm particularly fond of making Cattern Cakes. Something like a scone crossed with a biscuit, these are simple to make and well-suited to breakfast or dessert.

This recipe is adapted from one my favorite traditional recipe books, *The Festive Table*, by Jane Pettigrew.

INGREDIENTS

3/4 Cup butter, softened

1 Cup sugar - rapadura, succanat, raw

2 Cups flour

1 Cup almond meal

1 tsp. baking powder

1/2 tsp. allspice

1/2 tsp. nutmeg

1 egg, beaten

2/3 Cup currants or raisins - if you don't use currants, try to make your own raisins

INSTRUCTIONS

Preheat oven to 400°/200°C

Cream the butter and sugar together until fluffy in your stand mixer (or by hand).

In a separate bowl, whisk together flour, baking powder, spices and almond meal.

Add the dry mixture to the creamed butter and begin to mix slowly.

Add enough egg (usually one will do the trick) to form a stiff dough.

On a floured surface, roll out the dough to about 3/8 inch thick.

Sprinkle currants over the dough and roll up the dough, pinching the ends. If you've ever made cinnamon rolls, this part is similar.

Cut into slices about 1/2-inch-thick and lay them flat on a buttered baking pan.

Bake for 15 minutes until golden.

Let the tray of cakes cool for a few minutes before you move them too cooling racks. This allows the almond meal to set up and you'll get less breakage.

Butter and serve warm; though, they're equally good cold!

"Cleanliness is not next to godliness. It isn't even in the same neighborhood. No one has ever gotten a religious experience out of removing burned-on cheese from the grill of the toaster oven."

— Erma Bombeck, Author

For the Homesteadaholic: Chore Systems

Sometimes I get overwhelmed by the homestead activities that consume our days and I lose track of all the other work that needs to be done around here! I used to think I could organize and accomplish it if I just had the right chore chart. Well, I'm here to tell you that I've grown out of that notion and so have my kids. If you love your current chore system, then skip this section altogether. But if you're looking for something easy that works, read on.

Ditch the Chore Chart

A few years ago, after my fourth baby was born, I realized something important to my mothering, homemaking and homesteading life. I'm just not a chore chart type of person. Chore charts fill me with a sense of dread because they're a visual reminder that my life is chaotic, and nothing ever goes as planned. So, I went into my school room and tore every chore chart from the wall, determined to find a better way to do things. Having no plan and no organization was, of course, no better. With a home, homeschool and homestead to run, plus five children, there has to be some form of strategy! There's indoor work, outdoor work and school work that must be done every day. Every. Day. So, I prayerfully set out to find something better that would suit my personality and the way I run my home.

WHAT TO USE INSTEAD

Enter, Kid-of-the-Day. I can't claim to have thought up this idea. It was taught to me by my friend, Diann, who heads up something called the Family Centered Education Course (see the information in the resources at the end). There sat Diann teaching me how to inspire our children to own their responsibilities without being disorganized or, worse, harsh in our demands. When she mentioned her Kid-of-the-Day rotation plan for moving multiple children around to varied chores something in my brain clicked. This might just work for you as it has for me! If you have multiple children, plunk them in as you need, doubling up if you have more than seven children and/or more than six and take a Sabbath day of rest from assigned work.

Here's how it works: each day of the week has a child assigned to it and that child has both privileges and responsibilities. To decide which privileges and responsibilities, I sat down with my children and we talked about what they'd like to be responsible for when it was their turn to be Kid-of-the-Day. The following is what we came up with, but you should create your own list. Kid-of-the-Day will:

1. Pick their chores from the daily chore list first. Then, next kid of the day will pick, and so on.
2. Be the first to choose where they sit at the table and in the car. Once they have chosen, the next Kid-of-the-Day chooses and so on down the line until we run out of children.
3. Help with dinner preparation and setting the table.
4. Get two votes when we vote for family movie or activity on Family Night.
5. Be special helper to the youngest child for the day.
6. Get to pick which book we read first during family reading time.
7. Help things go right in the house and be a friend to all who need it that day.

This simple concept has saved our family from countless fights about where to sit in the car and who gets to go first with anything. There's nothing to fight about. Who is Kid-of-the-Day? That's all you have to ask. It became apparent, though, that this would also solve my chore dilemma, too.

I can't keep track of a chore chart, I can't remember to give cute stickers when something is completed, and I can't, for the life of me, keep up with the ever-evolving personalities of my children. I wouldn't want to wash the dishes every day for an entire year just because my mother decided it was my chore, so how can I expect my children to be happy with that arrangement either?

Varietal Chore List

Now, instead of a chore chart, we have a daily chore list that is new and spontaneous every day. It's penned with whatever writing implement I can find in the rush of morning breakfast preparations and is written on any handy scrap of paper. This often means that our chore list is penned with a broken crayon on the back of a grocery receipt. Whatever. I stopped being picky three kids ago.

Each day I write the things that need to be done only that day. I don't need my toilets scrubbed every day and I don't need laundry folded every day. We do those things once a week. However, I do need the kitchen floor swept and the kitchen scraps taken out to the chickens every day. Since the chores are different each morning, everything eventually gets clean. Each person gets to pick what they want to do from what's available that day so there's always variety. We use our Kid-of-the-Day order so that no one person is getting stuck with the less desirable chores every day. With my six-year-old, I still have to jump in and help with some things, but they're all accountable to me to do their best work. This arrangement has kept the chores more fun and easier to track.

A DAY IN THE LIFE

Just to give you an idea of what our time often looks like, here's how our days usually go during the week:

- **Sunday,** we take off entirely to observe our Sabbath except that everyone helps with animal care, food preparation/dishes as unassigned family service.
- **Monday** is a big cleanup day (also laundry day) and the kids have about five chores each inside and outside the house.
- **Tuesday** through **Friday** each person gets two to three chores, depending on if we must do some extra pick up for a school activity hosted at our house or if the livestock need something special during those days.
- **Saturday** is a family work day with Daddy home and our biggest outdoor homestead chores get done that day, so our house chores are usually limited to two tasks.

I have five children but only four of them are old enough to be included on the daily chore list. My toddler helps put away the silverware and feeds the cat, but we don't keep track of that on a chore list, nor do we hold her accountable yet if those things don't get done by her. My other children get extra chores if their daily expectations are not met. Today (I'm writing this on a Wednesday), here's what was on our chore list:

- Put away previously cleaned dishes
- Clean up dinner, wipe counters
- Take down one load of laundry from the line and fold
- Wash the kids' bathroom sink and tub
- Wash newly accumulated dishes
- Tidy and vacuum the family room for book club tomorrow
- Tidy and vacuum the basement for school group tomorrow
- Sweep the kitchen and dining room
- Take out scraps for the chickens and pigs
- Take out compost
- Clean the chicken waterer
- Help with the goats

If I feel we need a morale boost, I'll also add items like:

- Make monkey noises
- Tell a knock-knock joke
- Do a funky dance
- Use the word "spork" in a sentence

That kind of thing.

JUST TO CLARIFY:

Since I'm also my children's school teacher, I assign daily schoolwork that is to be done, as well as family learning and group activities with other homeschoolers. My children are involved with some sort of learning activity about eight hours every day. Onto this, we pile house and homestead chores as well as service, family time and, of course, sleeping and eating. Eight hours of work, eight hours of recreation, and eight hours of rest—a healthy formula that's negotiable for each person.

I don't pretend to have it all figured out; I guess the first fifty years of parenting are the hardest and I'm only twelve years in. However, I'm happy to use other people's wisdom and good ideas to manage the enormous task of keeping a home, homeschool and homestead functioning. Plus, anything that introduces variety into the several mundane tasks we must do every day is truly an asset.

If you love your chore chart and it works for you, keep it and rock on. If you find the chore chart confining, ditch it. Try a Kid-of-the-Day rotation, a simple list of daily tasks and let the children choose what kind of work they'll do. We can't choose to abstain from the work, so we may as well make it fun!

TIME AND REWARD – A LITTLE REMINDER

Each household is different and you're going to have your own priorities. Just remember, however, as you're writing your list, to put down only what needs to be done. This is especially important if your children are in public school, as your time with them during the evening hours is precious and occupied. The house and homestead must be organized and run efficiently, but not everything needs to be a chore. Spending time reading on the couch together is way more important than a sparkling-clean house. After de-junking my house and paring down to essentials, I'm far more comfortable with a "kinda" clean house than less time with my children.

Also, be sure to assign simple rewards to completed tasks. Our family doesn't reward with money very often, simply because we don't have much to spare. But there are plenty of other items and activities that a child can be trained to think of as a reward. Something as simple as earning free time to play will do as a treat, if that's how we teach them to think of it.

In Mildred Armstrong Kalish's very fine book, *Little Heathens: Hard Times and High Spirits on an Iowa Farm During the Great Depression*, she noted: "Without knowing it, the adults in our lives practiced a most productive kind of behavior modification. After our chores and household duties were done we were given 'permission' to read. In other words, our elders positioned reading as a privilege—a much sought-after prize, granted only to those good, hard workers who earned it. How clever of them."

ACTION ITEMS: In your homestead journal write down a list of five or ten basic chores that need to be done every day. Then make another list of chores that can be done once a week and once a month. Now, kick your in-laws, your nosy neighbors and any other hyper-critical person you know right out of your head—even if that person is you. Really look at your list and decide which of these chores are truly necessary enough to use up your and your children's precious time each day, week and month.

Once you're happy with that list, pinpoint those chores that still need you involved to teach your children how to perform them correctly. Honey, this can take years with some kids. Decide how you want to provide that education. I find that doing the chores together for a while, with me right next to my kids, is the best way to teach them. Especially when it comes to the details. For example, I always teach my sweeping/mopping child to do the corners first. Why? Because if you first pay attention to the minute and hidden areas of a room, the gunk in the middle is less likely to go unnoticed and it will all get clean eventually. When I'm present, I can show them exactly what I mean by "clean the corners first."

If appropriate for your family, create a Kid-of-the-Day rotation and apply it to your first chore list.

DIY: Take the list of necessary chores to your next family meeting and talk about starting a Kid-of-the-Day rotation—you can come up with a snappier name, if you like. Decide which responsibilities and which privileges Kid-of-the-Day will have as a family, with parents getting the final editing power. Type up your results on a nice piece of paper, laminate and display it somewhere prominent. I have a cheap-o laminator I bought years ago that still works just fine, and I use it all the time for stuff like this. You can change your list as you move forward, if you need to. Remember the list is a tool, not sacred text, and refining it is absolutely allowed and encouraged.

BUILD COMMUNITY: The most important community you'll ever build is your family community, and every successful family is built on love, forgiveness and work. Chores are an integral part of creating successful children who grow into hard working adults. As parents, too many of us expect either too little or too much from our children. Kids aren't indentured servants, and neither are they hapless morons; they're young and inexperienced, but very much able to perform as high as our realistic expectations. To build community here, analyze which kind of parent you are: the *do-too-mucher* or the *do-too-littler*.

About two kids back, I realized that I wasn't aging my first children in my head. I still saw those three as being little and unable to do very much. As my fourth child began to grow and our homesteading ventures expanded, I realized that I needed to snap out of it, organize myself, and start expecting my older and very capable children to shoulder more of the responsibility. This also gifted them the rewards of grown-up work and accomplishment around the home and land. I was often a *do-too-mucher* parent. I would delegate, but never enough, and then too often I would micro-manage their tasks. I didn't realize that, although they valued their free and play times, my children were right there waiting to be constructively taught how to work in their individual ways to become real assets in our home.

The *do-too-littler* parent is just as hindered, constantly expecting their children to do everything around the home and land without any time to be what they are: children. Education is the main "job" of childhood, not chores. Chores should be an important part of the education of every child, but they shouldn't be the child's only form of learning valuable lessons. I believe a lot of what they need to know they learn from good, old-fashioned playing—free time, during which they can use their imaginations and make their own decisions. No one likes to be told what to do every hour of every day or to be forced to do the "right thing" all the time. Rebellion follows a loss or misuse of free will, so don't be a *do-too-littler* parent who seeks to educate through force and unrealistic expectations but doesn't engage in much of the work yourself.

Plato said,

> "Do not train a child to learn by force or harshness; but direct them to it by what amuses their minds, so that you may be better able to discover with accuracy the peculiar bent of the genius of each."

Being a boss and being bossy are not the same thing, and it can be hard to find the fine line in between the two.

Locate where your weakness is concerning assigning meaningful chores and delegating responsibilities. Set concrete goals with your spouse, when applicable, to determine ways you can improve how you delegate, teach, and require ownership of the chores in your home. Your children will look back and thank you for teaching them to work hard, play hard, learn well and be responsible.

BOOKS TO READ: *The Life Changing Magic of Tidying Up*, by Marie Kondo is a book that has nothing to do with chore assignments or even children. This book is about de-cluttering and de-toxing your home of stuff. The accumulation of stuff has poisoned many of us—our stuff is everywhere and has been there for so long that we think of it as normal. Drowning in stuff is not *normal*. In fact, it prohibits us from living a meaningful, healthy life with balanced time for work, recreation and rest. Our stuff begins to own us until we're in debt paying for it, spending way too much time maintaining it and hardly ever able to use it. Read this book and kindly let it go. For more on this little book and/or de-cluttering, please see the *Downsize and De-Clutter* section in the Green the Homestead chapter.

Joy Berry's *"Help Me Be Good"* series is for children and their parents. They're small books with quality cartoon illustrations that beautifully explain the "why" of being polite, respectful, honest, hardworking and so many other values. We have all these books and I really appreciate how Ms. Berry makes it make sense to be a good kid. Or adult.

WEBSITES: The "Family Centered Education Course" (www.ensignpeakacademy.com/family-centered-education-program/) taught online by Diann Jeppson, as I mentioned above, is paradigm-shifting. I can't recommend this program highly enough! It is a LOT of work, and you will be tired by the time you're through, but you will gain a clear vision for your family and concrete goals to pursue together. Quality families don't happen by accident!!

HOMESTEAD LADY SPEAKS:
I hang out with my children more than I do with any other group of people; they really are the best friends I have. When they're young, having them work with me means that my tasks take longer. But that effort bears sweet rewards as they age. Apart from a few daily chores, most of our work is done together as a family. Take the chicken coop: I haven't cleaned that thing by myself in ten years because, from the time my oldest was two, she was out there with her child-sized broom and shovel helping Mama. Some people will be uncomfortable involving their children quite so much in the poo and the muck, and that's fine. However, I encourage you to find some meaningful task for your children to do. Children need to feel important and useful, just like you do. They need to have ownership in their homesteading

lifestyle and accountability to do their best work because learning to always do our best makes us happy.

The best way for our homestead kids to learn and value this lifestyle is for them to be at our side in everything we do; they don't have to love it all, they just need to know the basics of how to do it all. More than anything our children need to know that they're valued members of our family team. We needn't worry too much if the tasks aren't accomplished exactly the way we would accomplish them—the important thing is that our children share in these experiences with us. These are the times when we very naturally communicate our love and respect for nature. Our children will pick up on that and it will become an easy part of our family culture—something that will carry on into our grandchildren and great-grandchildren's lives.

"The most important work you and I will ever do
will be within the walls of our own homes."

- Harold B. Lee, Religious Leader and Humanitarian

For the Homesteaded: Family Work

There's an old Quaker proverb that I love:

Thee lift me and I'll lift thee and we'll ascend together.

What an incredible way to look at family life! Families are so different from house to house and situations on the homestead vary so broadly, but I think this little saying reflects the true goal of the homestead family. We struggle sometimes in our families and on our homesteads but, still, there is work to be done and family is the best way to get that work done.

Consider this section an expansion of the last one, which was far more basic in presentation. This segment presents lists of ideas covering five main topics. These lists are meant to give you inspiration, so take only what's of value to you. Some of you may be such veteran parents and homesteaders that you will have a long list of suggestions for me to include. (See the resources area at the end for a way to share your tips.)

To get our ducks in a row, let us start with a few definitions.

WORK

Work is that thing that must be accomplished on the homestead, whether indoors or on the land. Honestly, some days just getting out of bed is a chore, but we're going to limit our discussion to the work of the homestead. There are typically five areas of labor on the homestead in no particular order:

1. Animal Husbandry
2. Gardening
3. Homemaking
4. Cooking/Food Preservation
5. Provident Living

We could come up with more, I'm sure—especially as we break these categories down further. But, for the purposes of our discussion, these will do.

FAMILY

What is family? That answer can be complicated in our modern times. Simply and ideally put, family is a group of people whose hearts surround ours in loving protection and support. Adults with children under their care certainly qualify, but families don't need to be connected by biology to fulfill our homestead definition. Roommates, good friends and like-minded people can all form a family on a homestead if they're connected by common vision, service, hard work and affection. Most people find they come to love the land and the people they serve with their whole hearts. That's family to me.

Please know that I'm painfully aware that not all of us are homesteading in an ideal family situation. Some of us have family members or close friends who are constantly trying to dissuade us from even trying. Or, who are withholding their praise for our efforts or their labor to assist us. Sometimes these issues are symptoms of deeper demons in our interpersonal relationships but, regardless of their source, they make it even harder to do the work of homesteading with a hopeful and happy heart.

I don't have a magical solution for you except to encourage you to keep on keeping on. There's something therapeutic in all the work; a healing that comes when your hands are in the dirt. The service you render to your family by striving for self-sufficiency in their welfare will eventually be a balm to the wounds that might currently be painful. Keep your chin up. Things have a way of working out. If those seem like platitudes, just give it time—much of the bitterness will dissipate as you persist in your endeavors. Be sure to actively seek out like-minded homesteaders and make them part of your "family"—lean on them, work with them and learn from them. Don't feel alone because you're not. Hope is only lost when you stop looking for it. If nothing else, you've got one homestead cheerleader in your corner, and that's me. Go team!

When Life Happens

In large measure the size of our homesteads and the size of the family living in or near our home, along with their ages, will determine the amount of work to be done on our individual homestead. For example, if your homestead is your condo where you live with your husband

and your cat, both of whom are enjoying retirement, then your needs may be more moderate as far as getting the help you need to accomplish the work that there is to be done. It's perfectly possible that you, your husband and that freeloading cat can adequately take care of the patio garden, preserve a year's worth of food, see to your preparedness provisions, cook, clean and otherwise care for your homestead as the needs arise. As age advances, it will become necessary to employ the assistance of your son, daughter, nephew, church friend or neighbor to help with moving heavy containers in the garden and lifting full-brimmed canning pots in the kitchen.

Even though you can kill yourself on five acres the same way you can on fifty acres, typically the more land you have, the more there is to do. Add animals and a growing family onto that and you'll be plenty occupied. When children are still young, their need level is high and their contribution level to the homestead is conceptually low. If you're just starting out on your homestead journey, you can plan for the homestead to grow as your family ages into the work that needs to be done. If you're already living on an active homestead and just beginning your family, certain realities will need to be assessed head on. In other words, the quality care of the family comes first; which may leave the pigs unattended some years and the garden less than productive. A wise farm mother once advised me to "take care of everything with feelings first." Everything else is negotiable, as you have time.

It's in my nature to work—I'm extremely task-oriented. I get things done, and you can either get on board or get out of my way. The two entities that have taught me the most about knowing my limits, the poor wisdom of burning the candle at both ends and the folly of the notion that I am irreplaceable, have been my children and my homestead. Guess what? I've lost what amounts to years' worth of work on the various places we've homesteaded while I was pregnant or caring for a newborn. I was physically forced to slow down, not do certain things, and delegate for long periods of time. There were times when things simply went undone because I couldn't do them. Guess what else? We all survived just fine.

Each taxing pregnancy produced a wonderful blessing and asset to our home and homestead. Each infant that needed constant care, every hour I've spent educating in my home, every evening spent reading classic books, instead of turning the compost, have been precious moments that I wouldn't trade away, even if my younger self was sometimes chomping at the bit to jump back into the fray of the homestead work. And now, as they're aging, even with a toddler in tow, those children have been raised up to the homestead life. It's their normal and they don't question the need for the work, even if they still grumble about it, because we've involved them in every decision, included them in every planning session and treated them as equals to us in producing on the homestead.

I guess what I'm saying is that, if you keep your family first, everything else will work itself out for your good. I know that, when you have young children (and my oldest is only twelve), it feels like they'll always be young; like you'll never be done with the littles phase. It passes, and the new phases that come are just as challenging. However, if you've managed to share your passion with your little farm sprouts, they'll be primed to be full participants in your homesteading way of life. Even if they don't share your passion, at least they'll get where you're coming from and expect to have to do their part in running the home and homestead.

Please don't undervalue their emotional and spiritual support, either. Young children keep things fun, new and exciting in our everyday work with their endless supply of fascination. Teenagers and adult children can be such a physical help when it comes to all the manual labor to be done, but their unique perspectives and ideas can be a boon to our world-weary minds, too. In short, children, whatever their age, remind us just what exactly it is we're striving so hard to accomplish on the homestead. They are the point of all the work, after all.

Tips for Involving Others

So, let's take our homestead work list and match it to potential ways our family can help. Below, you'll notice suggestions for different ages of children, as well as grandparents and elderly friends. It may go without saying that very small children will require a great deal of supervision, but I've especially noted a few areas where I've experienced the need for direct oversight—as in, you're right there the whole time the task is being accomplished and can step in when needed.

Apart from that, you know your own family and the personality types of each person so try to honor each one individually and suit your supervision to their needs. (If you feel like you need to profile your family member's personalities better, I suggest you read It's Just My Nature, by Carol Tuttle. This book completely changed how I home educate and assign tasks on the homestead.)

INVITE

Don't hesitate to invite others outside your family circle to participate in the special work of the homestead. Never underestimate the power of quality relationships built with neighbors, like minded homesteaders, your congregation and even your Elk's Club. At our core, apart from a few jerks, something in us cries out to serve our fellow man. More often we need to give people the opportunity to feel helpful and useful by asking them to help us on our homestead when we truly need assistance. In our house we have a little catch phrase we sometimes use: "Helping makes us happy and smart!" Give your social circles the chance to be happy and smart as they help you on the homestead.

TAKE TIME TO TRAIN

This is a quick note on properly teaching each chore to children, friends, and visitors. It is imperative that you take the time to train each person in the tasks for which they'll be responsible. It's a fine line between allowing people to learn from their own mistakes and setting them up to fail. So be smart about your expectations and be willing to spend a lot of your personal time training others. Learning to delegate is an imperative skill to master as you mentor the people in your circle of homestead stewardship.

Don't expect to do all this well all the time, especially at first. Instead, expect to mess up quite a bit and eventually learn what you're doing gradually and all together. Plan to give it all you've got, but don't agonize if you experience a learning curve. Make a whole effort, not a flawless one.

DON'T FORGET THE FUN

Another idea to pay special attention to is that of recreation and fun. No one, not even the task oriented, are happy working all day, every day without a real break. Look at your week and schedule time to just be together as a family with no work. As was mentioned in the *Traditions* section of this chapter, even daily rituals like bathing and getting ready for bed can become special times for family.

Designate a family reading time. Find one night a week for pizza and a movie; you don't have to go outside your home or spend money, just be together. Even while working, you're going to want to schedule consistent breaks—the younger the child you're working with, the more breaks you'll need to take. Young children do well with patterns, for the most part, but their attention spans are short. Try twenty minutes of work and ten minutes of play for any child under the age of eight. This has the bonus of never allowing you to become bored, either, so enjoy it.

Five Areas of Family Work on the Homestead

These are in no particular order; everything is equally important. When I say "younger children" I mean any age from toddler to about eight years old. The label "older children" generally means age nine and up. However, this is completely relative to your family. I have a nine-year-old whom I could trust to run my entire house, including meal preparation, child care and cleaning. Yet I sometimes worry that the oldest will set the house on fire with one of her experiments. What's your family like?

1. ANIMAL HUSBANDRY

Oi, animals can be exhausting. One more thing to feed, manage waste from, and keep housed. Some days, after seeing to the temporal needs of five kids, I'm over it. Fortunately for the animals, I don't homestead here alone; my children love to be with the animals! Each animal has very specific needs so, understand that these ideas are meant to simply inspire you to come up with your own plans. When it comes to having my children work on the homestead, I'm guilty far too often of underestimating them; they really are so capable, and they want to help.

For young children who are too small to be around even medium sized livestock alone, you'll need to tote water buckets for them and place feed dishes so that they're accessible without the child entering the animal's pen. A goat head-butt can be a dangerous thing for a small person, so be smart about your expectations and set up animal housekeeping so that even the young children can successfully participate.

FOR YOUNGER CHILDREN:
- Check for eggs
- Daily observe broody hens and those with new chicks
- Put down new straw or other bedding in stalls
- Feed and water smaller stock like rabbits and vermicomposting worms
- Pass out kitchen scraps to smaller stock
- Love on animals, especially babies, to get them used to people (be sure to teach them how to properly handle each animal)
- Help harvest vegetables, fruits and herbs meant for livestock during the growing season; preserve any items for the winter
- Act as helpers during animal harvesting sessions, either to bring supplies or play with the babies so the grown-ups can get the messy work done

FOR OLDER CHILDREN:
- Move larger stock/movable housing from place to place
- Install or move fencing
- Help with birthing and doctoring
- Assist with harvesting animals for food
- Feed and water larger stock
- Assist with preparing shelters and equipment for winter weather
- Assist with milking, including cleaning equipment
- Check on baby animals daily and make sure their pens are predator proof
- Help maintain the bee hives during the year and assist in bringing in the honey harvest

FOR GRANDPARENTS AND ELDERLY FRIENDS:
- Observe the animals' moods, personalities and natures; be aware when things in the flocks or herds are peculiar or "off"
- Help with birthing and doctoring
- Play with young children during intensely busy times on the homestead, especially spring birthing and fall meat harvests
- Be the voice of reason when decisions are to be made about how many and what kind of animals are to be raised on the homestead

- Remind everyone to sit down together at the beginning of each year to make plans and calculations about meat animal harvests and birthing calendars
- Help with set up and clean-up of the honey harvest

MIND THE GAP

In her fine book, *Beyond Basics with Natural Yeast*, Melissa Richardson tells of an interaction she had with an elderly woman after one of Melissa's sourdough classes. The woman shared with Melissa her sadness at all that she had been unable to pass on to her progeny in the way of vintage wisdom and practical instruction. Melissa relates, "…then [she] grasped my arm gently, and with sadness said, 'I am amazed at what has been lost in only two generations.'"

Melissa goes on to note with regret that the elders of our society, who had been regarded as the guardians of all that vintage wisdom and practical instruction, were eventually replaced by search engines. She says, "And as they, their children, and their grandchildren died, so died all of the ancestral skills unvalued by rising generations."

In the Mind the Gap sidebar, (page 88) of this same book, Melissa discusses the great loss our culture has endured as families have moved away from traditional techniques of food preparation in favor of faster methods and cheaper products. She talks about the training that used to take place in the kitchen as systems and information were passed from one generation to the other. As she points out, cooking was a family endeavor not a hurried, solitary pursuit. Specifically, and in keeping with the topic of her book, Melissa encourages us to have patience with ourselves and our sourdough (or natural yeast) starters as we learn to use them, pointing out that those who use natural yeast are resurrecting a nearly extinct skill. She says:

"We are bridging an experience gap spanning entire generations. Like it or not, there will be a learning curve, but the distance is not impossible to bridge. Any gap can be closed with planning and hard work."

Keeping in mind all of Melissa's wise words, I say to the seasoned-citizen readers: my dear friends, your work is far from finished. Regardless of what your background is, how exhausted or inadequate you may feel, or how busy you think you are, it's time to shake the dust off your shoes and get to work mentoring the younger people around you. If your biological family members are unreceptive to what you have to share, find a different individual or group to mentor. Don't imagine that you don't know enough to be of value on the homestead. Even if all you have to share is a favorite family recipe, a tool that's been passed down or your simple memories from childhood, these things will serve as instruction and inspiration. Join in, speak up, and help out.

See if you can alleviate some of the hard work and planning of today's homestead learning curve by not just minding the generation gap but by bridging it. This work falls to you because you are capable.

Please help us.

2. GARDENING

A garden can be a very fluid concept, changing each year in composition, success and even location. There is ALWAYS something to be done in the garden so it's an easy place to make assignments for all ages and skill levels.

FOR YOUNGER CHILDREN:
- Help plant seeds with adult supervision
- Dig small holes for transplants
- Harvest extra weed plants and herbs for livestock
- Help build a scarecrow, or tie bird tape on berry bushes
- Harvest edible weeds and other small items with their own special basket or bucket
- Pull up weeds and lay on the soil surface for mulch
- With help from an older sibling, write up and place plant labels

FOR OLDER CHILDREN:
- Turn the compost pile and wet it down, if necessary
- Assist with starting plants from seed for the spring and fall garden
- Set up and maintain cold frames at the beginning and end of the season, even over winter
- Harvest green manures by chopping and dropping, composting or turning into the soil
- Stake and prune tomatoes, or any other vining plant you wish to grow vertically
- Harvest, clean, and sort seeds for saving
- Arrange bird netting over seedlings
- Prune spring fruit trees and fall perennials with your training
- During harvest season, one or two daily walk-thrus of the garden to harvest what's available
- Clean up and compost spent plants
- Clean garden equipment and store in a dry, designated space
- Sharpen and clean tools with appropriate supervision in the winter

FOR GRANDPARENTS AND ELDERLY FRIENDS:
- Help place manure and mulch
- Grade and sort the harvest as it comes in for immediate use and preservation
- Harvest rose hips, herbs, and other specialty items
- Help design and maintain plants for pollinators and other beneficial insects
- While instructing older children, service larger equipment and prepare for winter storage

- Along with the children, plan and maintain varieties for seed saving and oversee special pollination steps and harvest
- Maintain sprouts, microgreens and indoor plants

A FAMILY CULTURE OF WORK

As Chara (www.pantryparatus.com) says, you might be a homesteader if: "Your kids start sentences with phrases like: 'When I eviscerate a chicken I like to start with the crop'... or, 'I helped pressure can 40 quarts of broth last week.'"

I sometimes stop and ask myself why I'm so committed to this homesteading lifestyle--how did I get here?! I was raised in a city and have lived in one my whole life. While I've always gardened and kept bees, those are relatively tame hobbies (well, the bees are a little exotic, but in a relatively "city-chic" way). What made me take the leap and cross over to the other side with chickens? And then goats? And then ridiculously large gardens and acres?

I can't guess as to what made you take the plunge or what will eventually propel you off into the new adventure of homesteading, but for me, it was my children. This is an uncertain world and so many things that affect our lives are simply outside of our control. I vote, but I can't control the outcomes. I participate in my community, but I can't control how people think. I save my pennies, but I can't bring about worldwide market stability. I grow a garden, but I can't control the weather or the water. I stay active and useful, but I can't control what might happen to me because of others' actions while I drive down the road. Literally anything can happen.

Given that, I decided to trust in God and my own two hands, realizing that a lot of good in this world can be accomplished with old-fashioned work. It's not flashy, it's not sexy but it's something I'm good at—and something I intend to teach my children as well. I homestead because it's the best way I know to teach my children how to flourish in life, both spiritually and physically, through the value and purpose of work.

The difference between a hobby and a homestead, in my opinion, is motivation. If a garden is a hobby, then it can be taken up and put down as time allows. Too busy this spring? Ah well, no garden, no biggie. Homesteading, on the other hand, becomes a lifestyle. No garden? Disaster! How will we afford to eat our vegetables if we're not growing them? How will we ensure our meat is clean if we aren't raising it? How will we be able to stock our herbal medicines if we don't cultivate and harvest them?

Wherever you are on your homesteading journey (and that's a totally personal goal, not to be compared to other families), that place becomes sacred ground for you. You stay committed to the work because the work is meaningful. The work serves your family and gives you some security. The work also passes on much of your belief system, both as a person and as a parent. In short, the work becomes a definition of you. What you truly believe and how you really feel is simply evident for your children to see and absorb. (By the way, you don't have to have given birth to children for them to be "yours" and for you to be an important mentor in their lives.)

> So, whether you homestead to prepare because you're frugal or you're getting ready for Rapture or to survive a future zombie attack or because it's what you do to save the planet...just keep the faith! Hold to the work and it will pay off. Watch as your children and others you mentor accomplish with ease what probably took you decades to learn. The homesteading lifestyle will be their "new normal" and the next generation of homesteaders will be even better prepared for a life of meaningful work.

3. HOMEMAKING

Without the work that takes place inside the home, none of the work that needs to be done outside the home and on the land will ever get done. The humans must be maintained well, not just the animals and plants. During the height of spring planting and fall harvest many of our usual house chores are put on hold because there simply isn't time to sweep and dust every day. However, establishing quality rhythms of work and maintenance in the home will ensure that the homestead year runs smoothly. For a few ideas on chore systems, please see the previous section.

FOR YOUNGER CHILDREN:

- Make and display decorations for each holiday and special event
- Keep the family's books tidy on the shelves
- Keep personal space tidy and do assigned daily house chores
- Tell your parents and siblings you love them every day
- Do your school work daily and ask questions when you have them
- With your family, plan and produce surprises for neighbors and sick friends

FOR OLDER CHILDREN:

- Be home for dinner and help prepare it when it's your turn
- Make necessary batches of soap and personal hygiene supplies like toothpaste
- Keep personal space tidy and do assigned daily house chores
- Maintain personal finances well and provide for your own luxuries when possible
- Cheerfully do what needs to be done and express your gratitude to your parents
- See to your studies, remembering to ask for help when you need it
- Keep a journal
- Assign yourself one or more younger siblings to especially look out for and help when your parents need an extra set of hands

FOR GRANDPARENTS AND ELDERLY FRIENDS:

- Make large batches of laundry soap and cleaning products several times a year
- Help plan and pull off holidays and special events
- With the older children, plan and prepare the herbal supplements for the year

- Let them catch you with a book in your hand and a song on your lips
- Help plan family service projects for the year and see them carried out
- End the day with kind words, regardless of previous squabbles; share the lessons in patience and forgiveness the years have taught you
- Take your turn preparing meals and doing chores if you live on site
- Have monthly mentoring sessions with each child on a topic of common interest
- Tell your stories and write them down for your family to read later

ON THE SAME PAGE

You and your significant other need to make sure you're equally yoked when it comes to dreams and realities on the homestead. This lifestyle is a joint venture and you're going to need each other in ways that, perhaps, you never have before. Don't make assumptions about each other's needs, goals, expectations or limits. Have A LOT of conversations about your homesteading plans before you start the process, buy the farm or whatever it is you're going to do to take the plunge. Do your best to make good compromises and anticipate where there might be sticky situations moving forward. Keep conversing because attitudes can change over time.

As George Nash and Jane Waterman remind us in their book, *Homesteading in the 21st Century*,

"It may take only a single season, or it make take several years, but if you don't wear each other smooth, you'll eventually rub each other raw. The time to discuss your goals and dreams is before you make the move, not after, when you might discover, to your dismay, that your dream is your partner's nightmare."

4. COOKING AND FOOD PRESERVATION

Everything begins and ends in the kitchen. Food is needed to sustain life and the bulk of what we do on the homestead is to ensure that our family has constant access to healthy, nourishing food throughout the year. About 85% of my time is spent dealing with food at some point along the line of its production, use or preservation. This is absolutely an area where many hands make light work.

FOR YOUNGER CHILDREN:
- Wash and prepare produce for both cooking and preserving
- Stir pots and knead dough
- Be a dinner helper at least once a week
- Roll groats for oatmeal and learn to make granola
- Learn to make at least one breakfast on your own, such as scrambled eggs, oatmeal or smoothies

FOR OLDER CHILDREN:
- Make the jams, jellies, and juices for each fruit crop
- Keep a weekly sauerkraut or other vegetable ferment going
- Make weekly batches of yogurt
- Learn to make bone broth and consult the dinner menus to keep it stocked in the fridge and freezer
- Make dinner at least once a week
- Learn to make at least one favorite snack food a week, such as granola bars, crackers or chips
- Feed the sourdough starter, kefir and other ferments daily or as needed

FOR GRANDPARENTS AND ELDERLY FRIENDS:
- Keep the family in sourdough breads, flatbreads and muffins each week
- Plan and produce hard cheeses for the family for the year
- Help create a canning calendar for the year; be sure to plan around seasonal harvests, including meat
- Create a root cellar space for yearly cold storage of root and cole crops like cabbage
- Dehydrate onions and garlic as they're harvested and cured—don't forget the herbs

5. PROVIDENT LIVING

This is the area of all the leftover "stuff" of homesteading—the energy goals, the preparedness ideas, the waste management systems and the overall philosophy of self-sufficiency. You don't have to be religious to be grateful for abundance and you don't have to be poor to find ways to preserve and repurpose items. If you are of a spiritual bent, give thanks to the Providence that gifts you with skills and passion to make your homestead all it can be.

FOR YOUNGER CHILDREN:
- Turn off lights every time you leave a room without being asked
- Shut off the water in between brushing and rinsing your teeth; look for other ways to save water
- Learn fire safety willingly and remind your family to do their monthly fire drills
- Make fire-starters from toilet paper tubes and other recycled, flammable material once a month
- Have a homestead harvest party for your friends; be sure to include an activity where you teach them something about the homestead

FOR OLDER CHILDREN:
- Help build water barrels and manage them throughout the year

- With your family, organize a semi-annual candle making activity
- Learn to set and use a propane cook stove, rocket stove or Kelly Kettle-type low fuel stove
- Host a trading fair for your friends where everyone creates product with their own hands and gather together to trade with each other, instead of using money; see the DIY Homeschool Unit Study for more details on trading faires (www.homesteadlady.com/shop).

FOR GRANDPARENTS AND ELDERLY FRIENDS:
- Help plan and schedule preparedness drills for the family throughout the year
- Sew Wonder Ovens for the family and plan to use them a few times a week
- Help design and build an outdoor fire pit, keeping food preparation in mind
- Set up a canning or other homestead-related topic workshop for the young people in your neighborhood or congregation
- Research and practice methods of upcycling recyclables, especially plastic; how can these items be repurposed on the homestead?

Special Needs Homesteading

If you're homesteading with a special needs child, you're already aware that there are some specific considerations and amazing opportunities to be had when including your sweet kiddo in the work. Many parents of special needs children choose to home educate, and the homestead provides a learning-rich environment with so many opportunities for personal growth. The following are some insights and experiences from homesteading moms of special needs children.

Kathyrn is the voice behind the blog Farming My Backyard (www.farmingmybackyard.com), an urban homestead where she lives with her family that includes her eight-year-old daughter who is autistic and her three-year-old daughter who has sensory processing differences. She suggests involving special needs children in a homestead chore that:

> "…if it's not done perfectly, nothing will suffer. If they are responsible for feeding animals, there needs to be someone keeping an eye out to make sure that it happens at the right time. Other chores like planting, watering, or collecting eggs that have flexible timing or that can be split into small increments, are other good choices…My daughters really love collecting eggs and feeding the rabbits and the cats. I need to be right there helping them still, which is okay, because I am doing my own chores nearby. Some of the awesome things about homesteading is that you are outside and active…. It's beneficial to get that sunshine and movement each day. Another bonus is that there are lots of hands-on ways to spend time together, with or without a lot of talking.

> "Really look at what your child's strengths are and find ways that they can feel included and be a part of the work but still excel at the same time. They may not want to be involved in certain chores but, by including them in the things they are interested in,

you can really use homesteading to strengthen your relationship instead of just another list of chores and time commitments."

Tammy Trayer of Trayer Wilderness (www.trayerwilderness.com) and her thirteen-year-old son, Austin of Mountain Boy Journals (www.mountainboyjournals.com), have made their journey through autism and off-grid homesteading together with their husband and father. Tammy reports that homesteading, homeschooling and a healthier diet have given Austin the opportunity to grow, learn and thrive. Tammy incorporates her homeschooling into their homesteading and says that:

> "Austin is responsible for caring for our animals. We have meat rabbits, laying chickens and milk goats. He takes his jobs very seriously and his animals have played a role in helping him to come out of that autism shell. Last year we bred our milk goats for the first time and when two sets of twin goats were born in January, he was very proud because he cared for them.

> "When teaching a child or adult with autism, it is essential to have their undivided attention, and to give them only one task at a time. When you provide them with too much information at once, they will hear either the first thing you said and be heavily focused on it, missing everything else, or they will hear only the last thing you said. Writing their chores down on a piece of paper or, as we did, on the side of the chicken coop, will help your child remember all the steps involved in their chores. Typically, with autism comes auditory processing delays, so it is essential to learn how your child thinks and what their mental triggers are; you may need to put a spin on things to help them learn in unique ways. By doing this, you will help your child in overcoming some of their struggles while teaching them coping mechanisms for life. One example is the habit of carrying a notebook and pencil in their pocket to keep them from forgetting things.

> "Set chores that you feel your child can succeed in at first to build their confidence. Then, add harder chores as they age, or when you feel it's appropriate. Many people choose to coddle their special needs child—in some cases that may be necessary, so they do not hurt themselves. I chose not to do that but, instead, to push my son out of his comfort zone as often as I could in supervised situations. I taught him that the sky really is the limit, and nothing is impossible—all he has to do is try. Because of this mentality, I feel we have helped him overcome SO much. If I wouldn't have pushed him, he would not have tried and succeeded. When he failed, he experienced real life. I think that by not allowing a child to fail, we are setting them up for failure down the road. Giving a child a chance and showing them, you have faith in them goes a long way toward helping them succeed!"

As everyone settles into their roles and learns to provide the labor necessary to keep the homestead running, kinks will work themselves out. Each year the tasks will become habit even as they evolve and as children, special needs or otherwise, age. Circumstances change, the work of the homestead will constantly need review and revision. This is a healthy process and one that is of benefit to both the family and the homestead. There will be plenty of days where plans go awry, people are ill, the weather misbehaves, and distractions sidetrack our homestead goals but with a plan in place and the work divided up, we're far more likely to enjoy success.

I love this observation from George Nash and Jane Waterman in their book *Homesteading in the 21st Century*:

> "The goal of 'homesteading' is to live a more satisfying life, insofar as that can be defined as freeing yourself from the burden of useless possessions and obsessions, raising kids that grow up curious and hopeful, forging strong family ties, and gathering together around the dinner table to celebrate good food and good living."

MY KINGDOM FOR A FARMHAND

If you find you can't get the work done on your own, even with your family helping, considering hiring a farm hand. Homeschooled teenagers are usually a great fit for this work since, not only are they often self-motivated, they have more flexible schedules than their public schooled counterparts. For example, a farm hand can be quite convenient to have when you must go out of town but have a small dairy herd that needs to be milked. Or when you just have more manure to haul than you can deal with on your own.

If you need more steady help and would like to act as a mentor for future homesteaders and small farmers, you might investigate the organization Worldwide Opportunities on Organic Farms (visit The Homestead Family chapter of our links page for a link: www.homesteadlady.com/homestead-links-page). Their stated mission is to connect people who want to learn about homesteading/farming with those who are doing it. Typically, you trade room and board for their work. Though they're only obligated to work a minimal number of hours per day, some find this sufficient. Visit their website to learn more.

ACTION ITEMS: In your homestead journal, create your own list of family homestead work ideas. Make sure each person has a job, right down to the youngest. Aside from babies and the infirm, there's something that everyone can do, and the work is both enabling and ennobling.

DIY: Take five lengths of butcher paper and lay them out on the floor, weighing them down so they don't roll back up. Label each paper with one of the titles: Animal Husbandry, Gardening, Homemaking, Cooking/Food Preservation and Provident Living. First, have your family members brainstorm work that needs to be done in each category—you can use the information in this section as a guideline if you'd like.

After you have a good list going, assign each family member a colored pen and have them put their name next to each item that interests them or that they'd like to learn more about. You may want to specify that writing their name down doesn't mean they'll HAVE to do that job, just that they're interested in knowing more about it and might like to try it. Roll up your large paper lists and take them out again at several family meetings in a row until you're ready to start making some assignments.

If you're new to homesteading at this level, you're going to need time to do some research and try out systems that will need tweaking over time. If you've been homesteading for a good long time, you may have been stuck working the pigs for so long, you just don't enjoy it anymore and might like to do something else for a while.

Compile your initial notes into your homestead journal and make several tentative schedules, one for each family member under your stewardship. I don't advise that you tell your older family members what they need to be doing, but I would advise that you counsel with them and get their ideas and input. You're allowed to boss around your kids, though—with respect, of course.

BUILD COMMUNITY: I've said it before and I'll say it again, the most important community you'll ever build is your own family, in all its uniqueness and imperfection. Use the ideas above and make specific plans with yourself about how you can educate and encourage the family around you to help them accomplish the work they have before them. What will you do to lighten their burden or assist in their success? Try to anticipate weaknesses and potential problems for each family member and determine what you can do to bolster weak areas. You'll find a lot more purpose and pleasure in your own work if each family member is able to execute theirs effectively.

BOOKS TO READ: *Things My Grandmother Taught Me About Organized Living*, by Gloria Barry, is a fantastic little e-book. The book is short and a very simple read but it's packed full of some of the best common-sense advice I've read on homemaking and organizing. I'm going to read it with my kids as they age into more of the business of helping to run our home. I SO wish I'd had this book when I first got married and started having kids because it would have saved me a bunch of time and many headaches.

It's Just My Nature or *Child Whisperer*, by Carol Tuttle for learning what makes you and each of your family member's tick. This book goes beyond personalities and right down into the soul of every person you'll ever know, especially yourself. I'm a much better teacher, parent and person for reading these books.

Also wonderful for family homesteading inspiration is Teri Page's book, *Family Homesteading*.

WEBSITES:

For a mingling of family and homestead life, be sure to read:

- Reformation Acres (www.reformationacres.com)
- A Farmish Kind of Life (www.farmishkindoflife.com)
- Homesteading on Grace (www.homesteadingongrace.com)

- Flip Flop Barnyard (www.flipflopbarnyard.com)

For a bit of seasonal pleasure and learning as a family, check out the Waldorf-inspired, seasonal guides for families at Whole Family Rhythms (www.wholefamilyrhythms.com). These will take you through each season with suggested learning activities, crafts and explorations. This site is not about chores, more about what to do when you take a break from all the work—remember the fun!

HOMESTEAD LADY SPEAKS:

When I was growing up, every Saturday morning was cleaning time. With a single mom working full-time outside our home, busy school schedules and a Sunday Sabbath to keep holy, we just didn't have any other time but Saturday morning to clean the house and do the gardening. It was a lot of work, but the thing I remember most about those mornings is that there was always music. We had a record player (yes, I am that old) that would hold up to five records, each one descending to be played after the previous one had finished. Simon and Garfunkel, The Beatles, Jim Croce, Bing Crosby crooning out Christmas as soon as the weather turned even slightly chilly—their voices all blend into the sound track of my childhood. There was a pleasant rhythm to the work, an expectation of setting things in order, even if we'd rather be sleeping in and eating donuts.

We worked just long enough to get the tasks done and then my mom would call a halt so that we could throw some air-popped popcorn into paper bags and toss a few drinks into a cooler. Off we'd go to the drive-in movie theater, kick back in our little Toyota and relax to the sounds of a family movie coming in over the scratchy speaker hanging from our window. Sometimes the fun would be a breakfast picnic before the work began or a pizza night at Grandma's.

There's so much work to be done but life is very, very short. Learning to be happy as we accomplish our tasks is an important skill to develop. Learning to rest and take time for each other is equally as vital. Only together, as a family, can you find the balance that works for you. Find joy in that journey...as you sweat and bleed and wear out your lives together with work. No one promises that the homesteading life will be easy, just that it will be worth it.

CHAPTER 7

The Homestead Community

What does our community have to do with our homesteading ventures? Plenty, believe me. Consider this scenario: you've just lost an entire batch of home-canned broth because the seals on the lids didn't set for some reason, and you nearly burned your arm off removing the lid. You're ready to pitch your pressure canner against the wall and collapse in an exhausted, weeping heap. Then you remember that your canning-guru neighbor once told you what to do when the seals fail, so you give her a call. She assures you that she's done that herself dozens of times and you just have to move on. She relates the story of the time she added four cups of salt to her jam, instead of sugar, because her toddler was providing loads of distraction in the kitchen that day. By the end of the phone call, you're both laughing and you're ready to try canning broth again.

In this book, we talk a lot about forming community wherever we are so that we can support each other. All of us need to be prepared for the amount of work on the homestead, at whatever level we perform it. Sometimes we're taken by surprise that living the deliberate lifestyle of homesteading is a lot of work! We use each other as a resource while we learn, and that means venturing out into our communities, both to receive mentors and to eventually become mentors. The homestead itself is a community of plants, animals, people, and dirt that all work together toward a single goal of self-sufficiency. Learning to both contribute and receive from the homesteading community is a big key to becoming successful overall.

So, how do we participate in, form and contribute to the homesteading community? As with everything, we begin from where we are.

Here's what we'll be discussing in this chapter on communities for the modern homesteader:

HOMESTARTER level: Forming the homestead book club may turn out to be the one thing that keeps you motivated to dive into all those gardening, chicken-keeping, and preparedness books you have on hold at the library. We become accountable to our book group to at least try. Although they don't have to be fancy or labor intensive, book groups provide live discussion and an exchange of ideas. Which are two fantastic learning methods. Here we provide a sample itinerary for your group and some things to consider as you begin.

HOMESTEADISH level: Move out into your community as you take and teach classes. Yes, teach classes—don't freak out. We'll show you some simple things you can do here to be a better student and to pass on what you're learning to your homesteading and wannabe-homesteading friends. Being an active student, and eventually reaching out to teach, really are the best ways to learn!

HOMESTEADAHOLIC level: Attending a seed swap can be such a beneficial learning experience, even if you're not ready to learn to save your own seeds yet. If you've already got a big seed collection and want to acquire new and proven seeds for your area, a seed swap can be a gold mine. Hosting a seed swap doesn't have to be complicated; just follow the suggestions in this section.

HOMESTEADED level: Discovering how to share your personal experiences in meaningful ways is the goal of this section. A mentor is a true guide, continually pointing and sharing until the student can engage in the mentored activity on their own. Here we share some suggestions on how to become a homestead mentor. Because what we send out comes back to us again.

As with every chapter in this book, be sure to follow along at the end of each section as you reach the Action Items, DIY Projects, ways to Build Community, resource recommendations and parting advice.

"We are transformed—magically—into the literary society each time we pass
a book along, each time we ask a question about it, each time we say,
'If you liked that, I bet you'd like this.'
Whenever we are willing to be delighted and share our delight..."

— Mary Ann Shaffer, from *The Guernsey Literary and Potato Peel Pie Society*

For the Homestarter: Homestead Book Club

A homestead book club could be just the thing to motivate you, and others, to get through all those chicken titles tottering dangerously by your bed. And the ones stashed in the bathroom, and the car. Reading those books is only the beginning, too. After you and your friends have discussed the book on making cheese, you may just find yourself huddled around a group batch of mozzarella afterward. What was a simple book club just became a homestead mentoring group.

You don't have to run the whole shebang by yourself, either. Each person can contribute on the topic about which they're the most passionate and with which they have experience. Everyone doesn't have to know EVERYTHING, just a little bit about a lot of things. Plan to share the burden of planning, leading discussions, purchasing materials for activities and hosting in your home. Personally, commit to contributing in some way; you'll only get out what you put in from a group like this one.

Also, please consider that the homesteading community in any town is pretty close knit so if you commit to do something, do it, so as not to give yourself a poor reputation as someone who doesn't follow through. Life happens to everyone and no one should expect any member of your group to sacrifice exorbitant amounts of time and energy but thinking of it as your monthly community service will help you stay engaged at appropriate levels. You know yourself; volunteer to do what you will do and then do it.

These discussions and experiences will expose you to a wide range of homestead topics and you will find, as they all sift together, that there are things you really want to do. You'll also find the things you're not interested in doing at all. This is one of the most important things you can learn from others' experiences. Learning as part of a community will end up saving you time as your own homestead grows because, although you'll be making plenty of mistakes, you can make them as a team and avoid some of the problems other members have experienced. Just you, and your homestead book club, against the world.

Nuts and Bolts

The first thing you're going to need to start a book club is a list of books and some people. Below is a sample schedule of books for a whole year of book club meetings, if you'd like to use that. The people, I'll have to leave to you. The next item of importance is a place to meet. Obviously, if your group is small, you could simply use a member's house. This has the added benefit of a working kitchen, and even a yard, should you choose to have an activity after the discussion. You may also find that your local library or community garden center has space. There will be specific rules of use for each site, so be sure you're aware of them. You will need to select one or two contact people if you're using an outside facility, so be sure to determine who will serve as liaison for your group.

It can be handy to have a designated leader or president, but it's not necessary. Having even a touch of formality can help prevent hard feelings later; if everyone knows at the start who is in charge, who makes the phone calls, who handles finding speakers, who reserves the books at the library, and so on, then everyone can move forward with their assignments in clarity. Some groups prefer to keep it loosey-goosey, and that's perfectly fine. Just try not to step on each other's toes when there is organizational work to be done. The hope is that this group will last a long time as you all grow and mature as homesteaders together.

If you do have guest speakers and outside mentors come, be sure to send them a card of thanks with everyone's signature. If you take field trips together, plan to do the same with the people who host you. If you have a secretary designated, this is a great job for him to take over. Be sure to include, in your note, a query as to whether they have friends with similar interests who would like to come speak to your group. Some books clubs are just about the book and some good talk, so don't feel like you must bring in guest speakers or take field trips at all. Asking for everyone's input and even taking a vote on ideas like these will ensure that everyone's opinions are heard.

Sample Schedule

I encourage you to discuss a reading schedule with your group and create plan for the year. Remember to check with your local library to be sure they have the titles you select. Or, locate a source for discount purchasing of books for your group. Always remember the value of used books in good repair!

The following is just a sample schedule to inspire you. You can download a nicely formatted version of this schedule by visiting our links page under The Homestead Community chapter.

JANUARY

Book: *Week by Week Vegetable Gardener's Handbook* by Ron and Jennifer Kujawski

Activity: Download, print and begin to pencil in your garden plan for the year. Garden planners are available online for free, or you can purchase a very thorough one from Schneider Peeps (go to our links page at www.homesteadlady.com/homestead-links-page). In your group, look at each other's plans and decide who's going to grow what from seed and plan to share around baby plant starts as the season progresses.

FEBRUARY

Book: *Natural Bee Keeping*, by Ross Conrad

Activity: Invite a bee keeper to give a brief demo of equipment and how to get started—make plans to order bees and get equipment before Valentine's Day (many bee breeders sell out by then).

MARCH

Book: *The Backyard Homestead*, by Carleen Madigan

Activity: Make a list of five new things you each want to do on your homestead this year. Have a guest speaker come lecture on foraging or making maple syrup. If weather permits for doing either activity in person, go on a field trip.

APRIL

Book: *The Homemade Pantry*, by Alana Chernila

Activity: Make graham crackers and marshmallows from scratch together and improvise some S'mores for a treat!

MAY

Book: Rosemary Gladstar's *Medicinal Herbs: A Beginner's Guide: 33 Healing Herbs to Know, Grow and Use*, by Rosemary Gladstar

Activity: Have a guest speaker walk you through making a tincture; also, prepare a simple lip balm together.

JUNE

No book club; group service project*

JULY

Book: *Independance Days: A Guide to Sustainable Food Storage and Production*, by Sharon Astyk

Activity: Can jam or pickles together. Make a canning schedule for those who wish to get together during the rest of the season to can/preserve food as a group. Plan to share produce, equipment and ingredients.

AUGUST

Book: *Mrs. Sharp's Traditions: Reviving Victorian Family Celebrations of Comfort and Joy*, by Sarah Ban Breathnach.

Activity: Go to group prepared to share a story from your family (or other inspirational group) about the value of wholesome traditions. Brainstorm ideas for your family from this book; help inspire each other with personal stories. Make a "holiday tree" for each family (see instructions in the book).

SEPTEMBER

Book: *Butchering*, by Adam Daforth

Activity: Have an experienced homesteader walk you through the humane harvest of a batch of meat chickens. With your group, help them with their processing to learn every detail.

OCTOBER

Book: *Vegetable Gardening the Colonial Williamsburg Way: 18th Century Methods for Today's Organic Gardeners*, by Wesley Green (or *Founding Gardeners*, by Andrea Wulf)

Activity: Make a wicker cloche by using a basket making kit in the right shape and size. This project can be started that evening and finished at home.

NOVEMBER

Book: *Cooking with Fire*, by Paula Marcoux

Activity: Have a guest speaker guide you through the preparation of an actual, over-the-flame meal—NO CHARCOAL BRIQUETTES! Don't forget dessert and don't be afraid to fail. Invite families to this event.

DECEMBER

No book club; group service project*

*Be sure to take time to serve in your community together as a club. Selfless service creates feelings of community and true charity which will strengthen not only those you serve but your homesteading group, as well.

> **SAMPLE BOOK CLUB ITINERARY:**
>
> Welcome
>
> Business—include book for next month and place of meeting, questions from last month's activity
>
> Person in charge present book for the evening with short summation
>
> Person in charge read aloud suggested discussion questions and begin a guided, structured discussion
>
> Pros and cons of book
>
> Break for hands on activity
>
> To download a nicely formatted sample schedule and/or an itinerary sheet with a pretty graphic, in full color or black and white, please visit our links page (or look in your copy of *The Do It Yourself Homestead Journal*): www.homesteadlady.com/homestead-links-page.

BOOK CLUB REALITIES

Let's chat about what a book club is and what it isn't.

A book club is a consistent group of people meeting at a consistent time to discuss a pre-arranged topic (a book you've all agreed to read). This discussion leads to an exchange of ideas, which leads to wisdom and experience.

A book club is not a party. If you're the month's hostess, you don't have to go all out with food and decorations and hours of cleaning your house (unless that brings you joy). Don't waste your time or energy on superfluous tasks and don't show off for other club members, thereby making the club more about the presentation and less about what is being learned from these great books.

A book club is a place to share ideas on a common topic and a motivation to read books you might not otherwise get around to reading. However, a book club isn't a place where you all have to agree or even understand everything you've read. It does provide food for thought and a jumping off place for learning more.

Let it be what it is and don't force it to be something it's not. Just relax and read and be with like-minded people at ease with each other. And be sure to laugh together as much as possible. Taking ourselves too seriously is the death of every worthy self-sufficient goal.

> "...There is nothing in the world so irresistibly contagious as laughter and good humor."
>
> — Charles Dickens, from *A Christmas Carol*

ACTION ITEMS: Take out your homestead journal and make a short list of people you'd like to invite to be part of your book group. Then, make a quick list of titles you'd like to read together, and possible activities. Keep both, as I say, short to begin with.

Think about your schedule. Don't take on too much and remember that you still have a life beyond your homesteading efforts. A book club is supposed to be fun and inspiring, not overwhelming and soul-crushing.

- Do you want to meet once a month or once every other month?
- What can you reasonably and realistically accomplish this coming year?

Go back to your journal and write down all the wild ideas you have for your book group, as well as your concerns; just free write, without worrying about what you're getting down on paper. Put it aside for a few days and come back to your notes after you've had time to let the idea rumble around a bit.

 DIY: Set up a Facebook or Google group for your book club to make it easier for this tech savvy world to find you and stay updated with what you're doing. Even if your book group stays small, you'll want a central, online place to share questions, inspiration and plans. Email lists are great for quick updates and plans but the messaging back and forth can get cumbersome with a group of people. Social media provides a tidy place to keep track of what's going on with your group. If you've never set up one of these groups, just search online to find step by step tutorials.

To learn how to set up a Facebook Group, simply visit our links page under The Homestead Community chapter: www.homesteadlady.com/homestead-links-page. Believe me, if I can figure out how to set up a Facebook group, anyone can.

 BUILD COMMUNITY: Contact the people on your short list of possible participants this coming week to brainstorm the idea. Your list will most likely be comprised of people you know relatively well, so feel free to bounce ideas off each other. Either in person, or via email, make a tentative schedule of books and meeting dates. Figure out how you're going to get the word out about your group. Do you want to keep the group small or get as many people together as possible? Will you have activities or simply do a book discussion? What venue is available for your meetings?

If you're online but having a hard time finding like-minded people in person, try sites like Meet Up (www.meetup.com) that offer online connections for your local community and specific area of interest. Always exercise caution when sharing personal information online, but you may just find there are more people like you in your town than you originally thought.

 BOOKS TO READ: There are ten suggested books in the text for this segment so read one or all. You can check the Resources section in the back of this book for more ideas.

If you're looking for community agriculture inspiration, try Forrest Pritchard's book, Growing Tomorrow: A Farm-to-Table Journey in Photos and Recipes: Behind the Scenes with 18 Extraordinary Sustainable Farmers Who Are Changing the Way We Eat. Talk about a community building topic!

WEBSITES: My two favorite places to read book reviews from a wide variety of readers are Amazon (www.amazon.com) and Good Reads (www.goodreads.com). Just type in the title you're interested in learning more about and click customer reviews to read what people have to say about that book. I usually read the lowest ratings first to sort out the trolls, the vendettas, and then those with honest, constructive criticism. Then, I read on up through the ratings until I hit the five-star ones. Even in the five-star ratings, you need to read several and sort out the people who have only skimmed the book, or who are commenting on topics like the packaging.

Whenever possible, leave a review of whatever book you read, especially if there aren't that many there yet, to help readers coming up behind you. Try to say more than, "I really liked it" or "I really hated it." If you read a review like that, your first question would be, "Well, why?" Amazon requires at least 20 words to leave a review so use those words to let readers know what you think. Even if you only want to tell us how we can improve the next edition, *The Do It Yourself Homestead* is a great title to start with—wink, wink.

HOMESTEAD LADY SPEAKS:
I remember when I read my first homesteading book and how I connected with the ideas presented there. The title was *Back to Basics*, A Reader's Digest compilation. I was a high school student and realized, as I was reading and learning, that I wasn't the only person who was interested in these topics, even if my peers thought I was weird! Furthermore, I saw that there was so much more to learn and explore than I had originally thought of on my own. Each homestead title I've read since has built on that feeling of wonder and excitement. Through books, I've been inspired and instructed to expand and flesh out my homesteading efforts in ways I never could have without them. Books are among my greatest homestead mentors.

If you have a hard time holding still long enough to read or find it difficult to make the time, rest assured that most of the books you'll be using for your homestead education are formatted typically for non-fiction. You'll be able to pick them up and put them down as you like, reading sections of information without having to worry too much about losing a narrative. Unlike your favorite novel, your homestead books rarely need to be read from start to finish and in order. Also, audio books are becoming increasingly popular for the busy information-seeker on the go. Keep that homestead journal close by so that you can keep a running tally of new ideas for the year as you explore great books.

"Once you begin, you'll discover that there's somehow always a mentor somewhere along the way… . Eventually, you will become that mentor for somebody else."

— George Nash and Jane Waterman from *Homesteading in the 21st Century*

For the Homesteadish: Taking and Teaching Classes

Some of us feel like we just don't have time to cram one more thing into our schedule so the idea of taking classes on homesteading topics feels overwhelming. For others of us, we just want to stay quietly on our homesteads and not have to mess with people. Please believe me when I tell you that there's no substitute for getting out there and getting involved, face to face, with like-minded people in the setting of a classroom. These classes don't have to be long, difficult, or expensive to be worthwhile and edifying.

Remember, too, that the more we feel close to other homesteaders the more often we end up getting together. When some of your neighbors and friends think of you as the weirdo chicken-keeper, it's nice to know you're not alone. The homesteading community is worth building up, wherever you are.

Taking Classes

Let's start with attending community homesteading or gardening classes and workshops. You may not think that attending a class as a student really qualifies as community outreach on your part. Aren't you going there to be fed, not to feed? Nope. You're there to reach out and connect, and that connection is a service to your homesteading community. Learning is just an incredible side benefit.

Instead of simply walking in and sitting down when you attend classes, the first thing you can do is to greet the teacher. A friendly handshake and a hello will connect you with the instructor and help their words be more memorable. Ask them how long they've been involved with their topic, if they enjoy teaching about it, and if there are further resources you should seek out after you've enjoyed their class. All that information may be answered on their handout (so make sure you get one) but having a conversation with the person prepared to mentor you for an hour or two will help you connect even more with their message. When you do take your seat, don't be afraid to talk to the person next to you—yes, even you introverts. Have conversations during the class break with other students, check out any items that might be for sale at the class and, of course, ask questions during the instruction.

Being an active, engaged student requires taking good notes, too. Be attentive, even if the topic isn't the thing you're most passionate about. You just never know when you'll come across something that makes your homesteading ventures more worthwhile; today's boring class is tomorrow's vital need. So, pay attention. As an example, as a master gardener I attended a class at our water conservancy on watering systems for home gardens. The class was well done but it was a lot of technical detail that I tend to not absorb very easily and a lot of numbers and statistics that just wouldn't stay in my head. I doggedly kept taking notes and trying to pay attention. (Did I mention I was also pregnant and kept dozing off?)

Then, all of sudden, I sat up and started listening to the instructor talk about grass, and the pros and cons of having a lawn. This was pertinent to me because we had just changed out the grass in our front yard for a herbal/edible garden. We still had a lot of grass in the back, though, and each year I would take out more and more. It was difficult to know just how much grass was useful for the family, and how much was just taking up space. The instructor said, "If the only time you walk on your grass is when you mow it, then maybe it's time to rethink that space." Genius! That became our measuring stick for grass maintenance. (Plus, once I was home, my engineer husband was able to make total sense of all those pipe measurements and design ideas. I like having an engineer on staff.)

If you are physically limited or simply live too far from any potential class venues, online webinars and courses are also a wonderful way to learn. While face to face interaction is useful, personal connections can be made through the communities that are often built up around these online educational outlets. From Facebook groups to student participation during live courses, these online communities can be a great source of feedback and inspiration. Often you will be able to make connections with like-minded people that turn into true friendships, which can be just the thing to boost your spirits if you're feeling alone in your community or isolated on your homestead.

> "'The best thing for being sad,' replied Merlin, beginning to puff and blow, 'is to learn something. That's the only thing that never fails. You may grow old and trembling in your anatomies, you may lie awake at night listening to the disorder of your veins, you may miss your only love, you may see the world about you devastated by evil lunatics or know your honour trampled in the sewers of baser minds.

> There is only one thing for it then—to learn. Learn why the world wags and what wags it. That is the only thing which the mind can never exhaust, never alienate, never be tortured by, never fear or distrust, and never dream of regretting. Learning is the only thing for you. Look what a lot of things there are to learn.'"
>
> — T.H. White, from *The Once and Future King*

Teaching Classes

You may not think you know enough to teach a class on anything but hang with me while we go through this. First, let me share with you something I once read from a homeschooling mentor of mine named Diane Hopkins. She noted, as any homeschool parent would, that often people will say, about homeschooling their own children, "Oh, I could never start doing that. I don't know enough!" Diane points out that you homeschooled your kids from the time they were born to the time they left for Kindergarten. You taught them incredibly hard skills like eating with a fork and walking. You also taught the difficult concepts of manners and how to use a toilet. The real question is not "Can you teach?" but, "What can't you teach?" Seriously, if you can potty train, you can teach a class on baking bread if that's your thing. Everyone has a "thing." There is something else you know how to do that not everyone around you knows how to do. Even if they have the same skill, they don't have your unique voice.

At first, you may not feel comfortable charging any money for classes—you may decide never to charge money for various reasons, including a desire that all might attend without having to worry about paying. A lot of homesteaders feel compelled to share what they know simply because they feel strongly that the information they'd like to share will be useful to people. My husband calls my homestead ramblings, both written and verbal, my missionary work. I've come to feel like that's really what it is; I'm preaching the word of self-sufficiency and skill and you can, too.

I would suggest that you find a motivation for sharing your knowledge that doesn't factor in personal gain at first. There is absolutely nothing wrong with expecting to be compensated for your time and hard work and I will be the first to say so. After all, you purchased this book, right? However, finding your higher purpose at the beginning of teaching will carry you through all the times you end up instructing a class with the flu, or after an all-nighter with your newborn or your teenager. The commitment you have on a personal level to share your experience with others is what will keep you smiling through the comments of that one guy who has to know more than you and has done it all better; it will keep you laughing when your demonstration cheese doesn't curdle, or your bread doesn't rise. I've been there, and I promise you that finding your mission in the homestead field is more important than, but akin to, finding your niche. Ponder and journal until you figure out what is at the core of what drives you as a homestead teacher.

After you've done that, you're ready to start considering which topics you'd like to teach and in what format. I suggest the first thing you do is pull out a piece of paper and brainstorm everything you've done in your life so far—education, employment, hobbies, side work, community service, religious service, clubs and groups, homemaking skills, etc. Write it all down and then start branching off each topic with some specific details. For example, I could note that I've kept bees for around twenty years. Cool, bees are of interest to a lot of people these days.

Let's get more specific. I could branch off from the basic idea of bees and create classes like:

- How to get started with bee keeping
- How to involve kids in bee keeping
- How to capture a swarm of bees
- How to fail at bee keeping (Yeah, I've done that, too)
- Ten must-have pieces of bee keeping equipment
- Best books for bee keeping

These topics could be covered in local University Extension forums like Master Gardening classes or as simple classes in your kitchen or backyard. You can lecture for your book group (see the *Homestead Book Club* section in this chapter) or ladies' church auxiliary. You can find summits or conferences, either in person or online, which are in your niche and pitch class ideas to them. You could start a blog or write an e-book, or print book, or both. You could learn to write magazine articles. There's no limit to the number of places you can share your knowledge and passion. There are too many possible directions to send you off in to be of much use giving you specifics for this level of teaching, but as you are ready, your mentors in these venues will appear. For now, just kick around the idea of teaching in your home or local library.

> "When I was 26, about 7 years ago, I met a woman at a local farmer's market who would completely change my world. We struck up an innocent conversation, which led to her sharing with me about her milk cow, her garden, and some of the things she was doing to feed her 5 children with very little money to spend. I was completely intrigued! I honestly didn't know people still lived the way her family did. Over the next several months I would spend a lot of time with Ms. Addy and her family, learning everything I could about homesteading and simple living. She ignited a passion in me which has revolutionized the way I think and the way my own family is now living."
>
> - Kendra Lynne, Homesteader

 ACTION ITEMS: Get out your homestead journal and do the suggested work under "teaching classes" by brainstorming what you are interested in teaching. List your passions and don't be afraid to list something that isn't seemingly homestead related. Remember, homesteading is home-building, and that is a very wide umbrella under which many topics can reside.

Next, list your life experience—education, employment, hobbies, side work, community service, religious service, clubs and groups, homemaking skills, etc. If you are comfortable working digitally, there are quality mind-mapping and brainstorming tools online to help organize your thoughts and ideas. Mindjet is a popular paid mind-mapping aide and Coggle is a free Google tool. Whatever you use, be sure to get everything down on paper, from your first job to church service, from hobbies to community work.

You will naturally discover a common theme of strengths and passions emerging and this will give you a sense of direction. Each skill or passion you list, come up with at least five class topics you could cover. Keep your class topics short and focused on simple points. Less is more in a sixty-minute class, so avoid broad topics. Break down ideas into smaller themes that you feel will really answer a need in the newbie homesteading community.

DIY: If you are not very tech savvy (you're in good company), or simply need to fine tune your skills, find software or online tool to become familiar with that will help you create class fliers, invitations and handouts. Both Picmonkey (www.picmonkey.com) and Canva (www.canva.com) offer free photo editing and document creation online. If you have Microsoft Word, you can use the various templates that come with that—calendars, newsletters, handouts and more can all be edited to present the information you need. Microsoft Office allows you to access PowerPoint whereby you can learn to create visual presentations. Both tools are good ones to master and Microsoft has tutorials that you can access. Your public library may also provide classes on these topics from time to time, as part of their adult education programs.

Your students expect to be provided with a certain amount of written information that can be taken home and studied further. Play around with handouts and decide how much information you want to provide, especially if the class is free. One of my herb mentors would provide and host her classes for free and allow you to use the extensive handouts she'd prepared during class, if you refrained from taking notes on them. If you decided to make notes on the handouts during class or decided that the provided material was beneficial enough to purchase, she'd charge a reasonable fee for them at the end of class. This meant that she didn't need to provide and print information that you weren't going to find useful and you could vet the information before purchasing.

However you choose to work it, learning to create pleasing documents and digital invitations will provide incentive for your students to come, and to keep coming back. These needn't be fancy or expensive to create. By fiddling around a bit with online tools, you can start to become familiar enough with them to do your own design and formatting work for the digital and printed materials for your classes.

Don't let those fancy terms scare you off, either. You CAN teach a stellar class on cheese making with nothing but your wits, a pot, and some cultured milk. However, you may find that by providing a clear course description with your initial invitation, and a few printed notes on your presentation, you will avoid answering the same questions repeatedly, and your students will be more engaged and prepared for your class.

BUILD COMMUNITY: Teaching is all about creating community and there are so many communities from which to choose that need your assistance. One of my favorite communities is kids. Teaching adults is great fun but if you really want to change the world for good and make a difference, resolve to teach a child something useful. To that end, the challenge for this section is to create a short class to share at your neighborhood elementary school, library, or church group geared towards kids. You can also choose to share your efforts at rescue missions, hospitals, or orphanages.

Wherever you decide you're going to reach out, be sure to call ahead and let them know what you're thinking. Inquire as to whether they have special needs for which your skills may be useful. Ask about their in-house rules and any other regulations that might govern what you're able to do, especially if you're working with a state facility or a hospital. Your questions will NOT annoy them; they really, really love community support and involvement but just need to make sure that all the rules are followed for everyone's safety. If you're willing, please share your experiences with me via my contact information at the back of the book; I geek out on this community stuff and would enjoy knowing what you're up to.

BOOKS TO READ: *The Good Food Revolution*, by Will Allen. This is just an inspiring read for foodies, growers and homesteaders who are community minded, or who would like to be. From start to finish, this book will teach you the power of one. We're all just normal people living our lives but, sometimes and in some places, our small contribution can lead to great change. Every bonfire starts with one spark.

WEBSITES: Go to the Resources section in the back of this book and pick any one of the number of websites you see listed there. You will be well served, I promise.

HOMESTEAD LADY SPEAKS:
When I think back to my favorite teachers in school, they're all English/Literature and History teachers. Not because they created in me a love of good books, writing or history but because they fanned my already present sparks into great flames. They involved me in the process of learning more about topics I naturally found fascinating. They shared their own stories, their favorite books and people, and their own passions with their students. Without fancy equipment, a large budget or special effects, they mentored me and helped me to love learning.

If they could do it, the take-away lesson is that so can you and I. We can all be better students and generous teachers if we will only take the challenge and begin today.

Thank you to Mrs. Henry, Mrs. Sood, Ms. Starr, Mrs. Paylor and Ms. Robison (and the countless other teachers I'm forgetting in my advancing age)—I'm still enjoying the sparks you ignited!

"And God said, Behold, I have given you every herb bearing seed, which is upon the face of all the earth, and every tree, in the which is the fruit of a tree yielding seed; to you it shall be for meat."

- King James *Holy Bible*, Genesis 1:29

For the Homesteadaholic: Seed Swap

For the gardener, seeds can become a kind of addiction. I have boxes of seeds that I lovingly organize and categorize, even though I usually make a mess of my organizational efforts during the year. My love of seeds, and my rather thin pocketbook, are eventually what led me to learn to save my own seeds each year from the garden. A seed swap is a beneficial community with which you can get involved, no matter your level of experience—even if you have NO experience.

Start attending the meetings of your local swap and pay attention during any classes or presentations they offer. Some areas have huge, regional swaps with a whole day of classes and lectures, swapping and purchasing. Even if you're a beginner gardener, I encourage you not to be intimidated and to attend as many of these as you can. Ask questions of everyone. Many times, swaps will have seeds for sale or just to give away, so you needn't worry that you've brought nothing to swap. Gardeners are some of the most generous people I have ever known. Whatever you do, don't stop dreaming about the gardens to come from these little seeds.

As Forrest Pritchard wrote in his book, *Gaining Ground*:

"All farms require a resident dreamer, someone to thumb through seed catalogs in the cold days of late January, imagining summer fields of squash and cucumbers, tomatoes and sunflowers. Fall harvests are the reward of winter dreams."

Begin a Seed Swap

Full disclaimer—I began one in my very suburban town and it flopped, despite all my best efforts. Yet a similar group in a neighboring, but more agriculturally inclined county, is still thriving and I'm so glad! I'm sure the leaders of that group are amazing and did lots of great stuff that I didn't even think of, but my first piece of advice is to know your town. If there's only a small group of you who are interested in seed saving, then keep it small; don't waste your energy trying to get more people involved, they'll come over time and as they're interested. If you just can't find anyone else close to you to swap seeds with, branch out a bit to the next town or anywhere in your state. You can also join online seed saving groups and swap via the post.

KNOW YOUR WHO

Besides knowing your area and what the interest level is in seed swapping, the next thing you need to decide is how you want the group to run. We started ours using Robert's Rules (parliamentary procedure) and a board, hoping that the structure would provide permanence for the group, since we knew we were moving soon and didn't want it to collapse when we left. Our goal wasn't realized, but I did learn a ton about parliamentary procedure, so I'll be running for public office soon. Ha, ha.

Don't be intimidated by parliamentary procedure and setting up a board, if you feel like that's what you'd like to do; believe me, if I can learn it, you can learn it. If you work in the business world you're most likely already familiar with it. However, you can run your group in a much more informal way, as well. If you're the leader, be prepared to make a lot of the decisions for the group and to do a lot of the work. But, you're a cool homesteady kind of person and work doesn't even phase you.

> **FOR THE SEED NERDS**
>
> I am such a seed nerd. Getting seed catalogs in the mail every season is my grown-up version of Christmas morning—I'm so excited to see them come in! I'm not as excited to pay for the seeds each year, though, and it was that rather base motivation, the desire to save money, that led me to learn more about seed saving, and join a seed swap. Getting together with other seed nerds turned out to be such a help to me, and my gardening efforts, too.
>
> There are other reasons to swap seeds, above wanting to save money on seed purchases each year and wanting to share with your neighbors. Exchanging seeds local to your area, with other gardeners local to your area, increases the seed strength in that area. Seeds are like people and they adapt to their environment over time, creating with each new generation, plants that are best suited to their regional quirks and climate. The more we share quality seed with each other, the stronger and more vigorous everyone's backyard food supply becomes.

> Plus, getting together with other seed nerds will teach you something new every time. Each seed saver develops their own specialties. Dale Thurber of Delectation of Tomatoes participated in our local seed swap and his specialty was giant tomatoes. You know, those massive ones, nearly as big as your head, that are all heirlooms and so tasty? (To find his seed, go visit our links page under The Homestead Community chapter: www.homestead-lady.com/homestead-links-page). Another gentlemen grew amazing melons and was so generous with this seed and his growing tips. I stink at growing melons and so every time he started talking, I started listening. Don't let a fear of not knowing enough or not knowing how to do everything perfectly keep you from joining a seed group and really diving into participating.

Marketing and Managing

If your seed group stays small, so will your management headaches. However, in truth, you really want as many people saving seeds in your area as possible. Right?! For that reason, prepare for growth by laying an excellent foundation with the following tips.

FIND YOUR WHERE

The next item of business is a venue, some place to have your swaps. You'll need to double check that whatever venue you choose has enough space and that it will allow you to meet as often you decide you want to. Meeting monthly is great, if you can, and places like libraries and churches often have their calendars available for scheduling far in advance.

You can also meet in someone's home. I like meeting in homes at least sometimes because you can do garden tours and see a homesteader in their element to find out how they do things. Gardens are never boring to me and I learn something from each one I'm privileged to visit. You might be able to find space at your local nursery, if they have a conference room; likewise, your local water or land conservancy or public garden. Be sure to find out what the venue's rules are about live plants in dirt (should you choose to swap those, too) and selling seeds on site (in case anyone wants to do that).

GET THE WORD OUT

Those places I suggested for possible venues also make great places to advertise your seed group. Call around and ask at local nurseries, gardens, master gardener classes and even real foods groceries and health food restaurants if you can post a flier about your group. These places often have boards dedicated to community meet-ups. I also highly recommend you start a Facebook/Twitter page (Instagram and Pinterest might be useful to you, too) for your group. You could even start a blog to share information on seed saving and to have a central place where everyone can find out the latest on swaps and classes. Yahoo groups are still useful for this, too, and Meet Up is becoming very helpful at connecting people, as well. (See *The Business of Homesteading* section in the Homestead Finances chapter for a few more tips about establishing yourself online.)

After you've decided on a place, time and schedule, you need to start meeting together. You may discover that it would be handy to have the following moving forward:

1. **Seed bank:** a group box of collected seed that one person hangs on to and brings to every meeting.
2. **Seed bank check out form:** you may want to have people check out their seed like they would a book from a library, filling out a form that says what they've taken, when, and when they anticipate growing out more seed to return to the bank.
3. **Seed saving book list:** someone can oversee keeping a master list of any seed saving books the group members own and are willing to share with other members.
4. **Contact list:** you'll want a way to get in touch with everyone in the group, whether via email, text, or social media
5. **Schedule:** people are busy and you're more likely to get consistent attendance if you stick to the same day every month and people know what you'll be discussing
6. Job descriptions: if you've decided to go with a structured group where certain people have certain assignments, you'll most likely want to clearly delineate, in writing, what each job entails so that people can fulfill their obligations to the group
7. **Guest speakers:** it's a great idea to include outside mentors and teachers—we had wonderful mentors teach our swap on various topics from mushroom growing, to tomato seed saving, to growing year-round.

> "Might I...have a bit of earth?
> To plant seeds in—to make things grow—to see them come alive..."
> - Francis Hodgsen Burnett, from *The Secret Garden*

TEACH TECHNIQUE

One of the most important things to stress as you meet is proper seed saving techniques. Everyone must start somewhere, and you'll get members with lots of enthusiasm but little experience. No worries, that's what the swap is there for. Make sure you're covering, even if it's only for five minutes at the beginning, how to properly grow, save and store seed. You could feature one vegetable or herb each month to take from seed planting to seed saving, or you could have a quarterly class on the side covering various seed saving techniques.

Teach, teach, teach and move forward with your own efforts remembering that there's no such thing as a dumb question. If you need extra guidance, Seed Savers Exchange offers free webinars (and archives past webinars) on seed saving techniques on their site; you can find a link on our links page under The Homestead Community chapter: www.homesteadlady.com/homestead-links-page.

Lastly, be sure to include other garden items besides seeds in your swapping. Roots, plant starts, bulbs, tubers, tools and surplus harvest are all items that can and should be swapped in a seed group. A harvest festival at the end of the season is a great way to celebrate with swap members' families by getting together and sharing the excess bounty from the season. Hold an old-fashioned Harvest Home, complete with a Queen of the Harvest and some delicious, home-cooked food. Celebrate the miracle of the seed together. These seed and garden nerds are your people; have fun together as you make connections and friendships that will serve you in good stead.

> **SEEDS FOR GENERATIONS**
>
> Seeds for Generations is a small, family-run seed company (visit our links page under The Homestead Community chapter for more information). Jason Matyas and his family simply love growing their own food and wanted to provide that opportunity to others through selling heirloom, non-GMO seeds. They also wanted a business they could run as a family. Jason says,
>
> "I've been gardening my entire life. Literally. Before I could walk, my parents had me out in the garden with them, and once I could walk, the garden became my pasture for grazing. They planted important seeds in me—love for God's creation, enjoyment of working with my hands, and an appreciation for the thought and effort that goes into growing food.
>
> "I'm doing similar work with my seven children, working to pass on the legacy of gardening to them. We've done this for years now—my oldest daughter has been in the garden helping since she could walk. I'd been looking for a long time for an opportunity to have my kids work alongside me and to work with each other, and one winter I had an epiphany. Our family loves heirloom seeds and vegetables, why not share them with others? So, I decided to start a small cottage industry as a way to give my children an opportunity to work together with me in a family business."

ACTION ITEMS: In your homestead journal, write down anything you already know about seed saving. If you can't think of much, choose one of the books below to begin studying. Make a list of at least five seeds you'd like to learn to save this year. If it's the right time of year, allow one of your garden veggies to go to seed—lettuce is probably the easiest and quickest.

1. Observe how and when the flower stalk emerges; watch how the flowers fade and the seed pods appear.
2. Compare this process to what's written in your seed saving book, or favorite online resource, to decide when you might harvest the seed from your chosen plant.
3. Take notes in your journal about what you're doing and the questions that pop up.
4. Next, look around your neighborhood and wild spaces to see if you can find this seed setting process going on with other plants; see if you can identify which plants are going to seed using a field guide for your area.

Whether you already know how to save seeds or not, it's time to seek out your local seed swap group. If you can't find one close enough to you, you'll need to start one. This group does not need to be large, especially at first. Talk to your neighbors and associates, especially the ones that have gardens, and see if any of them know how to save seed or if they'd like to learn. Ask your local extension agent if there's a master gardener who could teach a class on the basics of seed saving to the small group you've gathered. Be sure to keep in contact with those who are interested in learning more and organize another time to meet.

A group will form out of your first attempts because like-minded people eventually find each other. Be sure to keep all your notes, about who you've contacted and what you might do next, written down in your journal because you'll need those when it comes time to organize a formal group.

DIY: Using your list of five plants, from which you'd like to learn to save seed, begin to plan your seed saving garden for the coming growing season. You most likely chose plants you already enjoy eating from or growing so perhaps you already have some experience with their care. The first thing to do while planning to include them in this year's garden is to determine if the variety you've chosen is a non-hybrid seed. Hybrid seeds are not used for seed saving because their genetics become unpredictable. Plan to use only non-hybrid or heirloom seeds in your seed saving garden. Also, try to find a seed saving friend for this year's planting, if you don't have a group yet. You plant some seed saving plants, and your friend can plant some others, then you can share seeds at the end of the season.

The next thing to do is to pick a location for your seed saving plants. To keep the genetics strong and plan against any losses, be sure to plant as many of each type of seed saving plant as possible in the garden. For example, instead of planning to save seed from just one tomato plant, grow at least five of that kind of tomato just in case you lose some during the season and to create a nice mix of seed tomatoes from which to choose. The more the merrier when it comes to seed genetics. Also, using your references, double check if these plants will require some sort of isolation from plants botanically similar to them to prevent cross-pollination—you don't want to end up saving a pumpkin that tastes like cucumber just because the bees messed up your plans with their efficient pollinating efforts. You'll also want to be sure to note how long it takes for each plant to mature the seed you're looking to save; lettuce is a lot faster than melon, for example.

If this is all new to you, don't worry too much about trying to get EVERYTHING right this first year. If five plants are just too many to plan for, plant only one type of plant for seed saving. This is your garden; you pace yourself the way in which you feel comfortable. Like I said, it's best to have a seed saving buddy so you don't neglect the steps outlined here. If you can find a gardening friend or group of friends to plan and experiment with, you'll have way more fun and you'll be able to compare results.

- **BUILD COMMUNITY:** This whole section of the book is about building community so all I'll say here is to not be intimidated by this process. If it's in your nature to be shy around large numbers of people, just find a few seed-saving friends to hang out with. If you do better with larger groups, don't stop gathering until you have a sufficient pool from which to draw. Know yourself, and don't try to work so far outside your comfort zone that you fail. Have some fun.

If you're curious to know what other seed savers are doing, consider a subscription to Jerre Gettle's magazine *Heirloom Gardener* (You can find a link on our links page under The Homestead Community chapter: www.homesteadlady.com/homestead-links-page). Incidentally, I don't work for them, I'm just a fan.

- **BOOKS TO READ:** *Seed to Seed*, by Suzanne Ashworth, has long been considered the Bible of seed saving and it is a fabulous book. My favorite part about it is the simplicity with which the material is presented. Anyone can understand what she's written and adapt it to their own efforts.

Another great one is *The Complete Guide to Saving Seeds*, by Robert Gough, which has cool graphics and covers 322 different plants. There are others and, as always, I encourage you to read every single one over time because you'll learn something new with each book you read. Books are cool that way.

- **WEBSITES:** Seed Savers Exchange (www.seedsavers.org) is a great online resource. Don't confuse that with Seed Save (Seedsave.org), which is another good one.

You can also search individual plants online for seed saving instructions from various garden experts and bloggers. It can be interesting to read the methods of several different people to learn their tricks. It's even fun to learn about their flops since you're going to have those, too.

- **HOMESTEAD LADY SPEAKS:**
I've been to many seed swaps and I never get tired of them. Nor do I get bored reading about the adventures of other seed lovers. A fun book to read is *The Heirloom Life Gardener*, by Jere and Emilee Gettle. Not only will this book teach you how to grow plants from seed, but it will also teach you how best to cultivate them for seed saving. Each plant is profiled beautifully, lovingly even.

Mr. Gettle has been called the Indiana Jones of seed saving and that seems to be accurate given his apparent devotion to hunting out the purest, most fascinating seeds from all over the world. You can purchase his heirloom and non-hybrid seeds through his family's company, Baker Creek Seed, at www.rareseeds.com (they also own Comstock Ferre & Co in Connecticut and the Seed Bank in California). Just reading the description on each seed profile is an education!

If you want to do more than just read a book or make an online purchase, I invite you to join us in the great state of Missouri at Baker Creek's headquarters in Mansfield for their annual Spring Planting Festival, celebrating all things heirloom. They also hold free Heritage Festivals the first Sunday of each month, March through October. If you're near California, join them for the Annual National Heirloom Exposition in the fall, which is a huge gathering of gardeners, growers and chefs from across the globe to celebrate and educate on pure food. You northeastern gardeners aren't to be left out, as you can join the Annual Harvest Fair at Comstock Ferre in historic Wethersfield, Connecticut in the autumn. Nor are these the only seed saving events that take place across the globe—there are many!

From books to groups to fairs, there are so many ways to fall in love with seeds and the community of people who save them.

"The mediocre teacher tells. The good teacher explains.
The superior teacher demonstrates. The great teacher inspires."

— William Arthur Ward, Author

For the Homesteaded: Mentoring

There have been so many people who have truly inspired and educated me on my homesteading journey, from teachers to gardeners to groups of friends just getting together and sharing what we've learned. Once you've benefited from that kind of open sharing, you can more readily understand the urgent need for more of it in our homestead communities. Many people in our times are looking to break away from an entitled, consumerism lifestyle to create a more stable, more hands-on approach to day to day living. While this is a great goal, the reality is that very few of us grew up this way; the lifestyle of self-sufficiency that was normal for my great-grandparents was in many ways lost by the time my grandparents were raising their children in the 1950's.

That's not to say that the "greatest generation" hadn't earned a rest and a bit of ease after fighting two terrible world wars, and then other conflicts that followed hard on their heels. However, the skill sets that were vital during the Great Depression and the war years were replaced by store-bought bread and bottled ketchup. Some of us desperately desire to revive homemade bread and home-grown ketchup but we just don't know where to begin. Enter you, the homestead mentor.

WHO, ME?!

The first thing you need to wad up and throw in the trash bin right now, is the idea that you don't have anything worth sharing. We touched on this idea in the *Taking and Teaching* section but, believe me, at this "Homesteaded" phase of your progress, you've got plenty to share.

Even when you were back in the "Homestarter" phase, I guarantee there was something you mastered that you could have shared with others.

One day, I was talking along with someone about butchering my hens for stewing once they'd aged out of egg production. The lady's eyes got big and she said, "Is THAT what you do with them?! I was wondering! So, how do you do that? Where did you learn?" Her questions made me stop and think: what did I used to think happened to aged-out hens? Where did I learn to butcher a chicken? When did I start thinking of that as a normal part of my year?

I realized that, while I certainly don't know everything there is to know about butchering a chicken, I certainly knew enough to teach her where to begin and how to learn more. I was happy to witness her excitement and thrilled to think of how much time I was saving her by telling her about my favorite resources for chicken education and offering to show her in a hands-on way when she was ready to try harvesting a chicken herself. The truth is, my husband and I had to learn that part on our own. We just practiced until we knew more than when we started out. We would have loved to have someone show us what to do!

> "I always like to encourage people that it doesn't take land or a lot of money to get started homesteading. Oftentimes people get discouraged thinking that they'll never be able to live their dream due to their current circumstances. Believe me when I say you really can homestead wherever you are, right now! You don't have to wait. So much can be done indoors, without a square foot of land. You just have to get creative, be resourceful, and think outside of the box. I encourage you all to pick one self-sufficiency skill to learn and start there. Homesteading isn't an easy life, but it's one of the most rewarding things you'll ever do!"
>
> - Kendra Lynne, Homesteader

What is a Mentor?

A mentor is an experienced and trusted adviser; a trainer, tutor, or coach. The best mentors expect their students to do the work, but the mentor is there to guide. Good mentors will show their students what they've learned through their own experience, including their mistakes. Remember my family-building mentor, Diann that I mentioned in the *Chore Systems* section of the Family Times chapter? She expected us, her students, to work hard but she was personally involved in that process and shared her own experiences with us. When I think of powerful mentors, her smiling face pops into my head.

If you desire to be an effective mentor, keep these things in mind:

1. The student does the hard work of learning. At no time should a mentor assume responsibility for the student's work, nor his success or failure.
2. A good mentor is involved, passionate and personable but it is up to the student to pursue his own vision and success.

3. A good mentor shares her own experience. I still remember many of the stories Diann shared from her own family life and the lessons she drew from them; so much of what she shared has become part of our family culture just because it was personal and pertinent.
4. A quality mentor is not aloof and untouchable; rather, she's right there with you being honest about her successes as well as her failures. Our homestead flops make some of the best lessons we could ever pass on as homestead mentors! *...Then there was that time I forgot dinner in the solar ovens during that rain storm and we didn't eat until 10pm...or there was the time I neglected to water 100 tomato transplants and they all died. Yay me, a real-life homesteader!*

Mentor the Next Generation

Every time we share ourselves, every time we take the time, it makes a difference for good. To that end, the following are some ideas for getting started on your journey to becoming a homestead mentor.

TOURS FOR ASPIRING HOMESTEADERS

One of the easiest ways to lend a hand to aspiring homesteaders is to give garden and/or homestead tours. One of the first myths we need to dispel for the newbie homesteader is that homesteading looks like *Better Homes and Gardens*. It doesn't. Homesteading involves dirt and poop and hands-on work. While it's important to maintain our homesteads with healthy systems and rotations to control smell, mess and poor health, the reality is that most of us are homesteading while living busy, parallel lives that don't allow for a lot of fluff on the homestead.

Plus, "cutesy" requires its own budget and some of us don't have the funds to make our homesteads look just so. Others of us aren't the fancy type. I plant daffodils with the herbs to deter deer and figure I've done my fashionable landscaping for the year. Don't wait for your homestead to look picturesque before you invite people to share in it as a work in progress. I always tell people that if they want to feel better about their own gardens, they can come visit mine anytime! Let's not show off for each other because that's a waste of time and is stupidly prideful.

Providing scheduled tours does one more important thing in that it provides people an *appropriate* time and place to contact you and view your homestead. Once you have a homestead or small farm, many people will prevail upon you to give them tours or to allow them to stop by and visit at any time that's convenient for them. Most people really aren't trying to be rude, they just don't have a realistic understanding of how busy you are on the land and in your home. When people who want to drop by, especially in the spring when baby animals are being born, you can politely direct them to your tour schedule and explain that you are unavailable outside of those times.

It would be helpful if you had some way for them to sign up easily, either with a link to a Google document, a Facebook group or page, or a website. (To learn more about resources for setting up a website, see *The Business of Homesteading* section in the Homestead Finances chapter.) If you prefer more organic communication, you can print old-fashioned calling cards with your contact information for people interested in visiting. I suggest you get an email address that's dedicated to homestead contacts so that sign ups and questions don't get lost in your personal inbox. Be judicious about how often and to whom you open up your homestead. In short, be generous, but be safe.

TOURS FOR CHILDREN

Along that line, I suggest you open your homestead to school groups or other groups of children. Children have a natural curiosity and interest in exploration but, in many ways, our culture dulls their senses and leaves them uninspired. Children aren't unintelligent, they're simply young. Provide them with a true homesteading experience, away from personal devices and personal indulgences, by allowing them to visit a place that produces items of importance. I always create a scavenger hunt as the tour portion of a school group's visit because we end up following the children's lead as to what we talk about. I like to finish up with a craft and a homestead snack.

When appropriate, have each child try their hand at a homestead chore like gathering eggs, digging a hole, planting a seed, kneading bread dough, harvesting herbs or making an herbal preparation. Make sure they leave with something to take home as this will help spark family conversation later. The item you gift them doesn't have to be large or expensive. How about a packet of seeds, some flowers from the garden, a home-baked treat, or a tuft of wool?

As I've mentioned before, there are some logistics to think about with tours. You may want to create a liability waiver form for all guests to sign. You might also want to speak with your insurance agent and/or your lawyer to work out any other liability details for your specific property. Create a list of rules for adults and children to follow and send them electronically with other details of the visit ahead of the scheduled tour time, so that everyone knows how to behave beforehand. If you want a certain number of adults per child during a children's's tour, make that very clear as you're working out details when setting up the tour.

Make sure everyone knows where the bathrooms are, and which entrances/exits are available for their use if you don't want them ranging all over your house. Be sensible and be clear with your instructions and everything will work out fine. Except when it doesn't, and you end up with some lazy adults and obnoxious children. It does happen sometimes but, more often than not, in my experience, everyone has a wonderful time. Life is give and take.

If you prefer not to have people come to you, go to them! Many schools, particularly charter and private schools, are building gardens and they love to have volunteers help with education and projects. For mentoring adults, be involved in your closest master gardener program, or any continuing education venue, and teach as often as you can. The more you do it, the better you'll get and the more natural it will feel. I've taught at big venues like regional conferences, as well as in smaller settings like book groups and garden clubs. Both have their challenges and rewards.

SWAPS, VOLUNTEERING AND MISCELLANY

Host a seed swap (see the *Seed Swap* section) or a food swap. Food swaps are a great way to offload extras from the harvest season and give you an opportunity to trade for the things you need. Be sure to use your homestead networks to invite everyone and pick a venue large enough for the people and their produce. Decide if you want to allow selling or if you'd just like to keep the experience trade-based. Get a crew together to set up and take down tables to hold everyone's wares. You can do a door prize raffle to encourage attendees to be on time, closing entry for the raffle five minutes after start time. If you'd like to make it a recurring event, be sure to set up a Facebook/Twitter page and email list to keep attendees abreast of new events throughout the year.

Volunteering your time for 4-H, living history, or any farm-focus educational experience in your area is not only a good way of sharing your knowledge, but it's just plain fun for you. Our family once lived in a city that had a "Farm Town" venue which exposed and educated children to the value of farm animals, how they behave and why they're raised. Also included was a garden area where children could participate in after school programs with hands-on, mentored gardening. Volunteering at such places only need take a few hours of your time but the rewards are hard to quantify as you end up touching so many people over the course of your volunteering.

Teaching and hosting classes is another simple way to provide outreach for aspiring homesteaders (see the *Taking and Teaching* section). You can also learn to teach online in the form of webinars if you prefer to do that. If you've ever used PowerPoint to give a presentation, you can teach a webinar because it's basically the same experience except that you record it. Even videos uploaded to YouTube will provide loads of help to people you've never even met. I sometimes wonder how we ever learned all those random things we need to know before YouTube. You don't have to have special equipment to shoot videos, either. The camera on your phone will usually work well enough and there are free video editing programs on the Internet; your phone may have an app for video editing.

Incidentally, should you wish to develop these as marketing tools later, you can monetize both without too much hassle. There's something to be said for simply donating your time and experience and there's also something to be said for being reimbursed for sharing your time and experience. Both are worthy.

> "The Internet has been an amazing resource in our journey. When we first got started homesteading in 2008, there really wasn't a lot on the Internet about self-sufficient living. This is what motivated me to start my blog, New Life on a Homestead, to share what

> we were learning and the mistakes we made so that others could glean from our experiences. I also found that although there wasn't a lot of information published online yet, there was an audience out there that was a wealth of knowledge. I learned so much from the people who helped answer my questions and troubleshoot my problems in those early years of stumbling along and trying to figure things out as I went. With the explosion of social media and blogs over the past few years, there is more information than ever online!"
>
> - Kendra Lynne, Homesteader

FEEL LIKE A REAL HOMESTEADER

I'd always had a garden and bees but the first time I ever really thought of myself as a homesteader and considered doing more than grow my own veggies and honey was when I took a series of classes on small farm livestock at Fickle Creek Farm in Efland, NC. The classes were simple but thorough, given in normal-person-speak for regular people who wanted to know more about animal husbandry. Each week we met at the farm and learned about brooding chicks, raising pigs and hoof rot in sheep. It wasn't all animals, though. We were also shown the realities of the profit margins on a small farm, the way a market garden is grown and even some ins and outs on running a bed and breakfast.

Ben Bergmann and Noah Ranells, who run Fickle Creek Farm, shared their knowledge, their experience and, most importantly, themselves with each of us, their students. They answered every question and spent their precious time helping us gain confidence to imagine we could go on and do what they were doing. I've never forgotten the experiences I had there, and they have, in large measure, influenced the decisions I've made and the way I've gone about learning and teaching in the ensuing years as we've considered ourselves "real" homesteaders.

If you're in the Piedmont area, contact Fickle Creek farm (wwwficklecreekfarm.com) for tours, farm store hours or to stay in the farm's bed and breakfast.

> **REAL HOMESTEAD MENTORS**
>
> I asked two homestead friends what they'd done to reach outside of themselves to help their community, despite how busy they are with their own homesteads. Samantha of Run Amuck Acres (www.runamuckacres.com) and Chris of Joybilee Farm (www.joybileefarm.com) both responded enthusiastically, having loved their mentoring experiences—so much so that they continue to pursue them in various capacities.
>
> Samantha says:
>
> "I founded the county beekeeping group, was elected president, teach the annual bee-school, and offer various workshops for beekeeping. I also initiated the local farmers' market, I am the market manager, vend at the market and offer a CSA program for my community."

> Chris says:
>
> "We started an artisan marketing group that included homesteaders, artists, and artisans, and planned an annual studio tour in the spring, a Canada Day artisan market on the grounds of the local art gallery, and a two-day Christmas market (all juried) so that quality artisans would have a local market for their work. We passed off the Studio Tour to a larger Arts organization who is continuing it and passed off oversight of the two markets to the Art Gallery, but I continue to oversee both markets and organize."

ACTION ITEMS: In your homestead journal, begin to make a list of ways you can share your homestead with the community. You may choose to interact and give back in different ways each year, so make a long list if you'd like. One year you could host a garden for homeschool kids; another year you could do a free pumpkin patch for families with loved ones at your local extended care facility. Maybe you'd like to host a class on butchering for the hard-core homesteaders you know, or a simple backyard chicken keeping class for the newbie homesteaders. What about an informal canning competition or bake-off with the elderly ladies in your congregation? Just brainstorm and don't worry too much about the logistics right now. Your homestead journal is a place to let your brain empty itself of all the neat ideas you have rumbling around in there.

DIY: If it's not on the Internet these days, it just doesn't exist. I don't use social media much for personal communication, but I use it all the time for the blog, books, and to connect with networks of homesteaders. If you're going to interact in the community and want to provide an easy way for people to get in touch with you, I suggest you get an email address for your homestead communication and set up a Facebook page for your homestead. This will require coming up with a name for your homestead, if you don't already have one. There are online tutorials for how to set up a Facebook page (visit our links page under The Homestead Community to learn how: www.homesteadlady.com/homestead-links-page) but there aren't any for naming your homestead, which can prove to be quite a challenge.

Be sure to gather your family together and get their ideas about what your homestead's name should be. Maybe it's the British in me but I strongly feel like naming your home gives it a personality and purpose. Who doesn't thrill when I mention "Pemberly", the impressive abode of Mr. Darcy from Jane Austen's *Pride and Prejudice*? What about a real-life house of Vanderbilt fame by the name of "Biltmore", nestled in the mountains of North Carolina? What about all the small and unassuming houses full of love and warmth that have born names we'll never know?

My homesteads have never been grand or fancy, but I have lovingly named them with an eye single to the vision of what they could be. The first house I named was a less than 800 square foot home in the middle of a Sacramento suburb. I had a tiny garden and a single bee hive, and I named my home "The Cottage," which made me feel cozy. Our next homestead was dubbed "Daffodil Hill" because we covered the entire slope of our property with daffodil bulbs to deter the deer that wandered through our North Carolina half acre. My next was "Pocket Farm," an acre homestead tucked into the middle of a Utah suburb. Our 20 acres in Missouri we call "WinterPast" to signify that the long wait for land in the country is over as we look forward to the harvests of many summers.

Whatever you decide, be sure to pick a name that is meaningful to you. If you have a plan in place to go further online using your homestead's name—like blogging or setting up a website to sell items from your homestead—then you'll want to make sure that the domain name hasn't already been used so that you can create your new site. There are several places online, like My Domain (www.mydomain.com), where you can determine if a name has already been used for a website and purchase your name if it's available. You buy the domain for a year, but you can easily let it go if you decide not to work with a website or blog in connection with your homesteading efforts. Either way, have a dedicated email or a Facebook page so that potential visitors can connect with you online.

BUILD COMMUNITY: I challenge you to carefully select your first group of homestead guests and prepare for their arrival by baking something sweet or preparing a homegrown harvest dish. Create a flier with information about your homestead, including lot size, what's grown every year and what your plans are for the future. Always include a resources list of your favorite books, blogs and supply houses. Your favorites will probably change from year to year but that's because you do, too, and that's wonderful! These small preparations might change the entire homestead course of a person's life; do NOT discount what you have to contribute as being unimportant.

BOOKS TO READ: I have an inspirational challenge for you! Read either, *Folks, This Ain't Normal*, by Joel Salatin or *Animal, Vegetable, Miracle*, by Barbara Kingsolver. If you want to make your eyes cross, your brain cramp and your heart rate double, read both together. They're both great homestead mentors but with totally different approaches and personalities. Even if you disagree with some of what they say or how they present it, you WILL learn new things and be inspired by their teachings. Both Salatin and Kingsolver make me want to go out into the world and do great things with tomatoes and chickens!

WEBSITES: To learn to use PowerPoint, please visit our links page under The Homestead Community chapter (www.homesteadlady.com/homestead-links-page) for a link to a YouTube video detailing an hour's worth of topic tutorials designed for the beginner. PowerPoint itself will have tutorials, but I often find great help from visual aids and love YouTube for that.

If you're the one in need of mentoring on all this homemaking and homesteading stuff, I invite you visit Untrained Housewife (www.untrainedhousewife.com) for a wide range of topics. Weed 'Em and Reap (www.weedemandreap.com) and Nitty Gritty Life (www.nittygrittylife.com) can provide inspiration along with humor—and we all could use a few more laughs.

HOMESTEAD LADY SPEAKS:
I was hiding in the bathroom with my book. I didn't even need to avail myself of the privy, I just needed to be alone with Forrest Pritchard. I had recently picked up his book *Gaining Ground*, the simple tale of how he doggedly restored his family farm against all odds. I couldn't put the thing down. It was almost aggravating how captivated I was when I had so much else to do. I mean, I had my own book to finish, children to homeschool, a home to run, and a homestead to see to. Yet there I was, squirreled away, just dying to know what happened next. I think I snorted I laughed so hard when he recounted his first attempts in the hay baler. I ached for him as the bills piled up and he gambled everything on grass-fed beef. I wanted his dad to approve of his efforts as much as Forrest did. As I was reading along over several days, it suddenly occurred to me that this was my story—this was every homesteader's and small farmer's story. All the work, the reward, the lean times, the family times.

Sometimes, the most effective way we can mentor, and be mentored by someone else, is to simply share our story. I'm beginning to think that there are only a handful of human tales and they just keep repeating themselves over the centuries, each generation convinced they're the first to feel and see the way they do. We become effective students and teachers when we find ourselves in the stories and pull the thread in each one that is like our own experience, listening to people with ears to hear and hearts that are willing to understand.

There's so much to learn, even in the bathroom with a book you don't technically have time to read. The truth is, I desperately needed some homestead inspiration just then and Mr. Pritchard came to my rescue with his broken-down tractor and his ramshackle farmer's market display. Lord love him.

What story do you have to share? Whom might you inspire today?

You just really never know until you start talking.

CHAPTER 8

The Prepared Homestead

Some modern mentalities are just weird. Take the cultural misunderstanding of the "preppers" out there. Media outlets use words like *bizarre*, *hoarder*, *fear-monger* and, my personal favorite, *fringe,* to describe people who prepare their homes and families against a time of hunger or emergency. Has it occurred to our culture that, historically speaking, the idea of NOT living a preparedness lifestyle is actually what's weird? There hasn't been a time in our nation's history, or even the history of the world, where storing extra food and other provisions for the winter, or other times of want, hasn't been the norm. For many, many cultures and peoples around the world, it still is. So, what's missing from our "advanced" culture that we no longer consider this important? Why are we so preparedness poor and complacence rich?

I can't speak for everyone, but I think the answer for some lies in a sense of ease. Not in actual ease or freedom from strife, just the feeling of it; an almost blind, tenacious clinging to the idea that we have everything under control. The more we come to find our place in the homesteading lifestyle, the more we see that there's so much we still must learn and do to be self-sufficient. We begin to recognize that self-sufficiency is a life-long quest and we settle down to the fact that we may not ever reach a place where we can say,

"That's it; I'm done now. Totally self-sustaining, that's me. Bring it on."

Because of that perspective, we learn to prepare our homes and families with a few basic provisions and supplies and slowly, but continually, build on them.

Aside from temporal provisions, preparing against a day of want provides us with an emotional and even spiritual strength that will stand us in good stead when hard times come. Losing a job and having to live off food storage for months doesn't just have tangible consequences. There's a lot of emotional fall-out from that experience, too. It can be frightening to be in that place of trepidation and uncertainty—to be so much in need. However, the more we practice a preparedness lifestyle, the more emotionally stable and experienced we'll be when true calamity strikes. And true calamity comes to us all in various ways. Without going crazy and spending thousands of dollars at once, or getting our families all in a dither, we can methodically and rationally prepare one step at a time.

No tin foil hats. No conspiracy theories. The aliens don't help us. It's just us, our families, and maybe a few chickens or pigs. That's what I think of as normal.

Here's what we'll be discussing in this chapter on the preparing for common place problems and special disasters on the homestead and in our communities:

HOMESTARTER level: The easiest place to start with your preparedness efforts is with the shortest-term preparation: 72 Hour Kits, or Bug Out Bags. Without using a bunch of chemical preparations or nutrition-less foods, you can build your own healthy BOBs to make sure your family is prepared with the basics for a short amount of time. We'll get you started in this section.

HOMESTEADISH level: Water is absolutely basic to life. Learning to store, reuse and conserve water on the homestead will go a long way towards making you prepared for any eventuality. Being smart with water is just plain being smart. In this section, we'll be covering some basic storage ideas, along with ways to cut back your water consumption and re-use gray water.

HOMESTEADAHOLIC level: Being prepared includes storing food for your family. There are many ways to preserve and store healthy food on the homestead and we list here just a few. You might be surprised at what you already have in your pantry that you just never thought of as *food storage*.

HOMESTEADED level: Having food is one thing. Knowing how to prepare food without the luxury of electricity or propane may be a new concept for us, though. Listed here are some off-grid cooking techniques presented for your consideration with my frank confessions that these methods are often tricky for me to learn. If I can do it, I promise, so can you! From open-fires to wood stoves to Wonder Ovens, there's something in this section for everyone.

As with every chapter in this book, be sure to follow along at the end of each section as you reach the Action Items, DIY Projects, ways to Build Community, resource recommendations and parting advice.

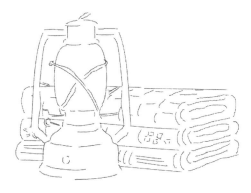

"'This earthly life is a battle,' said Ma. 'If it isn't one thing to contend with, it's another. It always has been so, and it always will be. The sooner you make up your mind to that, the better off you are, and the more thankful for your pleasures.'"

- Laura Ingalls Wilder, from *Little Town on the Prairie*

For the Homestarter: A Healthy Three Days of Preps

There is no shortage of information on the topic of 72-Hour Bug-Out-Bags on the Internet and in quality preparedness books (see the suggested resources at the end of this section). My goal here is not to list for you everything you should have in your BOB, as that would be reinventing the wheel. However, for those who are thinking that they really don't need a BOB, let's start there.

Why Do I Need a BOB?

A simple way to begin your disaster preparations is a 72-hour kit. They're also often called *bug-out-bags*, (or BOBs), from the popular Korean War term "bug-out," referring to an urgent retreat. The basic idea is that you collect enough items to keep you alive for about three days, or 72 hours. Here are some examples of situations in which you might need a BOB:

- Wildfire
- Flood
- Earthquake
- Ice or other winter storm
- Tornado
- Hurricane
- Gas leak in your neighborhood or apartment complex
- House fire or apartment fire; perhaps not even in your unit, but in your complex

- A broken pipe in the apartment above causes flooding – landlord evacuates the unit for repair
- Your car breaks down in the middle of nowhere (yes, you need a BOB for your car, too)
- You forget some hunting/camping gear, but you've got your BOB
- The power goes out in your house, but your BOBs are nearby, so the flashlights are, too
- Extended power outage in area

And that's just a few circumstances! There's just no knowing when hardship will strike, so we prepare ahead of time to minimize the trauma we may experience in the event of an emergency. This kind of precaution is especially important to take if we have small children or elderly people in our care; these two age groups do not typically adapt quickly, or necessarily well, to adverse, emergent situations. If we can make sure we've secured a few necessary supplies and even provided for a little fun with something simple like crayons, we will be much more likely to come through a difficult experience with less panic and pain.

I'VE GOT A BOB, NOW WHAT?

Once they're packed, they do require some maintenance. If you put your packs together yourself or purchased them ready-made, you should congratulate yourself! That's a big deal and a huge accomplishment. However, you still need to make sure that your BOBs stay up to date with fresh food and supplies. In our family, we rotate food and clothing out of our emergency packs every six months. That doesn't mean we replace everything in the bag; we just change out foods if they've expired, seasonal clothing and Play-Doh® that's dried out. (No kidding, it's fun to have those mini containers in their BOB, both for the kids and the grown-ups).

I am not one for details and the older I get, the worse that gets. Ask me my birthday and I'll have to stop and think. So, I need to associate important routines with special events and holidays to remember them—and make them fun. I do this with our BOBs and it might be helpful for you, too. For example, a Catholic could plan to rotate the contents of his BOB around Michaelmas and Easter every year. Jewish families could plan to do it at Sukkot and Passover. I'm LDS (a member of The Church of Jesus Christ of Latter-Days Saints) and so we restock them at our Biannual General Conferences, which are in October and April. Those who aren't religious might try Halloween and May Day. If you're pagan, try adding the BOB updates to your equinox celebrations. Whatever time of year, whatever the occasion, just make sure those BOBs are freshly stocked.

Keeping BOB Healthy

Sometimes, the things we might put into an emergency kit aren't necessarily what we would use at home, especially if we're accustomed to using herbal or wholistic wellness methods. The food, too, can be full of ingredients that we would never consume on a regular basis. Assuming that our bodies can suddenly cope with chemicals and other substances that they're not used to is not the wisest thing. In a difficult situation, the last thing we will want is mood swings, migraines, and constipation because our BOBs are full of products we're unaccustomed to—no matter how conveniently packaged.

If you're new to herbal wellness preparations, healthier foods, and a make-your-own mentality, I invite you to read on and determine if any of these suggestions are something you'd like to try for yourself. The key to having a useful BOB, that's well-suited to you, is to work with it a lot long before you might need it. We've slowly been filling our BOBs with items that will nourish and care for us in the ways to which we've become accustomed. But you may have different needs and priorities. I'm including a list of items that were once in my BOB, and those items that are now in my BOB having proved themselves to be healthier in some way.

Don't feel like you need to reproduce what you read in its entirety, or that you need to do everything right now. Organic, holistic, natural are words that can often translate into very expensive. If you already have a BOB put together (or if you have purchased a pre-packaged one), decide first what items you want to switch out of your pack. Then budget enough money each month to make a few changes at a time.

I encourage you to learn ways to cut costs by making your own items (like salves and tinctures). Other items (like freeze-dried prepackaged foods) can add costs quickly, so just save a little at a time and pay for these changes gradually. If you still need to put a BOB together, and are building it from scratch, good for you! Maybe we can save you a headache here and there. Please bear in mind that I have five small children so my pack, out of necessity, includes many things that a childless, single, or older person just wouldn't NEED to carry.

Bob Before and After

Here's a list of the original, commercial items that were in my first BOB:

MEDICAL SUPPLIES:

- Topical antibiotic salve
- Pain medication/anti-inflammatory
- Antacid (for indigestion/acid reflux)
- Anti-diarrheal
- Lip balm
- Toothpaste

- Name brand deodorants
- Sunblock, in various brands
- Mouthwash
- Bar soap
- Feminine powder
- Commercial shampoo/conditioner
- Bug repellent

COMMERICAL FOODS:

- Cracker and peanut butter packets
- Cheese crackers
- Various brands of granola bars
- Table salt and sugar
- Fruit roll ups
- Foil wrapped storage crackers and cookies
- Chocolate snacks
- Freeze dried food pouches

> You may still choose to keep these in your pack as a backup, to augment your holistic preparations or to share with others. For example, I still have a bottle of named brand pain medication in my pack. If you have prescription medication, be sure to have a week's supply in your BOB and do NOT make any changes to your regimen without consulting your holistic or conventional medical practitioner. (Human-speak: don't mess with your meds before you talk with your doc!)

Now here's a list of the healthier, holistic-y items that are now in my BOB:

HOLISTIC SUPPLIES:

- Mom's Stuff Salve® (www.momsstuffsalve.com) or homemade salves for wounds and skin care
- Styptic (coagulant) herbs like shepherd's purse, cayenne, plantain and yarrow
- Lavender oil and Arnica Montana (homeopathic) for pain balancing
- Cayenne tincture to promote balance during shock
- Antibiotic and immune support herbal preparations
- Exodus®** to help gut regulation and function
- Peppermint essential oil to aid with stomach upset and breathing
- Non-petroleum-based lip balms
- Tooth powder or another non-fluoride/sulfate tooth cleaner

- Bentonite Clay for cleaning teeth, the gut, and assuaging wounds
- Homemade deodorant, or other sulfate free deodorants
- Botanical sunscreen, homemade sunscreen, or plain coconut oil
- Castille soap without dyes or fragrance, or homemade bar soap
- Dry shampoo
- Essential oil mixes to repel insects, or homemade bug repellent

FOODIE FOODS:

- Whole grain cereal and homemade trail mix in locking plastic bags or vacuum-sealed
- Homemade dehydrated fruits like apple chips
- Organic, whole grain crackers and snacks, like Annie's®
- Gluten and refined sugar free snacks, like Lara® and Cliff® bars
- Sea salt and raw sugar***
- Organic, fair trade chocolate bars
- Homemade fruit roll ups
- Mary Jane's® Organic, freeze dried foods for backpackers. There are other companies with similarly cleaner products whose selection of Organic options is low but none of their products are genetically modified or contain MSG, according to their websites. Try: Valley Food Storage (www.valleyfoodstorage.com), Preppers Market (www.preppersmarket.com), Thrive (www.thrivelife.com)

** Exodus is a digestive aid created by Jonelle Francis as part of the Yeast Beaters Cleanse found at her website, Feel Good Foods (www.feelgoodfoods.com). I purchased her real foods, family cookbook, and the recipe for Exodus is in there. So, I now make my own with items I can easily find at my local health food store or online.

*** I have a backpacking kitchen bag in my BOB that has a cutting board, silverware, plastic bags, soap and small containers with whole grain flour, raw sugar, sea salt, pepper. The flour can be replaced with arrowroot powder to use as a thickener in campfire soups or to administer to bug bites if you prefer. This is NOT an essential part of my BOB and would be left behind first if my pack were too heavy.

FOR THE LADIES ONLY (Seriously, Gentlemen, this will be way too much information.)

I recommend "mama cloth," or similar cloth pads for everyday use, but I'm torn about menstrual cycle products to use in my BOB. Cloth menstrual pads and patterns to make your own can be purchased if they're something you'd like to try using. I've used a Diva Cup before and considered putting one in my BOB because it's small and very pack-friendly. However, to keep a menstrual cup sanitary you really need a good supply of water for each time you empty it, and water availability can be iffy in an emergency. You'll need water to wash the pads, too, of course, so they aren't a perfect solution either.

Even with the environmental impact, I would be willing to use only disposable pads in my BOB if they didn't give me such a horrible rash; apparently, I'm allergic to the chemicals or plastics in these products. As a compromise, I put a few disposables and some cloth pads in my BOB and will hope for the best. Let's face it girls, if we're menstruating when calamity strikes, it's going to be a mess. Oi.

(Back to our regularly schedule programming, Gentlemen.)

If you are new to a whole-foods/crunchy diet and/or a more holistic way of balancing your health and well-being, or you have yet to make the switch, please make changes gradually. It is unwise and unsafe to jump in if:

- your gut isn't balanced
- your system isn't accustomed to whole foods
- without knowing how to effectively use alternative wellness methods

You don't have to know everything (indeed, that's impossible) but you do need to know enough—and only you can know what that looks like for yourself.

Don't let anything come between you and the goal of getting your BOBs in place for your family this year. Just do it. Please don't let the task of making them healthy keep you from doing it, either. Believe me, if I can do it, YOU CAN DO IT!

> "By failing to prepare, you are preparing to fail."
>
> -Benjamin Franklin, Author and American Founding Father

ACTION ITEMS: In your homestead journal, make a list of what's already in your BOB that you might like to switch out for healthier options. Check the labels on the food in your BOBs. Is it up to your standards? Does it contain any allergens that will cause members of your family to become ill? Decide what you're willing to compromise on and what you're not. If you find items you'd like to switch out, set goals for doing it by specific dates. Believe me, if you don't make yourself date these goals, they'll never happen. Don't stress if you find you can't afford to achieve your goal by the set date. Just conscientiously set another goal, if necessary.

DIY: Dehydrate your own fruits, veggies and jerky this year to add to your BOBs. Learn to make fruit leather, trail mixes, jerky, fruit and vegetable combinations that can easily be added to your kits. Schedule a rotation of these contents every six to twelve months. Not only is this a great way to save money on packed food, but it will also enable you to control the ingredients in your snacks and trail food. If you don't own a dehydrator yet, use your conventional oven set down on its lowest setting. Or use a solar oven and leave the lid ajar to allow the moisture to escape and dry your foods.

To learn to dehydrate just about anything, read *The Ultimate Dehydrator Cookbook*, by Tammy and Steven Gangloff. Be sure to check out the Shelle Wells *Prepper's Dehydrator Handbook*, too – this one has some clever ideas and troubleshooting tips on top of solid dehydrating instruction.

BUILD COMMUNITY: Assembling your BOBs will be a fantastic thing to do as a group because you can buy in bulk and assemble en masse. Get your homestead book club or your ladies auxiliary from church together and go in on cost and shopping trips to find these healthier items. Get together one night with everyone and everything and just start putting together those kits. Remember to include special items for the kids like stickers, card games, and art material.

BOOKS TO READ: *Survival Savvy Family*, by Julie Sczerbinski is a wonderful, all-around preparedness read for families. She includes tips on BOBs, food and water storage, emergency drills and planning and natural disaster awareness. Her book is easy to read, sectioned well if you're looking to read by topic and includes useful charts for creating a preparedness binder for your family.

The Ultimate Survival Manual, by Rich Johnson. Outdoor Life has a series of these books by different authors and they all have merit. They're highly visual with a lot of graphics and set up to present a lot of varied information in short snippets. For example, this book by Mr. Johnson has 333 skills covered in just the one book. Clearly, nothing is very in depth, but for the survival dummy like me, they're a great place to start.

If you have children, a fun book is *Playful Preparedness*, by Tim Young. This book is full of simple games that everyone can play with their kids to get them prepared without a bunch of gimmick equipment or irrelevant crafts. There are age appropriate suggestions for each game and a bunch of bonus information at the end.

To learn how to make your own toothpaste, salves and just about anything else you might need, please read *The Beginner's Book of Essential Oils*, by Chris Dalziel. I wish I'd had this little gem as I began using herbs and oils to improve my family's health and well-being, not to mention keeping and cleaning my home in a more natural way. Also useful for DIY herbal preparations is Colleen Codekas's book, *Healing Herbal Infusions*.

To learn how to make your own soaps and other body products you can visit Majestic Mountain Sage's blog (www.blog.thesage.com) or read Jan Berry's eBook called *Natural Soap Making* (please visit our links page under The Prepared Homestead for a link to the book: www.homesteadlady.com/homestead-links-page).

 WEBSITES: For personal perspective and a lot of useful information, Graywolf Survival (www.graywolfsurvival.com) and Survival Mom (www.survivalmom.com) can give you a lot of advice on the topic of Bug Out Bags.

To access a basic checklist for a 72 Hour Kit from Fema.gov visit our links page under The Prepared Homestead chapter.

HOMESTEAD LADY SPEAKS:
As Scott, a former combat veteran and the voice behind Graywolf Survival, will tell you in his several articles on the topic of BOBs, be cautious about how much weight you add to yours. The temptation for a lot of us is to try and cover every eventuality, especially if we have younger or older people in our care. Right now, my pack is so heavy with items I feel I really need to carry that I can hardly lift it, let alone carry it for three days or more.

I have a goal this year to go back through my BOB and separate out items by type and bundle them together in smaller bags. If I need to, I can go through my pack and remove a few of my "non-essential" items. Even writing that causes me anxiety because all those items seem essential to me. But, the truth is, my family and I could survive on less. The other thing I need to do is switch over all the food in my pack to freeze-dried only – it's so much lighter!

It can be difficult to wrap your brain around the idea of surviving only on what you have in a backpack for a minimum of three days. I assure you, you'll be so happy you made the mental effort if there comes a time to use your BOB.

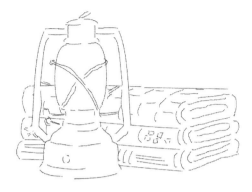

"Drink waters out of thine own cistern and running waters out of thine own well."

- King James *Holy Bible*, Proverbs 5:15

For the Homesteadish: Water—Storage and Conservation

Water is so fundamental to life…water is life! Regardless of whether you consider yourself a prepper or a full-fledged homesteader, we are all united in our basic need for water. There are many ways to store water, and it's important that we learn other ways to be water-savvy on the homestead, too.

STORE EMERGENCY WATER

It would be incredibly difficult for most families to store what might be considered a year's supply of water for each family member. Water is heavy, it takes up a lot of space and it needs to be kept cool, dark and rotated. Many families do strive, though, for a three-month's supply, which might be more doable. If that feels too daunting, aim for the FEMA (the U.S. Federal Emergency Management Agency) recommendation of at least a two-week supply. The standard recommendation for water storage is one gallon a day per person but that's just not enough for hygiene, cooking, animals, illnesses, a hot summer or climate, or nursing mothers/hard workers. Planning for two gallons per person seems more appropriate to me. Regardless, bottom line, store as much water as you can safely store, and don't forget to plan for pets and livestock if you have them.

> **WATER WHEN YOU'RE OUT AND ABOUT**
>
> Having water with you when you travel is important, even when there is no emergency or disaster. Most of us don't drink enough water each day, so carrying it with us throughout the day is a good way to remind ourselves to drink! Or, am I the only one who struggles with this?

To store water in a vehicle in spring, summer and fall, find an out-of-the-way place to tuck it in. Wrap your water in insulated material to keep it as cool as possible. When hot, plastic leaches chemicals into the water stored inside, so try to avoid exposing these containers to heat. If you live in a severe winter weather area you can either toss your water in the car each time you go somewhere, or you can under-fill your containers to allow for expansion if the water freezes.

For every-day use, we have Life Factory® (www.lifefactory.com) glass bottles with silicone sleeves that prevent breakage. I love these bottles for peace of mind, especially in the summer heat; I don't have to worry about plastic chemicals poisoning our water. They're heavy for hiking, though, so I recommend you have a Klean Kanteen® (www.kleankanteen.com) or other stainless-steel water bottle for carrying with you on long hauls. Camelbak® packs (www.camelbak.com) are also useful (but the reservoirs that hold the water are made of plastic). It's important that you remember to dry the reservoir every time you use it to prevent mold. Camelbaks® are light-weight and easy to carry and, for their many fans, the little bit of work to keep them clean and dry is totally worth it.

Containers

I'm not a big fan of plastic, but there are several very handy plastic storage containers these days that make keeping water a lot simpler. For example, many people use 55-gallon drums for home water storage; they're big enough to be worth your time and can be emptied with a hose and a pump that you can buy especially for these drums. You can buy the drums new online or at your local preparedness store (if you're lucky enough to have one). You can also purchase them used from local food processors.

Buying used containers means you'll need to clean them yourself, and you'll need to be extremely thorough to ensure that no bacteria-causing product is left behind. I can't advise using drums that have had cleaners or other chemical containing solutions stored in them because, no matter how you scrub, you run the risk of lingering contaminants. I can't heartily recommend the used ones at all, quite frankly. However, they are an option for us cash-strapped folks. There are plans online for building your own rack on which to store these drums laying on their sides, which is a space saver and makes them easier to use when it's time (although they have been known to leak on occasion, just FYI). Plan to rotate all water storage every six months, regardless of the container.

Another option for larger water storage containers are super-tanks that store anywhere from 250 to 500 gallons of water and are designed to be taller than they are wide. They look like a hot water tank. As opposed to removing water with a pump, these tanks typically use a gravity-fed system to deliver the water from a spigot. They aren't designed to be portable, of course, but they make excellent use of the floor to ceiling space. Plan to put them where you can readily access and rotate out the water from where they're stored. If you don't have the space for larger containers, there are smaller (five to seven gallon) containers that are easier to maneuver and lighter to stack. Often called "water bricks," these containers can be found online, as well as at your local camping store.

The easiest and most portable commercial water storage option is to purchase individual water bottles from big box stores in bulk, though the plastic used in these bottles is often not very durable. Whenever possible, your home water storage containers should be made of dark, thick, food-grade plastic. Such containers won't allow light to permeate through to the water which can reduce the growth of algae and bacteria.

You can create your own water storage containers by upcycling juice and soda containers. Save the lids, wash the containers and fill them with water, but be sure the containers are sparkling clean. Add a water-cleaning agent to your water (most experts recommend bleach) and keep them out of the light. Even with these efforts, you may still develop bacteria in these upcycled bottles. Be sure to check the integrity of your re-used containers often. Many juice and soda bottles are made from recycled plastic which can often mean that they just aren't that thick or strong. Because of this, I really recommend commercial, food-grade water storage containers, if you can financially manage it, for ease of use and safety. If you want to avoid plastic altogether, there are some stainless steel* containers to be had online, but they're double to quadruple the cost of plastic. It is what it is. Do your best.

*Be sure to store only non-chlorine and non-chloramine water in metal containers to avoid corrosion.

Water Everywhere

You can also choose to store foods with high water content in your home. Home-canned soups and stews, as well as fruits and veggies, are all preserved with water. I had a canning friend who would put water in her empty canning jars as she stored them to keep them useful while they were empty of food. Please remember to secure your stored items, especially glass jars, properly against earthquakes. You can't guard against every disaster, but you can properly strap down water containers and hold in glass jars with a locked cabinet door or a sturdy bar.

Here's one last home water storage suggestion, especially for those that live where severe weather events occur, such as hurricanes, ice storms, etc.: there are plastic liners you can buy for your bathtub that are form fitting and designed to be filled with water from the faucet in case your municipal water service is interrupted. A standard tub can hold between 80 and 100 gallons of water. This can be invaluable in an emergency! Be careful that children in your home don't play in, contaminate, or drown in the water.

Finally, be sure to put your home water storage on a rotation schedule, which should occur every six months. You do NOT want to access your water storage in an emergency, only to find that it's bright green and deadly. In rotating your supplies, remember that, the stale water can be used to wash people, clothes, and dishes, as well as to flush toilets and water the garden. I try to make sure I'm being careful with every drop of water I use or consume, but I still have so much work to do in this regard. Water is an amazingly pure and powerful resource and we are blessed to have access to it, clean and fresh each day. The best way to show our gratitude to nature and our Maker is to use it well and to store as possible against a day of need.

WATER ON THE HOMESTEAD

It's a simple truth that animals and plants need water to thrive, just as people do. One of the best ways to keep your garden thriving is to always use plants for your growing zone and be water-wise in your plant selections. For example, if you live in a desert, don't try to grow bananas. Bananas not only require a lot of water to flourish, but a certain humidity level in the air to bear fruit. If you have purely ornamental parts of your garden, you may consider "xeriscaping" those areas. Xeriscape is a type of garden design that optimizes irrigation needs through plant selection and ground cover choices. Having said that, however, most homesteads cultivate veggies and fruits, all of which are water hogs compared to other plants. See the *Growing Smarter* section in The Homestead Garden chapter for more ideas on water use and direction in the garden. Or read up on permaculture, most specifically using swales and mulches in the garden.

There are several options for water storage on the land and around the homestead. Natural or man-made ponds are an obvious source of water for livestock and land. If you live in a city and its legal, having a well dug can mean water independence. You can go one step further and run your well pump with a solar panel so that you aren't relying on the electrical grid to pump your water. Even if you have access to public water, a well can mean available water regardless of what else is going on in your local world.

If a well isn't feasible, the next largest container is a rainwater-harvesting cistern. Even the small ones are large, so find a local source to avoid high shipping costs. Cisterns can be placed in or above the ground, and are usually made of metal, plastic, or cement. You'll need to make decisions about placement and design, so be sure you do thorough research on what's available to you locally. On a smaller, and possibly more realistic, scale for small-space homesteaders are rain barrels. These rainwater catchment barrels are a wonderful way to preserve the water falling for free from the sky. Rainwater can be used in any number of ways on the homestead, most typically to water the garden. Check local ordinances, and the regulations of your HOA, to be sure rain barrels are legal where you are. Rain barrels have the added benefit of being something you could conceal behind a shrub. Plus, they're simple to make yourself!

Because of the enormity of the topic, this section is meant only to be a list of general ideas and suggestions. I'm hopeful you found something that sparked your interest. Each idea has its own list of pros and cons, so simply tackle one item at time. The most important lesson here is that you need to begin some kind of water storage program, and you need to do it *today*. Now, pardon me while I go check my own water barrels...ahem.

WATER-WISE WILLPOWER

Following are some water-wise ideas for the modern homesteader from Kendra Lynne (www.newlifeonahomestead.com). Kendra and her family have been transitioning their on-grid home to an off-grid one and Kendra shares that journey on her blog, along with so much more useful homesteading information. Kendra says:

"The low-flow faucet adapters have saved us a lot of water. It was a very simple and effective way to reduce our [water] usage right off the bat.

"We installed rain barrels to catch water for watering the garden instead of using the water hose. This reduces our need for using the well pump, whether we'd be pumping by hand or using an electric submersible pump.

"[Another suggestion, is to] turn running water off when not in use. For instance, when you're brushing your teeth, turn the water off between rinsing. When you're showering, turn the water off between lathering up and rinsing off.

"We actually pulled out the garden tub in our master bathroom to make better use of the space. It was so big and would have required a LOT of water to fill, which wouldn't be very efficient. Since the master bathroom is the coldest room in the house during the winter, the tub's old spot is a great place for the chest fridge and freezer because they won't have to run as often in the cooler room.

"I purchased a plunger washer, washboard, and a wringer for washing clothes by hand. Although we do still have our washing machine hooked up for larger items, such as quilts, sheets, and towels, I do enjoy the therapeutic, methodical rhythm of hand washing smaller everyday items.

"We installed a hand pump on our deep well. It's not solar, though we do have the option to upgrade to solar in the future if that's something we want to do. It's working well, though we only get 5 gallons of pressurized water to the house at a time due to the size of our pressure tank. We could opt to fill a larger holding tank to gravity feed to the house, but so far we haven't had a need to do that."

ACTION ITEMS: After doing some research, pick three areas of water management and storage that you would like to explore in the next six months. Write these down in your homestead journal. Be sure that one of the areas you chose concerns water storage in the home, the car, or on the homestead. Tally up the cost of containers and any other equipment and begin to budget for it with this month's paycheck. Calendar a schedule for purchasing, filling, storing or otherwise getting your water storage in place. By the end of the six months, complete these three goals and set three more for the next six months. Be sure to write your time-line on your family calendar so it's a visible reminder to make your goals part of your monthly activities.

DIY: There are two easy ways to be proactive with your water. First, reduce the amount you use right now. Read Kendra's suggestions in the side bar or come up with your own. Really, this is the simplest place to start.

Next, if it's legal where you live, start using water barrels. There are some quality barrels you can buy. Naturally, it's cheaper to make your own, especially if you need several. The project will take a few materials and a bit of know-how with tools (like drills and saws), but the process is no more than opening the top of a barrel and putting in a spout.

To learn to make and install your own rain barrel, please read the article from Betsy Matheson featured at Mother Earth News found on our links page under The Prepared Homestead: www.homesteadlady.com/homestead-links-page.

To follow these instructions, you will need:

- Barrel or trash can
- Drill with spade bit
- Jigsaw
- Hole saw
- Barb fitting with nut for overflow hose
- 1 1/2" sump drain hose for overflow
- 3/4" hose bibb or sillcock
- 3/4" male pipe coupling
- 3/4" bushing or bulkhead connector
- Channel-type pliers
- Fiberglass window screening
- Cargo strap with ratchet
- Teflon tape
- Silicone caulk

You can also read these instructions in Matheson's book, *DIY Projects for the Self-Sufficient Homeowner.*

BUILD COMMUNITY: If you decided to do the rain barrel project, try to get a group assembled to build several together. If you decide to use a 55-gallon drum for the barrel, there's typically a savings when you buy a higher volume. Turning this into a group project also means that, if you're not the tool-savviest person on your street, you might be able to share in a neighbor's knowledge. If you really aren't comfortable leading the way on this project, find a do-it-yourselfer to help the group with following the instructions. Sometimes, you can find mentor-lead classes on building your own rain barrel through your local agricultural extension or public works department.

BOOKS TO READ: *The Prepper's Water Survival Guide*, by Daisy Luther, covers finding, treating and storing water. It's short and to the point. *The Water Wise Home*, by Laura Allen, may also be of use to you.

The Prepper's Blueprint, by Tess Pennington is a great, all-around prepping book that has several chapters on water storage.

 WEBSITES: To learn more about storage water from their whole category on the subject, visit:

- Food Storage Moms (www.foodstoragemoms.com) for water storage, or any storage
- Ready Nutrition (www.readynutrition.com) and Survival Mom (www.survivalmom.com) are great for general preparedness
- For a link to an overview of water storage from Ready.gov, please visit our links page under The Prepared Homestead chapter: www.homesteadlady.com/homestead-links-page.

HOMESTEAD LADY SPEAKS:
There I was this morning, kneeling with the toddler in the church parking lot. I was fruitlessly trying to sop up the oatmeal barf from her Sunday dress with the only thing I had on hand: a roll of toilet paper from our car's smashed, but serviceable, emergency kit. The putrid oatmeal was just everywhere—in her sandals, the gathers of her petticoat and belts of her car seat. The toilet paper was unequal to the task. I suddenly remembered that we had glass water bottles still in the car from yesterday's trip to town. Our emergency kit water bottles had long since been used, and I'd neglected to replace them. But these others would certainly answer the need my poor two-year-old had, to come out from under the barf.

She stood there good-naturedly while I doused her dress and shoes, cleaning it all as best I could. As I dried her down with a towel from the kit and we went inside to watch her siblings perform in the special children's program that day, I was so grateful for that water. After the program, we put the smelly cutie pie back in the car and took her home. We wound our way along the roads that cut through a land alive with water in the lush, green fields and hidden ponds. I kept thinking back to the great relief I felt finding that small amount of water just when I needed it. My little one certainly wouldn't have died without that water to help us clean up. But, she was more comfortable than she would have been otherwise.

Though I've never been threatened with a sustained or fatal lack of it, I've had so many opportunities to be grateful for water and the blessing of having enough with which to wash and drink. I've always lived in cities and had access to municipal water. Even with its fluoride and chlorine levels, city water is there when you turn on the faucet. What a miracle that is. Being a native Californian, I became used to living a water conservation lifestyle because of the endless cycles of drought. However, even during those cycles, we always had enough water to sustain life. Just not enough to sustain lawns. My short stint in water-conscious England taught me to use one basin of rinse water for the dishes, instead of turning on a stream of water to wash each one. You do what you must do when water is scarce, and you're accustomed to consuming it in certain ways.

I remember the days of unexpected water loss when I lived in Russia. We'd wake up one morning and the water would be off for no one knew how long. Every time we'd ask about the cause people would just say, "They're cleaning the pipes." I was never sure who "they" were or what pipes were being referenced. One thing I discovered living in Russia was that knowing the why of things doesn't change reality. We simply learned to store water in every container we could find. Doing that we could to still brush our teeth and flush the toilet, and even drink the precious stuff.

I've never struggled through the daily chores of hauling water from nearby streams for drinking, cooking, and cleaning. So many people around the world do just that every day. The closest I've come is moving from the city to the country where I'm now learning how to use and maintain a well. As we look at this new contraption and all the electricity it takes to keep the well-pump working (without which we have no way to draw up the water except to rip open the top and use a bucket), we've become very aware of how many times a day we turn the handles of the faucets in our home.

I keep asking myself questions like: do I really need to wet my toothbrush before I use it? And: is there a better way to clean my hands throughout the day than just turning on the water each time to do it? These are good questions to ask. It's also clear that we need a good water storage program for this new home, and we need to get it up and running as soon as possible. Ice storm season is coming, and the electricity often goes out, so…

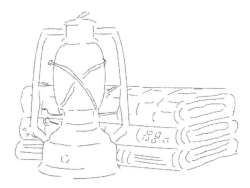

"We hear, 'Many more people could ride out the storm-tossed waves in their economic lives if they had their year's supply of food ... and were debt-free.' Today we find that many have followed this counsel in reverse: they have at least a year's supply of debt and are food-free."

- Thomas S. Monson, Religious Leader and Humanitarian

For the Homesteadaholic: Health-Wise Food Storage

There's food storage, and then there's healthy food storage. You can find cheap, highly processed commercially preserved foods in bulk to store for your family's needs. They're readily available, too, which makes building your food storage supply gradually a simple process. However, is the cheapest, most readily available food storage the best to have? Is it the healthiest option available? Is it possible to get healthy food storage?

WHAT IS HEALTHY FOOD STORAGE?

That, my friend, is what you call a loaded question. For those struggling with various health issues, food can be a hot button topic. Start extolling the virtues of wheat to someone who has found balance with a gluten-free diet and you might just lose your head. On the other hand, many people don't seem to struggle with the yeastie-beasties (yeast overgrowth in a broken gut exacerbated by grain consumption – see the *Ferments for Starters* section of The Homestead Kitchen chapter for more information). These people are equally protective of their whole grain breads and pastas. I can guarantee that you and I are going to be so busy building our own healthy food storage that we won't have time to worry about other peoples' dietary choices, so learn to relax regarding what others are doing.

Also, try to cut yourself some slack as you build a food storage pantry. Most of us would prefer to purchase storage foods that are 100% clean and pure as the driven snow, but the reality is that we just can't afford to purchase certified Organic everything. Give yourself a learning and budget curve as you strive to find the healthiest items you can afford, as you can afford them. Remember what Toby Hemenway said about our food growing efforts in *Gaia's Garden*?

"Overall, doing an imperfect something is better than doing a perfect nothing."

Start by building a week's supply of supplemental foods, then a month's, and then three months' worth. After that, going up to a year's worth of longer term storage items won't feel so overwhelming a goal. Organic brands are usually more expensive. So, if that label is important to you, make sure you build up your stores slowly, so as not to break the bank. For example, when you go to the store, just buy two cans instead of one. Always remember that, some food is better than no food. If the ultra-healthy, Organic brands are simply outside your budget, but you still want the healthier options, buy the green beans instead of the Gatorade.

There are really two kinds of food storage: *long*-term and *short*-term. We're going to cover a little of both here. The goal is to think about what we might include in our food storage to keep it as healthy as our every-day menus. Let's look at each approach individually first, because there are some significant differences between long- and short-term planning. Bear in mind, the more disposable income you have available to spend on your food storage, the "healthier" you can require your store-bought items to be. Alternative, Organic-labeled packaged foods are going to be generally more expensive than their conventionally produced counterparts. There's more to food storage than packaged foods, though, so let's begin.

WHY FOOD STORAGE?

If you're a Christian, stop me if you've heard this one; everybody else, I have a little story. If you turn to the 25th chapter of the book of Matthew (verses one to thirteen) in the Bible, you'll find a story told by Jesus that has a double, even triple or more, meaning. For those of you who are not that religious, don't worry—I'm not whomping up a sermon, I'm just making a point. Here's what Matthew recorded:

"Then shall the kingdom of heaven be likened unto ten virgins, which took their lamps, and went forth to meet the bridegroom. And five of them were wise, and five were foolish. They that were foolish took their lamps and took no oil with them: But the wise took oil in their vessels with their lamps.

"While the bridegroom tarried, they all slumbered and slept. And at midnight there was a cry made, Behold, the bridegroom cometh; go ye out to meet him. Then all those virgins arose and trimmed their lamps. And the foolish said unto the wise, Give us of your oil; for our lamps are gone out. But the wise answered, saying, Not so; lest there be not enough for us and you: but go ye rather to them that sell, and buy for yourselves.

"And while they went to buy, the bridegroom came; and they that were ready went in with him to the marriage: and the door was shut."

Now, there are about sixty-seven different Sunday School lessons and chats over lunch that we could pull out of this one story, but I'm going to focus on the most basic and that's this: stuff happens. Big stuff, life changing stuff, stuff we plan and stuff that just comes out of nowhere. There's no way we can have emergency essentials lined up for every potential personal, familial, political, professional or natural disaster. Even the coolest prepper-dude and the most conscientious mom are not going to be able to prepare for everything. BUT, we can all do something, as Matthew goes on to write, to "watch therefore."

The Bridegroom wasn't asking for the moon; He just needed the wise virgins to have oil in their lamps. Now, some might look at this story and think those wise virgins were a bunch of selfish brats. Why couldn't they just share their oil so that everyone would have some and they could all go party? My response to this is, to be honest, not seemingly very Christian. Want it?

Grow up. It's nobody else's job to take care of you, just like it's nobody else's job to take care of me. A certain amount of oil was needed to last the night; some brought enough, and some didn't. It's not brain surgery, it's oil. We prepare to take care of ourselves because nobody is obligated to do it for us. Period. The process isn't even that hard, it just requires some planning and sacrifice.

And look, it was The Christ who labeled five of them as *wise* and five of them as *foolish*. If someone has a problem with that, they can take it up with Him.

Short-Term Food Storage

Here are some basic ideas of short-term foodstuffs you can start acquiring for your home food storage program that will typically last between one and twelve months.

FRUIT AND VEGGIES
1. Home-grown, commercial, or Organic fresh and frozen fruits and veggies. In an emergency, these are eaten first and are gone quickly, but they count!
2. Fresh apples, winter squash, garlic, potatoes and other suitable "root cellar" crops. All of these can be stored in a cool, dark place for months, depending on conditions
3. Home-canned vegetables and fruits. You control quality when you can it yourself and the importance of these items cannot be overstated for keeping nutritional balance in your food storage menus.
4. Home-grown (or from a trusted, local grower) dehydrated produce. Tomatoes, zucchini, pumpkin, onions and much more can be dehydrated and stored in your pantry for at least a year. These items are so important to have to offset the grains and proteins you store. My body has a hard time digesting grains and I struggle with meat sometimes, so having fruits and veggies in my food storage is important for me and for my family.
5. Organic brands of canned vegetables and fruits if you can afford them.

6. Home-canned fruit juices. It's nice to have something sweet and cozy to drink; you can combine them with water kefir to make homemade, healthy soda pop. (Please see the *Ferment all the Things* section of The Homestead Kitchen chapter for more information on kefir.)
7. Home-canned ketchup and barbecue sauce. Remember that tomato-based canned goods tend to have a shorter shelf life, so plan to use them up first (at our house, they go first because they taste divine).

FATS AND PROTEINS

1. Homemade butter from fresh milk. Store this in your freezer or refrigerator for the longest shelf life because raw butter will continue to culture. For longevity, try making clarified butter, or ghee; some Indian families will keep a small pot of ghee on their shelf for several months.
2. Avocado, olive or other virgin oils. These won't last as long as the less healthy canola oil, so buy them in smaller, usable quantities. You can learn to cook with them and rotate these oils often through your kitchen, always keeping your storage supply fresh.
3. Frozen or dehydrated eggs. Be sure to keep these cool.
4. Bulk blocks of quality cheese. These cheeses be cut up and waxed for longer term storage. Cheese also continues to culture, so the longer you store it, the sharper it will get.
5. Home-canned bone broth or Organic store-bought canned soups and broths.
6. Organic peanut or other nut butters. You can also store whole nuts to make your own nut butters, but you will require a food processor or manual grinder to blend them.

GRAINS AND BEANS

1. Homemade breads stored in the freezer.
2. Homemade crackers, granola and fruit leather. Storing these items in glass is healthier, but make sure to earthquake-proof your food storage areas so you don't end up with broken glass in a seismic event.
3. Organic macaroni-and-cheese and other packaged foods you trust.
4. Flax seeds* for adding to just about anything to increase healthy fats in your food. Be sure to have a way to grind or masticate them for better digestion.
5. Boxed, Organic, whole-grain pastas. Or you can store the grains required to make your own pasta, and a way to grind them with and without electricity.
6. Rice cereal, cream-of-wheat, and whole grain flours. Oatmeal, flours and cereals are all made from grains that have been cracked, rolled or ground up; in this state,

they lose their nutritional value rapidly and they spoil quicker which makes them better for short term storage where they should be rotated often.
7. Organic grains like brown rice*.

Flax seed and brown rice can go rancid due to the high volume of (healthy!) oils contained in both, so they're recommended for shorter term storage.

MISCELLANY

1. Homegrown or Organically purchased herbs to be used as medicine and food.
2. Organic condiments and dressings. These aren't needed to sustain life, but if you've been living off your food storage grains and pastas for any length of time, they sure can make those bland dishes taste a little more interesting. (Most of us *experience*, not just eat, food.)
3. Organic chocolate or other favorite treats. Do NOT skimp on these, as you will want them. In stressful times, it truly helps to have as many familiar flavors as we can get, and treats are comforting.
4. A bottled water brand you trust. See the previous section in this chapter entitled *Water Storage and Conservation* for more information on water storage ideas.

Don't discount the food in your freezer, refrigerator, and root cellar. These items are great for a short-term storage plan because their shelf life is also shorter—you *must* eat them first. I don't worry too much over the shelf life of the items I grow and preserve myself because I'm pretty sure we'll consume them in one year's time. I still have pickles and jams from a few years ago, but all the sauces, broths, meats, fruits and veggies that I've put up myself have usually been consumed by the time the next harvest season rolls around.

My dehydrated items always go first because my children enjoy eating them as snacks. Fruit leather and raisins, "chips" made of apples and zucchini, and dehydrated tomato slices (that become quick spaghetti sauce when blended or an addition to salad when soaked in olive oil and garlic) all make wonderful, dehydrated snack foods.

> **MAKE CHOICES AHEAD OF TIME**
>
> If you live alone and are employed outside your home, you may need to purchase Organic options as you might not have the time to plant a garden yourself. The point is to get at least something stored, and to make sure it's as healthy an option as possible whether you grow it yourself or not. Simply do what fits your circumstances best. If you really need gluten-free and can only afford rice pasta as opposed to quinoa pasta, pick the rice. Or skip the pasta altogether and buy beans, because they're wholesome and cheaper. I encourage you, though not a millionaire, to still try and feed your family a whole foods diet, even with food storage.

> If you're planning on storing a lot of beans, make sure you get your system used to digesting them by eating them now. You can't just suddenly start eating beans three times a day and not compromise your gut, causing acute discomfort. Please also be particularly careful with how you plan your food storage if you follow a GAPS or Paleo diet—or any of the others that require strict prohibition of specific foods. I know firsthand that this can be hard, but if you plan now, you won't have occasion to be sorry in the event of an emergency or financial set back when you're eating your stores.
>
> We were recently living off our food storage and I discovered just how unaccustomed my tummy was to digesting so much grain. Due to various circumstances (one of which was moving and not having our substantial garden available), our food stores had become almost completely depleted of fruits, vegetables and protein. I carefully rationed what there was of those items, but the only way to keep us eating every day was to use a much larger quantity of grains than I would have normally. That financially tight time passed after a couple months and we were able to cautiously buy food again. You'd better believe the first things we picked up at the farmers market were fresh fruits and veggies.
>
> If you're a prayerful person, another thing to keep in mind is to ask a blessing on your food to make it as healthy as possible. We take special care to pray over every meal and ask for harmful elements to be removed and for those remaining to be healing to our bodies. It's hilarious to hear my kids ask for the food to taste good, too, so we can eat it—am I that bad a cook? Whatever Higher Power you believe in, petition for help here. You're doing your best to eat the healthiest food you can with the resources you possess. That Higher Power will understand and assist you.

Short-term food storage is a perfect way to build up a supply of the things you use all the time and would really miss if you couldn't access them. We've been financially challenged these past few months and I've learned what I pine for in my penny-pinching kitchen. Top of my list right now are cultured dairy and eggs. I didn't realize how often I utilize these items to replace the high quality, but expensive, protein of meat. While I love beans and so does my family, after a while, you're kind of "over it" and need other options. I'm planning on increasing the number of dairy products in our food storage program as soon as I can. Someday my lazy hens might keep us supplied with eggs!

Whatever you do, be sure to have an adequate variety in your food storage. In the Biblical story of the Exodus, manna (a bread-like substance that was miraculously provided each day) was a great blessing to the ancient Israelites when they were starving, but they soon grew bored with it; even when God sent them quail, they were still whining. While I'd like to think I'm way less picky than that, I'm not so sure I'm being honest with myself.

HELPFUL TOOLS TO HAVE FOR SHORT-TERM FOOD STORAGE:

- Water-bath canner
- Pressure canner

- Dehydrator
- Cheese press
- Steam juicer

You can also use a solar oven to dehydrate fruits and veggies by venting the lid. A solar dehydrator works wonderfully, too.

Each tool requires you gain new skills, and that can be difficult or require a financial investment. But, you can do it over time and with careful planning.

Long-Term Food Storage

Here are some items to consider for your longer-term home food storage program that will last anywhere from two to twenty-five years. Always remember that the fresher the item, the more nutrition it will have. Having said that, however, there are many foodstuffs that will last a long enough amount of time to make them very worthy of saving.

1. Organic freeze-dried produce, meats and dairy. Dehydrated produce can also be stored longer than a year, if processed properly, but freeze-dried items have a much longer shelf life.
2. Coconut oil is my favorite long-term, healthy food storage fat. I buy in bulk and use it within one year, but I've read that it can last several years. For more information, go to our links page under The Prepared Homestead and look for the Tropical Traditions link: www.homesteadlady.com/homestead-links-page.
3. Raw honey. This is used for both food and medicine, just like coconut oil. It might crystallize while it sits (you can reheat it and its fine), but it will last indefinitely—well, not at my house where it's devoured rapidly.
4. Sugars. Organic, low-processed cane (succanat and panela/rapadura), raw sugar crystals, coconut sugar, Grade B maple syrup are all good options. Even if you don't think you use much sugar anymore, you can still trade these items. Incidentally, my local cooperative grocery store will order me 50-pound bags of Wholesome Sweeteners raw sugar for a discounted price, which they'll do for any of their bulk foods; see below for other bulk foods providers for building up your stores.
5. Powdered fair trade cocoa. This will only last about two years so be sure to rotate it (Aw, poor you will have to find a way to rotate the cocoa.)
6. Sea Salt. Store this in whatever variety you feel is healthiest, which can be finely-ground or in large crystals. You do NOT want to run out of salt.
7. Dried natural leaven (aka natural yeast or sourdough). Once dried, this stuff will last indefinitely, meaning that you do NOT need to store commercial yeast. If you learn ahead of time how to use natural yeast, that is. (See our *Ferment All the Things* section in The Homestead Kitchen chapter for more information.)

8. Larger containers for water storage, like 50-gallon drums.
9. Organic dried beans, lentils and pea of all kinds.
10. Organic wheat, whole oats*, rye, barley, quinoa, amaranth.

*To make oatmeal from whole oats you'll need a grain roller/flaker.

I included several grains in that list but how long each will last varies quite a bit. Hard, red or white winter-grown wheat*, stored properly, can last 25 years, while quinoa can last about 3 years. Don't discount lesser known grains like quinoa, though! Quinoa is a great substitute for rice if your tummy has a hard time digesting the latter. A little tip for using quinoa: make sure you soak and then rinse the grains several times before you cook them up, or you'll taste the natural saponins (basically soap suds) on the grain. Sometimes I soak, sprout, rinse and dehydrate my quinoa to grind it into flour for cookies, pancakes and breads; it bakes up like fine corn meal. (To learn more about pre-soaking grains and seeds, please see the *Ferments for Starters* section of The Homestead Kitchen chapter.)

When determining how long a food will last for longer-term storage, you really must consider how it's being stored. A food item in a plastic zip bag just won't retain its nutritional or taste value as long as something stored in a vacuum-sealed bag.

*To learn a little more about hard wheats (information on other grains also available), please visit the Bob's Red Mill link on our links page under The Prepared Homestead: www.homestead-lady.com/homestead-links-page.

HELPFUL TOOLS FOR LONG-TERM FOOD STORAGE:

- Grain roller/flaker
- Grain grinder (both electric and manual)
- A food storage room that's cool and dark with good shelving

Yes, each one requires you gain a new skill and make a financial investment. That can be hard. But you can do hard things.

> ### GROW IT AND SAVE?
>
> Can you save money by keeping a garden and putting up your own food? If you are currently buying Organic brands in bulk for food storage then, YES! A garden can be a lifesaver for your food storage budget because certified Organic foods are usually double to triple the cost of their conventionally produced counterparts. If you're not growing a lot of your own food yet, buying bushels of local produce for preservation can also be a thrifty option.

> Similarly, Community Supported Agriculture programs, (or CSAs) and farmer's markets will most likely be able to save you some money when you buy large amounts of their high-quality product for preservation. Bear in mind, however, that if you do not yet own canning or other food preservation equipment, you will need to purchase those items. Such equipment will save you money every year you use it, and much of it can be purchased used, but be aware of that cost.
>
> If you're currently buying the cheapest items you can find at the grocery store, regardless of how they're grown or packaged, or how nutritional they are, then whether or not your garden can save you money becomes a matter of debate. The more homegrown foods and the less packaged foods you consume, the better your health will most likely be. The healthier you are, the less money you'll spend on health-care related purchases and the less work time you'll lose due to illness. Could the key to your grocery and health-care budget possibly be the humble, homegrown tomato?
>
> I suggest you start growing at least part of your own food, or find local growers to supply you, and experiment on the cost-effectiveness of DIY food preservation. Give yourself at least five years of a completely homegrown, handmade, hand-preserved diet. If you still think you were healthier eating conveniently packaged and processed foods, then I won't argue the money with you.

ACTION ITEMS: Take an inventory of your food storage in your homestead journal. Make a list of anything canned, bottled, boxed, or bucketed. Note the expiration date of each item; if you can't find an expiration date, try to remember when you purchased it and make a note of that on the item itself. If you have a commercially packaged product, it should have a "Best if used by" date somewhere on the packaging. You can also look online at the manufacturer's website for more information. Train yourself to write an expiration date on everything you purchase or preserve as part of your food storage program. I use a large, permanent marker to write on the sides of buckets and cans so I'm able to quickly spot aged-out products in my storage room.

Look up the recommended amounts of food for a year's supply, considering the size of your family. Note how much of each item you still need to purchase to have "enough." Bear in mind, these recommendations are for the *bare minimum* amounts of food to keep your family alive—you won't be dining in excess with these amounts. Don't freak out if the numbers you write down look big. You don't swallow a pan of lasagna whole, do you? Of course not! You cut it up into single-size servings, until the whole pan has been served. Apply this same principle with acquiring food storage; just do a little bit every shopping trip and learn to can, dehydrate and otherwise preserve everything you're able.

As beloved children's author, Maj Lindman wrote:

"One thing at a time and that done well is a good plan."

You can use a simple online food storage calculator from About.com OR and access a much fancier one from Food Storage Made Easy by going to our links page under The Prepared Homestead: www.homesteadlady.com/homestead-links-page.

DIY: Listed in the text of this section are some helpful tools to have and to learn to use. Here, I'll list them again: water bath canner, pressure canner, dehydrator, solar dehydrator, solar oven, cheese press, steam juicer, oat roller, grain grinder (both electric and manual) and a food storage room. Pick one of these, acquire it and learn to use it.

If you have the space in your home already, I suggest you set up a food storage area first. (If the idea of setting up a space for stored food feels too daunting right now, how about figuring out how to use a grain grinder? Or a canner?) You should choose the coolest room in your home for your food storage (basements work well, if you have one). You should also cover the window, if there is one, to avoid direct sunlight resting on your stored foods. You want enough room for your food storage, as well as enough room to walk around and access items as easily as possible. Sturdy shelving units come in handy in a food storage room, too.

If you already have a food storage room or you want to start your new one off on the right foot, make sure you earthquake-proof your storage items as much as possible. Here are five simple steps to get you started securing your food storage against earthquakes and other potential hazards:

1. Brace your shelving against a wall stud or the floor using proper hardware. If you have multiple shelving units, you may want to secure them to each other, as well. Do NOT skip this step!
2. Secure your glass-preserved produce individually. Having your hard work ruined by broken, glass canning jars would not only be an irritating waste of food, but also a real danger. The easiest way to secure your glass jars once they're full of produce is to simply put them back into the box in which you purchased them. These boxes have cardboard dividers inside that protect the jars from breakage. If you no longer have these boxes, fashioning your own cardboard dividers would be a simple matter once your jars are placed on their shelves. The thicker the cardboard, the more protection the dividers will supply. Use a box cutter to easily cut up salvaged cardboard boxes and place strips between your secured jars. You could also use individual sleeves of bubble or Styrofoam wrap, like what is used to pack and ship dishes, but this would be a more expensive option. For upcycled materials you could use old socks, the sleeves of shirts or the legs of pants cut into appropriate lengths to place over your jars but still allow them to sit flat against the shelf.
3. Secure jars on the shelves with straps, wire mesh, or bars to prevent them from sliding off in the event of an earthquake or other disturbance.

4. Secure stacks of boxes or buckets with cords or rope. You could also use a cargo net. You still must be able to access your food storage, so try a few different methods and decide which is less annoying to work around. You may not be able to prevent your stacks from toppling entirely, but you can break their fall with a few well-placed braces.
5. Place anti-slide, rubber shelf-liners on your food storage shelves, especially wherever you have glass bottles. This can't take the place of straps or bars on your shelves, but it can provide a bit of traction.

BUILD COMMUNITY: Bulk food orders are a great way to build your local food storage community. You can organize group buys and discounts with local friends and family using a simple email list. Collect the names and email addresses of people interested in participating in group food storage buys and then network with each other as you find great deals.

Companies like Azure Standard (www.azurestandard.com), Rainy Day Foods (www.rainydayfoods.com) and the LDS Bishops Storehouses (to learn more go to our links page under The Prepared Homestead: www.homesteadlady.com/homestead-links-page), and those already mentioned like Thrive, Prepper's Market and Valley Food Storage can all accommodate large, group orders and are happy to help you answer storage questions. Azure Standard can be particularly challenging to navigate, with their very specific drop off schedules, so be sure to educate yourself on their methods—they're always happy to help!

Sometimes the best way to assist the people you know to help themselves is to just offer them an opportunity to participate. Many people aren't unwilling to store healthy food items against a day of hunger or to save some money, but it doesn't occur to them to get organized and do it until you come along and invite them to participate in a monthly group order. These group buys work equally well with fresh produce from CSAs and even dairy orders from local farmers who offer such services. From wheat to beans to zucchini, there's something for everyone to store.

BOOKS TO READ: *The Pantry Primer* and *The Organic Canner*, by Daisy Luther, for simple, wholesome food storage recommendations and methods. Daisy shows that you don't need a lot of money or fancy equipment to stock your pantry with nourishing foods.

Getting Prepared, by Robin Egerton and Angela England, is great for the newbie to avoid becoming overwhelmed.

A generally good book for emergency preparedness, including basic food storage, is *Survival Mom: How to Prepare Your Family For Every Day Disasters and Worst Case Scenarios*, by Lisa Bedford.

WEBSITES: For frugal, healthy eating with storage friendly and handmade ingredients, please visit Don't Waste the Crumbs (www.dontwastethecrumbs.com). Tiffany has also designed a Frugal Real Foods Meal Plan program that is for sale on her site if you feel you need someone to help you get things set up at first (find a link on our links page under The Prepared Homestead). This is NOT a food storage program, but rather an organized way to learn the art of frugal, healthy eating with wholesome ingredients that will, hopefully, find their way into your food storage program.

For food storage advice and absolutely no-nonsense advice, I suggest The Organic Prepper (www.theorganicprepper.ca). If you're interested in learning how to cook with your food storage, among other things, My Food Storage Cookbook (www.myfoodstoragecookbook.com) is the place to go. To learn absolutely anything about grains, flours, beans and more, please visit the Bob's Red Mill on our links page: www.homesteadlady.com/homestead-links-page.

HOMESTEAD LADY SPEAKS:
Looking at the lists of various foods above, I realize that most of what I personally store would be considered shorter-term items, so I naturally rotate them often and easily. In part, this is because I typically have a large garden, as well as livestock, and preserve a lot of my own food, so shorter-term items are more consistently produced on my homestead. If you don't have the opportunity to grow and preserve your own produce and protein, the practice of food storage rotation is still a very important one. You don't want anything to get so old it no longer has nutritional value or, worse, spoils. Utilize your farmer's market and/or local farms for preserving fresh produce whenever it's available.

Rotating my long-term food storage is the easiest part of food storage since it mostly consists of big bags and buckets of grains and beans. They remain undisturbed in my food storage room until my smaller containers in the kitchen run dry. I leave the forty-five to fifty-pound buckets in the storage room so that I'm not lugging them around all the time, trying to find a place for them in my kitchen. Although, I do need to keep a running list of what I have in my storage room at all times, alas, I never do! Consequently, I end up thinking I'm out of items that are just shoved to the back of deep shelves. This isn't all bad because I end up buying extra storage and am, therefore, even more prepared...right? My lack of organization must have a bright side, I tell myself.

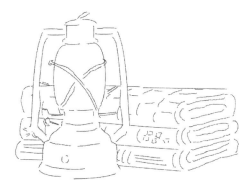

"Fire makes you adapt and tune in. Your creativity emerges from the limitations imposed by this primitive technique. Working with recipes and perfectly calibrated ovens—sometimes it's easy to lose our connection with the food itself."

- Russell Moore, *This is Camino*

For the Homesteaded: Off-Grid Cooking

As I said before in the *Energy on the Homestead* section of the Green Living chapter, "off-grid" doesn't have to be an all or nothing venture of complete independence from public utilities—especially at first. Even if your home is still connected to the power company, learning a few off-grid cooking techniques can save a great deal of energy and money on a homestead. Alternative cooking skills are also useful to have from a preparedness perspective; if the grid ever goes down unexpectedly, you'll still be able to prepare meals for your family beyond reconstituted powdered milk and the gummies stuck between the couch cushions.

Unless you grew up using these skills, they will take a certain amount of practice. So, I encourage you to begin today to figure out which ones you can manage without too much hassle. In fact, you may discover you truly enjoy using these methods of meal preparation. We're going to cover four methods in this section:

- Solar Ovens
- Wonder Ovens
- Open Flame
- Rocket Stoves, and the like

GRILLIN' – THE FIRST STEP IN OFF-GRID COOKING

No need to freak out at the idea of off-grid cooking because you can begin with something you probably already do—outdoor grillin'! For some of us, the best part about this option is that we already own and know how to use a propane grill. Or, if you have access to charcoal, a Webber Kettle grill will also work wonderfully. Grilling is surely not a new concept, but I think it bears mentioning here to make sure we've covered the basics. If you keep several propane tanks filled, or several bags of charcoal on hand, your grill can be a good emergency preparedness tool, for as long as the propane and charcoal last. Outdoor grilling can be a great place to start if you're new to the idea of cooking somewhere other than your kitchen. Once you learn to grill with these tools, grilling over an open flame won't seem that hard at all.

A fun dish on which to practice your grilling skills is kebabs (we say kabobs at our house, but we're just unhip that way). You can Google a myriad of marinade recipes to enhance their flavor, but kebabs are essentially food on a stick, cooked over a small flame. In Turkey, the traditional birth place of the kebab, the stick would most certainly have meat on it (typically lamb), but you can also make veggie, or even fruit, kebabs.

Maybe you grew up eating these wonderful things and are wondering why I'm bothering to talk about something so basic. I was apparently deprived of Turkish food as a child and hadn't eaten a kebab until I lived in Russia (the wonder of the kebab spread to the Caucuses). A Russian family treated us missionaries to homemade kebabs over an open, outdoor flame. They were simply delicious. (Random memory: their yard also had the first cistern I'd ever seen. This city girl asked if it was for swimming. The father laughed a bit but was very kind in explaining the idea of water conservation/redirection.) I guess I'd forgotten about kebabs until a few summers ago, when I was desperate for something easy that could be grilled quickly to satisfy the children who were nipping at my heels for their dinner. Quick and simple, kebabs are a tasty place to begin your off-grid cooking ventures.

Solar and Wonder Ovens

Whereas an outdoor grill uses fuel to cook, a solar oven is designed to capture and utilize the heat of the sun's rays. Set up properly, it will maintain an even temperature using only the power of the sun to cook your food. Solar ovens are kind of like slow cookers, but without requiring any electricity to run. I encourage you to research the several brands on the market or make your own to test-drive their use. For best results, plan to preheat your solar oven like you would a conventional oven. However, with a solar oven, that just means setting it out in the sun about a half hour to an hour before you need to use it. We've cooked all kinds of things in ours, from fish to soup and bread to "boiled" eggs.

A neat feature of this kind of cooking is that—unless you leave the food unattended for a *really* long time—you won't burn what you cook in a solar oven. My SOS® brand solar oven is light and easy to use, with only a few parts. The main body is a black box that has a plastic lid attached with six metal clasps so that it doesn't blow away in the wind, and so wayward kiddos don't get into a hot oven. There's also a sun shield that attaches with spring clasps to magnify the sun's heat and draw it into the oven. Occasionally, you need to move the solar oven, to follow the sun and keep the temperature even. However, the wide design of my SOS® means that I don't have to turn it as often. Some food items require only half an hour to cook, while others may take several hours.

I will buy another one, especially if I can find one on sale, but these are economical to begin with. There are other brands, of course. I have a friend who has a Sun Oven® and loves it. It's quite a bit sturdier and deeper than mine, it folds up for toting around and isn't too heavy. It has an internal thermometer, too, which is quite handy. I lost the thermometer that came with my SOS®—I needed to attach it somehow so that didn't happen. The SOS® isn't wider, though, and my biggest pet peeve with a commercial solar oven is that they will hold my 9"x13" casserole dish, but nothing bigger. I feed seven people daily so, I need a bigger oven, especially for off grid-cooking. What I'll have to do is buy or build two to three more of them so that I can make a main dish, side dish, bread and dessert all at once. Extra solar ovens could also be used to dehydrate food during the harvest season, which is another function of these already highly useful off-grid cooking ovens.

THE WONDER OVEN

I love my Wonder Oven (aka Wonder Box) for off-grid cooking. Like a powerless Crockpot, Wonder Ovens are basically super-insulated bags with a top piece that acts like a lid. These ovens were developed from the Wonder Boxes designed to help African moms cook a pot of food using a minimal amount of fuel. The idea, which is not a historically new one, is to bring a pot of food to boil and then shut off the stove (or bank the coals of the cook fire). Next, you place a lid on the pot and put the whole lidded pot into the center of the box's bottom. Finally, you put the lid on the box and let the pot sit for as long as it takes to cook the food. For example, brown rice will usually cook in a little over an hour.

Although sewing isn't really my thing, I decided to make my own Wonder Oven, since I strive to be frugal—which is why off-grid cooking appeals to me in the first place. The boxes are made from cotton fabric, or any other natural fiber like wool, and stuffed with polystyrene pellets to insulate them. I'm not much of a seamstress so I went to a neighbor's house, where she could mentor me in putting them together. Another nice thing about sewing these with someone was that we were able to split the cost of the pellets. We got a price break for ordering them in bulk, but it would have been way too many pellets for just one person (unless we were each making five-or-so boxes). Thankfully, another friend suggested we stuff our boxes in the bathroom, better yet, in the tub to minimize the mess; there's major static electricity on the pellets and the little buggers can just go everywhere. And I mean, everywhere.

Food Storage Made Easy (www.foodstoragemadeeasy.net) has a great tutorial on making your own, which you can find in PDF form on our links page in the Prepared Homestead chapter: www.homesteadlady.com/homestead-links-page. Visit Meghan at My Food Storage Cookbook (www.myfoodstoragecookbook.com) to learn how use a Wonder Oven. She's passionate about having them on-hand as a preparedness tool and shared the following with me:

> "I know for myself how critical owning a Wonder Oven is to being able to cook with your food storage. There's no tool out there to compare it to. The conservation of fuel saved by using a wonder oven, worst case scenario, could be what literally saves your family from running out of fuel (and starving) before running out of the food you have stored.
>
> "Let's talk about that for a minute. Most people might have plenty of food put aside, but the thing they're often not thinking about is the amount of fuel they'll need to cook the food they're storing. And storing fuel is, in many ways, more difficult than storing food. And just consider for a moment the types of foods we store in consideration of how much "cook time" they require. We many times store whole grains, legumes, foods that in their "whole" form naturally require more time to cook. And yet, how are we planning to cook them? Without a thermal or other "free" way to cook (Sun Oven cooking of course falls under this category as well, so long as the sky is clear), you're sure to use up your fuel reserves far before using all your food!
>
> "How would it feel to have boxes and boxes of stored food sitting neatly stacked in front of you, but no way to cook it? It's essential that you have a plan to be able to conserve the fuel you do have in living without power otherwise you'll rapidly find yourself in that situation."

For off-grid cooking you can use the Wonder Ovens for just about anything like preparing grains, veggies and soups. With the right baking pan, you can even bake bread in a Wonder Oven. We use ours constantly to transport and keep warm meals for the hours we're at dance and music classes. I mellow soap batches and culture yogurt in my Wonder Ovens, too. The only drawback I've discovered is that they do take up quite a bit of space. You can't compact them to store them because it will also compact the insulating pellets, making them less effective. I only have two of these ovens but would like at least two more. We store ours in two soft-side canvas hampers, so they'll be safe from compaction, rips, and soiling. The hampers also make it simple to transport the Wonder Ovens when I need to. Taking good care of them will allow them to reward me with years of simple, off-grid cooking.

Outdoor Fire Options

Let's start with the basics.

OPEN-FLAME COOKING

We've been trying to improve our open-flame cooking skills but, to be honest, I have never been very good at keeping a fire going. Very good?! Ha—I'm like the anti-fire! I stink at it. It is time to un-stink at it because I may very well need to feed my family using open flame of some kind. If that happens, I don't want us to starve. My biggest accomplishment last year was to learn to build and maintain a fire for my open flame cooking experiments. You may laugh but, for me, this was a big deal!

If you've never cooked on an open flame, my first piece of advice is to practice a lot because it's part art and part science. You're going to burn some food as you learn what you're doing. I also suggest you get a good book on the subject, or a willing mentor, because both will be full of great advice on the subtle nuances of cooking times, flame size and coals, etc.

The other thing I suggest you do as you're learning is to keep it simple. One of our favorite foods to prepare in the coals is a potato (sweet or brown). I wanted a way to bake potatoes in the fire without aluminum foil, and—voila! God naturally jacketed the potato, I'm sure, just so we could cook it over coals. For those to whom cooking over an outdoor fire is old hat, I'll wager the idea of roasting a mere potato will seem overly simplistic. And I thank you for your indulgence as I revel in my new found culinary project. For those learning with me, you can also roast an onion with its layers of skin still on—über yummy, would be my very scientific report on that technique. Cooking time varies from fire to fire, but when you insert a knife (watch you don't get burned!) and it pushes right through, you can know your onion is done.

The flavor of a roasted onion prepared this way is amazing. All the sugars in the onion do a happy fusion-dance in your mouth, and the onion comes out hot, steamy and smooth. I just cut it up and eat it with whatever we've prepared because it makes everything else taste delicious. You can toss some salad dressing on top, or some apple cider vinegar with spices, and eat it that way, too. Similarly, all food cooked over open flame has a unique flavor and appeal.

We don't use charcoal briquettes when we cook over an open flame because that's cheating (and I'm always doing things the hard way). Practically speaking, charcoal is expensive and won't be around in an emergency. However, if you want to use it when you're first learning how to flame-cook a meal, you won't hear about it from me. With all these methods, I just say do what you gotta do. Only, do it today because they all take practice.

ROCKETS, THE IRISH AND MORE!

There are other flame options besides a fire pit. Three others are EcoZoom, the Rocket Stove and Kelly Kettle. These are all low fuel, high flame apparatuses that we have tried and really enjoyed using. These stoves are designed to use small amounts of fuel, while still producing a flame strong enough to boil water and cook food. They're particularly well suited to camping, backpacking and, especially, emergency situations where wood to burn can be scarce.

You can do some Internet searches to determine which ones you might be interested in, and then ask around amongst your friends to see if they have one that you can try out. You can build any of these highly efficient, little stoves or you can purchase them at camping stores and websites. Again, practice with them because they're very different from your indoor stove or even your outdoor grill.

You can also learn to dig what's called a Dakota Fire Hole, which is basically a Rocket Stove you build in the dirt. To learn how to build one, you can find a link from Graywolf Survival on our links page: www.homesteadlady.com/homestead-links-page.

If you're in your cozy home this winter, already using your wonderfully fuel-efficient wood burning stove for heat, why not use it for cooking as well? You can use the flat top, or even learn to cook inside the wood box. There are little tent-like apparatuses you can buy, and tricks you can employ, to use the wood stove as an oven. Online searches will connect you with various peoples' experiences and advice. There are also brands of wood stove you can buy that are built to cook in and on—just like your great-great grandmother may have used.

It does take work to learn a new way of doing things in the kitchen, but the rewards are well worth it, I believe. We've not only gained useful skills, but something unexpected happened when we started to make off-grid cooking a part of our normal food preparation: we grew closer as a family. We do the bulk of our off-grid cooking together outdoors, during the warmer months (although you can use all these methods year-round). Especially with open-flame cooking, I need all hands-on deck to build and maintain the fire while I prepare the various parts of the meal. This simple exercise brings our family together.

We've discovered that my oldest daughter is quite skilled at fire-building and that the others are very happy to help gather kindling and wood. Further, we've also realized how much we enjoy sitting around talking while we get the fire going. Cooking our food outside has meant that all of us are out there, more often, preparing and eating meals together. We put up a table on the patio for summer dining because eating outside is so enjoyable. Plus, we needed to be close to the cook fire to tend our meals. We didn't eat outside that often before we took some of our food preparation outside and off-grid. Sometimes the rewards of a simpler lifestyle are a sweet surprise.

AN OUTDOOR KITCHEN

To really go off-grid with your cooking, you may decide to set up an outdoor kitchen area on your homestead. Pinterest can provide you with a myriad of fancy ideas and, if that's what floats your boat, feel free to experiment. However, you don't need a lot of expensive equipment or designs to get started. Teri Page of the off-grid blog Homestead Honey (www.homestead-honey.com) says that their first outdoor kitchen wasn't much more than a tarp, a rocket stove, and a propane burner. Since then, they've built their own shelves, put in a table and some seating. They've also included a cabinet and counter area, and even a sink piped with running water from rain-water catchment barrels that gets run through their Berkey filter if they're drinking or cooking with it.

Even if you're not completely off-grid like Teri's family is, you can still benefit from setting up an outdoor food preparation area. Teri points out,

"An effective homestead outdoor kitchen space could be as simple as setting up a few card tables under a shade structure, or as elaborate as you can imagine. I highly recommend making shade a priority, because in the hottest months, you will want to have a cool place to cook. Make your outdoor kitchen the most ideal spot to hang-out, and you will want to spend hours there preparing and enjoying homemade food!"

-From her book, *Creating Your Off-Grid Homestead*

ACTION ITEMS: Take out your homestead journal and write down just one of these off-grid cooking options. Pick the one that feels the most inviting to you. Hop online and see if you can learn more about it, particularly looking for articles that detail people's experience using that method. Write down a list of equipment you'll need for the method you chose. Now, look at your list and think, is this the method you want to try first?

If you still can't decide, go to YouTube and find videos of your method being used. Now what do you think? Next, talk to your family, since their enthusiasm, or lack thereof, might help you decide which method you'd like to try this season.

DIY: Make a Wonder Oven if you have even a little bit of sewing skill. You really only need to be able to sew a straight line. Like I said, you can use the tutorial from Food Storage Made Easy that's linked on our links page (www.homesteadlady.com/homestead-links-page); be sure to use a sturdy cotton fabric or broadcloth.

As I mentioned earlier, try to do this with a friend. Not just for moral support, but also so that you can buy the polystyrene pellets together. You'll need another set of hands to stuff the bags with the pellets anyway. Once completed, you'll want to find a sturdy box in which to store your oven, when it's not in use, so that the pellets don't get compressed. You also want to protect the ovens from people walking on them and from kids sitting on them—they do look a little like bean bag chairs, after all.

If you want to buy a kit to make sure that you're doing it correctly the first time, My Food Storage Cookbook sells. They also sell assembled Wonder Ovens, if you don't feel like making your own (which I totally understand!). Prepared in Every Way sells completed Wonder Ovens, too. You can find links for both sources on our links page under The Prepared Homestead chapter: www.homesteadlady.com/homestead-links-page.

If you stopped reading when you saw the word "sewing," how about making a rocket stove out of bricks using the easy-peasy tutorial from Mom With A Prep found on our links page? All it takes is 27 concrete bricks, a grate, and some metal screening. Plan to build this stove outside and start using it tonight!

BUILD COMMUNITY: There are meet-ups for recreational outdoor cooking that you can join. I had no idea that there were whole groups of people who get together to cook with Dutch ovens, but there are. There are similar solar oven groups. Join them, and you will eat well and learn much. Internet searches can help you find groups local to you.

BOOKS TO READ: *Cooking with Fire*, by Paula Marcoux and *Cook Wild*, by Susanne Fischer-Rizzi. Both books are excellent for open-flame cooking methods and recipes, but I like *Cooking with Fire* just a bit more for the beginner. Ms. Marcoux spends a fair amount of time explaining techniques that a newbie doesn't inherently know. The feel for open-flame cooking will come with time but, it's tricky to get a handle on it, at first. Solar oven companies all have cookbooks you can purchase, but most regular recipes can be prepared in a solar oven by simply providing a longer cooking time.

WEBSITE: Pinterest is your best friend on this topic because there are scads of posts with recipes and techniques. You can vet them on online and see what looks interesting to try this month. Then pick a new idea for next month. Good keyword searches to use on Pinterest are: outdoor cooking, outdoor cooking techniques, off grid cooking, camp food, fire pits and Dutch ovens. People have whole boards dedicated to nothing but these topics!

As has already been mentioned, Meghan from My Food Storage Cookbook (www.myfood-storagecookbook.com) has several good articles on how to operate a Wonder Oven, including recipes and advice on using them effectively.

Melissa K. Norris has a great podcast on using Dutch ovens which you can find a link for on our links page under The Prepared Homestead.

HOMESTEAD LADY SPEAKS:
Take it outside! Not only is outdoor cooking a handy emergency/preparedness skill to have, but it's also just plain fun. Everything tastes better cooked over a flame—it's like magic.

I will freely admit that this was one of my biggest challenges of the past year. We really wanted to learn to prepare food outside without electricity, but I was immediately daunted at the prospect. The reason? Historically, I'd always been the worst fire-builder and maintainer in my family. When I was a kid, we only camped once a year and my mom was the fire ace, so I always just left it to her. My mother's so fire-savvy she kept us heated one year with nothing but a cord of Eucalyptus and a small woodstove. She would sleep next to the woodstove and feed it through the night, banking the coals in the morning before we all left for school and work. It was the first experience I had with the reality of "keeping the home fires burning," and it was more impressive because she had to work away from home all day.

Because I was such a wuss about the fire, I started with my solar oven first and learned to use it well. Then I moved on to my Wonder Oven. Only after that did I turn my attention to cooking over an open fire. Most of what you read about Dutch oven cooking involves using charcoal briquettes, and they're great for beginners because you get consistent heating and results. However, like I said, I think they're cheating. Our pioneer ancestors didn't have charcoal briquettes and I can't afford to buy them. Those two realities spurred me on to figure out how to do this outdoor cooking thing with nothing but wood from my property and some matches.

Renewed disclosure, as I admitted earlier: I still mostly stink at it. I can build a fire and keep it going but it takes all my concentration. The fire experts in my family are my husband and my oldest daughter and we all still use matches. BUT, I learned the essentials and I'm so proud of myself. We've also gotten better at looking around our homestead for wood and have learned what makes the best kindling on our property. Additionally, we've managed to scrounge items on which to cook, like stones and planks, and had great discussions about fire safety and land stewardship. I've burned my share of potatoes, but we've totally aced the Homemade Solar Oven S'mores. Life is give and take.

SOLAR OVEN S'MORES

First, I recommend that you learn how to make homemade marshmallows. You can produce, without too much hassle, a much healthier, junk-free version of the humble marshmallow. No high fructose corn syrup (HFCS), no preservatives, no plastic packaging. Just water, honey (or other sugar) and gelatin. You can make them in ways that are friendly to dietary guidelines such as GAPS (Gut and Psychology Syndrome), Paleo and Weston Price. Homemade graham crackers are also worth learning to make.

INGREDIENTS:

Homemade marshmallows (see the *Make Your Own Stuffs* section in The Homestead Kitchen chapter for a recipe)

Homemade graham crackers

Chocolate bars of your favorite persuasion (these can also be homemade, if you want to show off)

You'll also need a 9"x13" casserole dish, a solar oven, a spatula, hot mitts and some utensils and plates. If you're into eating off plates, that is. My family just grabs forks and eats the hot S'mores out of the pan like piglets. Uh, do piglets use forks?

DIRECTIONS:

Set out your solar oven with the lid on in a sunny spot to preheat it a bit. Butter your casserole dish. Cover the bottom of the dish in graham cracker squares. Break the chocolate bars into bite sized pieces and put one or two on top of each cracker. Place one or two marshmallows on top of the chocolate—or however many it takes to fill the surface of the graham cracker with chocolate and marshmallows.

Now you face a choice, and a grave one at that. You can put another piece of chocolate on top of the marshmallow and another marshmallow over that, forming a kind of sandwich of happiness. Or, you can just leave one layer on each cracker. The decision is yours. But I caution you that the double layer will be doubly sweet and tempting, so bear that in mind.

> Regardless of how many layers of marshmallow and chocolate you decide to use, wait until your creation is fully baked and being served up before you put on the top graham cracker layer, especially if you like the inside melted all the way. The S'mores are a bit messy when they come out of the oven, and by waiting until the very last minute to put that top cracker on, you tidy up the presentation and reduce the mess when eating them. Again, my crew couldn't have cared less what they looked like, since it all tastes the same. But, if you're serving individuals who stick their pinkies out when they eat, presentation can be important.
>
> Once your dish is full, take it out to the solar oven and put it inside, re-securing the lid and leave it in the sun to melt both the chocolate and the marshmallows. How long that takes is entirely dependent on how mushy you want everything. If you just want all the ingredients warmed but not melted, be sure to check on your dish after about twenty minutes if it's a hot day. If you want it fully melted (the way we like it), give it a bit longer. If you're doing this in the fall or winter, give it even more time. Just FYI, a homemade marshmallow will melt a lot faster than a store bought one.
>
> Use a spatula to serve up the individual baked s'mores and place the final graham cracker layer on top, like a hat.

Solar oven cooking is always a bit of an adventure at my house. Between my five small children and our homestead, I get scatterbrained and can forget to check on my solar oven's contents regularly. With a roast chicken, it's not as big a deal. But with something as temperature sensitive as this, I do try to pay more attention. (Remember the fire building difficulty? Attention to details is a real chore for me.)

> And with that last recipe, you have my last word on preparedness for your homestead.
> In fact, you've made it nearly to the end of the book—how do you feel?
> Hungry for S'mores?

CONCLUSION

You Have Arrived

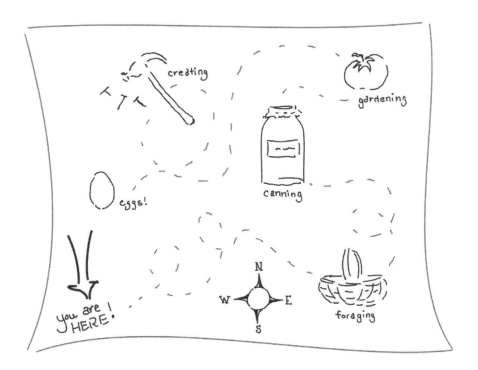

"You might be a homesteader if, like us, you suffer from Pasture Deficit Disorder every time you leave your front gate."

- Cheryl from Pasture Deficit Disorder.com

And here we are. At the end. I have one last question for you: have you ever felt what Cheryl describes above? I feel strongly, you'll know in your heart if you're a homesteader if you've ever had this feeling. The pasture doesn't have to be yours. But the feeling is.

If, when you go for a drive, and you spot an open space and your heart yearns to:

- Rip off your shoes and walk in the grass
- Dig your fingers into the dirt
- Grab a wheelbarrow and start gathering up organic material to build the soil, as if you really are back in Eden
- Bring in a few goats to clear the brush
- Plant an orchard to feed the future
- Gather friends and neighbors to start a community garden
- Set up an egg selling stand to share the bounty of the flock you've installed on that land.
- Or any number of other ideas you might have for that green, open space...

Honey, you might be a homesteader.

At the very outside, you've got a homesteader's heart and that's the most important part. Everything else in your head and your hands can be trained but you've got to have your heart in it or you'll peter out and die on the homesteading journey. Maybe you're being dragged into homesteading by a spouse, or maybe you see the merit of the idea, but you're just overwhelmed and even a little bit scared by the notion. I say, work on your heart first.

Sure, read all the books you can and talk to homesteaders everywhere. Get online and follow the blogs of the homesteaders you enjoy. But never neglect your heart. Try to emotionally connect with homesteader ideals and goals. Try to feel the great WHY of homesteading and remember, the root word of it is *home*. Homesteading is about home, about family—even if your family consists of you and your hound dog. Homesteading is about making our family's life as bounteous and self-sufficient as possible.

There's so much in the way of weather, war and wealth that we just can't control; homesteading gives us some peace of mind to face all of them calmly. As we seek self-sufficiency, our family bonds grow stronger, too. We know ourselves and our loved ones—we know our strengths and our weaknesses. Some of the greatest, most profound teaching moments of my mothering/homeschooling career have taken place over a pile of tomato seedlings at planting time. Or with a pitchfork in my hand as we muck out the barn.

> "But mere seeing from a distance did not satisfy her;
> she longed to go alone far into the fields and hear the birds singing,
> the brooks tinkling, and the wind rustling through the corn, as she had when a child.
> To smell things and touch things, warm earth and flowers and grasses,
> and to stand and gaze where no one could see her, drinking it all in."
>
> — Flora Thompson, *Lark Rise to Candleford*

Is homesteading going to be a lot of work? Well, of course it's work—anything worthwhile is going to require your best.

But

if the words here have resonated with you at all
if you found yourself nodding or laughing or even crying
if you realized you're not alone, or the only one doing or thinking of these things
if you felt inspired to give it a try, or try again, or keep going

...you just might be a homesteader.

Once you realize that, it's time to fully build, together with like-minded and supportive friends around you, your own Do It Yourself Homestead.

Congratulations, my friend, you have arrived.

Resources

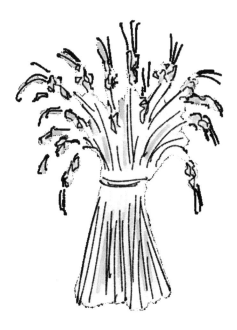

Whatever your level of homesteading, I'm hopeful these resources, together with the ones already provided, will prove helpful. Included below are titles already featured in the sections of each chapter, as well as a few others that have been of assistance to me. These lists should not be considered exhaustive. Please do your own reading and researching to find the information that's right for you. No matter how I try, I still haven't managed to read every word out there, so you shouldn't rely solely on me. I will say that the best resource homesteaders have access to is each other.

On that note, much of what you have read in this book has been gathered as a direct result of my interactions with many of the writers and homesteaders that feature in these lists, and throughout the book. They have my gratitude and respect.

Books, E-Books and More!

As often as possible, I always suggest vetting any print material through your local library before you buy it. If your local library doesn't carry a title you'd like to read, put in a request. This will tell our wonderful librarians that homesteading is a hot topic and that they need to seek out more books to cover the need. If you can't read before you buy, the reviews on Amazon.com and Goodreads.com are both a wealth of information when it comes to deciding whether a book is right for you to try. Previously owned print books will be cheaper than buying new, of course. If you enjoy reading books on your electronic device, bear in mind that e-books are typically cheaper than print, as well. I'm a sucker for a print book, especially with a book in which I'm going to want to take notes! I make a mess writing in a lot of my books, I must admit.

Whenever possible, be sure to leave your own reviews on one or both sites to help not only future readers, but the author as well. For example, if you have anything to share about this title, *The Do It Yourself Homestead*, we'd really appreciate and value your feedback at Amazon.com and Goodreads.com. Even if you'd like to explain how we can improve the next edition, we'd love to hear from you!

These materials are organized by chapter in alphabetical order. Some titles are listed more than once because they appear in more than one chapter.

GENERAL HOMESTEADING

*Backyard Farming On an Acre, More or Less, by Angela England
*Backyard Homestead, The, by Carleen Madigan
Back to Basics, a Reader's Digest publication
Encyclopedia of Country Living, The, by Carla Emery
Family Homesteading, by Teri Page
Homegrown and Handmade, by Deborah Neimann
Homesteading Handbook, The (and other titles), by Abigail Gehring
Homesteading in the 21st Century, by George Nash and Jane Waterman
Weekend Homesteader, The, by Anna Hess

DVD - The Homestead Blessings video series by the West family

IN THE HOMESTEAD KITCHEN

Against All Grain, by Danielle Walker
Art of Baking with Natural Yeast, The, by Melissa Richardson and Caleb Wornock
Art of Fermentation, The, by Sandor Katz
Art of Natural Cheesemaking, The, by David Asher
Art of Simple Food, The, by Alice Walker
Beyond Basics with Natural Yeast, by Melissa Richardson
Bountiful: Recipes Inspired by our Garden, by Todd Porter
Conquering Your Kitchen, by Annemarie Rossi
Elliot Homestead: From Scratch, The, by Shaye Elliot
Feel Good Cookbook, The, by Jonell Francis
Food in Jars, by Marisa McLellen

Gift it From Scratch (and other titles), by Kathie Lapcevic
Harvest, by Stacy Lynn Harris
Heal Your Gut Cookbook, The, by Hilary Boynton
Home Cheese Making, by Ricki Carroll
Homegrown Whole Grains, by Sara Pitzer
Homemade Pantry, by Alana Chernila
Independence Days: A Guide to Sustainable Food Storage and Preservation, by Sharon Aystk
It Starts With Food, by Dallas and Melissa Hartwig
Made From Scratch Life, The, by Melissa Norris
Maple on Tap, by Rich Finzer
Natural Beekeeping, by Ross Conrad
Nourished Kitchen, The, by Jennifer McGruther
Nourishing Traditions, by Sally Fallon Morell
Organic Canner, The, by Daisy Luther
Prepper's Dehydrating Handbook, by Shelle Wells
Putting Food By, by Hertzberg and Greene
Quinoa 365: The Everyday Super Food, by Greene and Hemming
Rawmazing Desserts, by Susan Powers
Real Food Fermentation, by Alex Lewin
Root Cellaring: Natural Cold Storage of Fruits and Vegetables, by Mike and Nancy Bubel
Saving Dinner Basics: How to Cook Even if You Don't Know How, by Leanne Ely
Things to Make With Roses, by Jan Berry
Traditional Kitchen Wisdom, edited by Andrea Chesman
Ultimate Dehydrator Cookbook, The, by Tammy Gangloff
Whole Foods For the Whole Family, by Roberta Johnson
Wild Fermentation, by Sandor Katz
Zero Waste Cooking, by April Lewis

DVD – "At Home Canning for Beginners and Beyond," by Kendra Lynne

IN THE HOMESTEAD GARDEN
All New Square Foot Gardening, by Mel Bartholomew
Backyard Apothecary, The, by Devon Yong
Backyard Orchardist, The, by Stella Otto
**Bountiful Container*, by Maggie Stuckey and Rose McGee
Carrots Love Tomatoes, by Louise Riotte
Complete Book of Herbs, The, by Lesley Bremness
Complete Guide to Growing and Using Sprouts, The, by Richard Helweg
**Complete Idiot's Guide to Small Space Gardening, The* (and other titles), Chris MacLaughlin

Edible Forest Gardens, Vol. 2, by Dave Jacke
Four Season Harvest (and other titles), by Eliot Coleman
Gaia's Garden, by Toby Hemenway
*Gardening Like a Ninja, by England and Piersall
Healing Herbal Infusions, by Colleen Codekas
Herbal Medicine-Maker's Handbook, The, by James Green
Indoor Kitchen Gardening, by Elizabeth Millard
Let it Rot: The Gardener's Guide to Composting, by Stu Campbell
Medicinal Herbs: A Beginner's Guide (and other titles), by Rosemary Gladstar
Plant Propagator's Bible, The, by Miranda Smith
Restoration Agriculture, by Mark Shephard
*Small-Space Container Gardens, by Fern Richardson
Straw Bale Gardens, by Joel Karsten
*Vertical Vegetable Gardening, by Chris McLaughlin
Week by Week Vegetable Gardner's Handbook, The, by Ron and Jennifer Kujawski
*You Grow Girl (and other titles), by Gayla Trail

GREEN THE HOMESTEAD

All That the Rain Promises and More, by David Arora
*Apartment Therapy—Complete and Happy Home, by Ryan and Laban
Art of Home, by Hannah Keeley
Craftcycle: 100 Earth Friendly Projects and Ideas for Everyday Living, by Heidi Boyd
Edible Wild Plants, by Elias and Dykeman
Encyclopedia of Country Living, by Carla Emery
Forager's Harvest, The by Samuel Thayer
Hannah's Art of Home, by Hannah Keeley
Humanure Handbook,The, by Joe Jenkins
Life Changing Magic of Tidying Up, The, by Marie Kondo
Little Britches (series), by Ralph Moody
Little House on the Prairie (series), by Laura Ingalls Wilder
Natures Art Box, by Laura Martin
Nesting Place, The, by Myquillyn Smith
Newman's Own Organics Guide to a Good Life, The, by Nell Newman
One Plastic Bag: Isatou Ceesay and the Recycling Women of the Gambia, by Miranda Paul
Sink Reflections, by Marla Ciley aka The Fly Lady
Slug Bread and Beheaded Thistles, by Ellen Sandbeck

LIVESTOCK WHEREVER YOU ARE

Accessible Pet, Equine and Livestock Herbal, The, by Katherine Drovdahl

Art of Natural Cheese Making, The, by David Asher
Backyard Chickens Naturally, by Alanna Moore
Backyard Guide: An Introductory Guide to Keeping and Enjoying Pet Goats, The, by Sue Weaver
*Backyard Homestead Guide to Raising Farm Animals, The (and other titles), by Gail Damerow
Butchering (either title), by Adam Danforth
Chickens From Scratch, by Janet Garmen
Chicken Health Handbook, The, (and other titles), by Gail Damerow
Chicken Hot Topics, by Jessica Lane
Complete Guide to Working with Worms, The, by Wendy Vincent
*Free Range Chicken Gardens (and other titles), by Jessi Bloom
G, is for Goat, by Patricia Polacco
Goats Produce, Too!, by Mary Jane Toth
Good Meat: The Complete Guide to Sourcing and Cooking Sustainable Meat, by Deborah Krasner
Gourmet Venison (and other titles), by Stacy Lynn Harris
Home Cheese Making, by Rikki Carroll
Homegrown Pork (also titles for dairy cows, chickens, donkeys and more), by Sue Weaver
Keeping a Family Cow, by Joann Grohman
Meat Goat Handbook, The, by Yvonne Zweede-Tucker
Meat Smoking and Smokehouse Design, by Stanley Marianski
Milk Cow Kitchen, by Mary Jane Butters
Milk Soapmaking, by Anne Watson
Nourishing Traditions, by Sally Fallon
Omnivore's Dilemma, by Michael Pollan
Raising Goats Naturally, by Deborah Niemann
Salad Bar Beef (and other titles), by Joel Salatin
Storey Publishing House titles are helpful and animal specific
The Trayer Wilderness Cookbook—Homesteading The Traditional Way, by Tammy Trayer

HOMESTEAD FINANCES

American Frugal Housewife, The, by Lydia Marie Francis Child
Backyard Beekeeper, by Kim Flottum
Christmas Trees, by Lewis Hill
Dave Ramsey's Complete Guide to Money, by Dave Ramsey
Earn It, Learn It, by Alisa Weinstein
Fleece and Fiber Sourcebook, The, by Ekarius and Robson
From Grit to Great, by Thaler and Koval
Graining Ground, by Forest Pritchard
Hope: Thriving While Unemployed, by Angi and Carl Schneider

Homemade For Sale, by Lisa Kivirist
How to Blog for Profit Without Selling Your Soul, by Ruth Soukup
Leadership and Self Deception, by The Arbinger Institute
Live Well on Less Than You Think, by Fred Brock
Market Gardener, The, by Fortier and Bilodeau
Mend it Better, by Kristin Roach
*Mini Farming, by Brett Markham
Money 911, by Jean Chatzky
One-Page Financial Plan, The, by Carl Richards
Organic Mushroom Farming and Mycoremediation, by Tradd Cotter
Ox-Cart Man, by Barbara Cooney
Seven Habits of Highly Effective People, The, by Stephen Covey
Smart Money, Smart Kids, by Dave Ramsey and Rachel Cruze
Starting an Etsy Business for Dummies, by Gatski and Shoup
Sustainable Market Farming, by Pam Dawling
Total Money Make Over, The, by Dave Ramsey

FAMILY TIMES ON THE HOMESTEAD

All Year Round, by Ann Druitt
A Time to Keep (and other titles), by Tasha Tudor
Bringing Back the Family Farm, by Forrest Pritchard
Children's Year, The, by Stephanie Cooper
Elliot Homestead: From Scratch, The, by Shaye Elliot
Festivals, Family and Food, by Diana Carey
Handmade Gatherings, by Ashley English
Heal Your Gut Cookbook, The, by Hilary Boynton
Help Me Be Good Series, by Joy Berry
Homespun Seasonal Living Workbook, The, by Kathie Lapcevic
It's Just My Nature (and other titles), by Carol Tuttle
Jane Austen Cookbook, The by Maggie Black
Kids Around the World Celebrate, by Lynda Jones
Little Heathens, by Mildred Armstrong Kalish
Little House Cookbook, The by Barbara Walker
Magic Art of Tidying Up, The by Marie Kondo
Mitford Cookbook and Kitchen Reader, by Jan Karon
Mrs. Sharp's Traditions, by Sarah Ban Breathnach
Nourishing Traditions, by Sally Fallon
Tasha Tudor Cookbook, The by Tasha Tudor
Things My Grandmother Taught Me About Organized Living, by Gloria Barry

Winnie the Pooh Cookbook, by Virginia Ellison (illustrated by E.H. Shepard)

THE HOMESTEAD COMMUNITY

Animal, Vegetable, Miracle, by Barbara Kingsolver
Complete Guide to Saving Seeds, The, by Robert Gough
Folks, This Ain't Normal, by Joel Salatin
Gaining Ground, by Forrest Pritchard
Good Food Revolution, The, by Will Allen
Growing Tomorrow: A Farm-to-Table Journey in Photos and Recipes, by Forrest Pritchard
Heirloom Life Gardener, The, by Jere and Emilee Gettle
Seed to Seed, by Suzanne Ashworth

THE PREPARED HOMESTEAD

Beginner's Book of Essential Oils, The, by Chris Dalziel
Cook Wild, by Susanne Fischer-Rizzi
Cooking with Fire, by Paula Marcoux
Creating Your Off-Grid Homestead, by Teri Page
DIY Projects for the Self Sufficient Homeowner, by Betsy Matheson
Getting Prepared, by Robin Egerton and Angela England
Healing Herbal Infusions, by Colleen Codekas
Natural Soapmaking, by Jan Berry
Organic Canner, The, by Daisy Luther
Pantry Primer, The, by Daisy Luther
Playful Preparedness, by Tim Young
Prepper's Dehydrator Handbook, The by Shelle Wells
Prepper's Blueprint, The, by Tess Penington
Prepper's Water Survival Guide, The, by Daisy Luther
Survival Mom: How to Prepare Your Family For Every Day Disasters..., by Lisa Bedford
Survival Savvy Family, by Julie Sczerbinski
Ultimate Dehydrator Cookbook, The, by Tammy and Steven Gangloff
Ultimate Survival Manual, The, by Rich Johnson
Water Wise Home, The, by Laura Allen
Beyond Basics, Reader's Digest

* May be especially helpful for smaller space homesteaders

Websites and Blogs

Just like anything else you read, always remember that just because a "fact" is on the Internet that doesn't make it true. Having said that, however, it's quite a blessing to have access to so much homesteading information online. I particularly enjoy reading personal experiences, triumphs, and failures along with good, solid facts. Take everything you read with a grain of salt and be willing to try new ideas on your own homestead.

Also bear in mind that, while every effort has been made throughout this book and here, in the resources section, to provide you with online links that I believe will be around for the duration of the Internet, websites and blogs close shop all the time, for various reasons. While each edition of the print book, e-book and other electronic materials will be checked repeatedly for broken links and dismantled websites, I make no guarantee that each recommended online resource will be available to you. That's why I encourage A LOT of print book reading and personal communication, along with your online education.

With that in mind, here are some of my favorite homestead-related websites and blogs:

A Chick and Her Garden – www.achickandhergarden.com
A Farm Girl in the Making - www.afarmgirlinthemaking.com
A Farmish Kind of Life - www.afarmishkindoflife.com
Against All Grain – www.againstallgrain.com
An American Homestead - www.anamericanhomestead.com
Apartment Prepper - www.apartmentprepper.com
Attainable Sustainable - www.attainable-sustainable.net
Backpacking Chef - www.backpackingchef.com
Backyard Chickens - www.backyardchickens.com
Backyard Chicken Project - www.backyardchickenproject.com
Backwoods Home - www.backwoodshome.com
Baker Creek Seeds – www.rareseeds.com
Becoming Minimalist – www.becomingminimalist.com
Book Design Templates – www.bookdesigntemplates.com
Brown Thumb Mama – www.brownthumbmama.com
Butter for All – www.butterforall.com
Cobble Hill Farm Apothecary – www.cobblehillfarmapothecary.com
Common Sense Homesteading - www.commonsensehome.com
Container Gardening - www.containergardening.about.com
Creative Vegetable Gardener – www.creativevegetablegardener.com
Cultures for Health - www.culturesforhealth.com
Curious Soapmaker - www.curious-soapmaker.com
Dave's Garden – www.davesgarden.com

Deep Roots at Home – www.deeprootsathome.com

Delicious Obsessions - www.deliciousobsessions.com

Don't Waste the Crumbs - www.dontwastethecrumbs.com

Edible Wild Food - www.ediblewildfood.com

Family Food Garden - www.familyfoodgarden.com

Farming My Backyard - www.farmingmybackyard.com

Farm to Consumer Legal Defense Fund - www.farmtoconsumer.org

Ferment Tools - www.fermentools.com/blog

Food Storage Made Easy - www.foodstoragemadeeasy.net

Food Storage Moms - www.foodstoragemoms.com

Fork in the Road – www.forkintheroad.co

Fiasco Farm - www.fiascofarm.com

Flip Flop Barnyard - www.flipflopbarnyard.com

Fly Lady – www.flylady.net

Frugal Mama – www.frugal-mama.com

Frugally Sustainable - www.frugallysustainable.com

Game and Garden - www.gameandgarden.com

Gnowfglins - www.gnowfglins.com

Graywolf Survival - www.graywolfsurvival.com

Grit - www.grit.com

Grow, Forage, Cook, Ferment - www.growforagecookferment.com

Harvest Right – www.harvestright.com

Healthy Green Savvy – www.healthygreensavvy.com

Herbal Academy – www.theherbalacademy.com

Herbal Prepper - www.herbalprepper.com

Hobby Farms - www.hobbyfarms.com

Holistic Squid - www.holisticsquid.com

Homespun Seasonal Living - www.homespunseasonalliving.com

Homestead Chronicles - www.homesteadchronicles.com

Homestead Honey - www.homestead-honey.com

Homesteading On Grace – www.homesteadingongrace.com

Homestead Lady - www.homesteadlady.com

Homesteadin' Mama – www.homesteadinmama.com

Home Talk – www.hometalk.com

How to Butcher a Chicken – www.howtobutcherachicken.blogspot.com

Imaginacres - www.imaginacres.com

Imagine Childhood – www.blog.imaginechildhood.com

J & J Acres - www.jandjacres.net

Joybilee Farm - www.joybileefarm.com

Just Plain Living – www.justplainmarie.ca
Learning and Yearning – www.learningandyearning.com
Learning Herbs – www.learningherbs.com
Life From Scratch - www.lifefromscratch.com
Lil' Suburban Homestead - www.lilsuburbanhomestead.com
Little House Living - www.littlehouseliving.com
Lovely Greens – www.lovelygreens.com
Majestic Mountain Sage's blog - www.blog.thesage.com
Melissa K Norris - www.melissaknorris.com
Mom With a Prep - www.momwithaprep.com
Money Saving Mom – www.moneysavingmom.com
Mother Earth News - www.motherearthnews.com
Mountain Rose Herb - www.mountainroseblog.com
Mr. and Mrs. Not Made of Money – www.notmadeofmoney.com
My Aqua Farm - www.myaquafarm.com
My Food Storage Cookbook - www.myfoodstoragecookbook.com
My Pet Chicken – www.blog.mypetchicken.com
My Repurposed Life – www.myrepurposedlife.com
Nerdy Farm Wife - www.nerdyfarmwife.com
Nitty Gritty Mama – www.nittygrittymama.com
North Country Farmer - www.northcountryfarmer.com
Nourished Kitchen - www.nourishedkitchen.com
Organic Life Guru - www.organiclifeguru.com
One Ash Farm and Dairy - www.oneashfarmanddairy.com
One Hundred Dollars a Month - www.onehundreddollarsamonth.com
Pantry Paratus - www.pantryparatus.com
Pasture Deficit Disorder - www.pasturedeficitdisorder.com
Penniless Parenting – www.pennilessparenting.com
Permaculture Research Institute - www.permaculturenews.org
Piwakawaka Valley - www.piwakawakavalley.com
Practical Self-Reliance – www.practicalselfreliance.com
Ready Nutrition - www.readynutrition.com
Real Milk – www.realmilk.com
Red Worm Composting - www.redwormcomposting.com
Reformation Acres - www.reformationacres.com
Rockin' W Homestead – www.rockinwhomestead.com
Rooted Childhood – www.rootedchildhood.com
Schneider Peeps - www.schneiderpeeps.com
Seeds for Generations – www.seedsforgenerations.com

Seed Save – www.seedsave.org
Seed Savers - www.seedsavers.org
Simple Life Mom - www.simplelifemom.com
Simply Canning - www.simplycanning.com
Soule Mama - www.soulemama.com
Sprout People - www.sproutpeople.org
Stoney Acres - www.ourstoneyacres.com
Strangers and Pilgrims on Earth - www.strangersandpilgrimsonearth.blogspot.com
Strictly Medicinal (formerly Horizon Seeds) – www.strictlymedicinal.com
Sun Calc – www.suncalc.net
Survival Mom - www.survivalmom.com
Temperate Climate Permaculture - www.tcpermaculture.com
Tenth Acre Farm - www.tenthacrefarm.com
The 104 Homestead - www.the104homestead.com
The Bread Geek - www.thebreadgeek.com
The Elliot Homestead - www.theelliothomestead.com
The Fewell Homestead – www.thefewellhomestead.com
The Free-Range Life – www.thefreerangelife.com
The Homesteading Hippy - www.thehomesteadinghippy.com
The Minimalist Mom - www.theminimalistmom.com/
The Organic Prepper - www.theorganicprepper.ca
The Pistachio Project – www.pistachioproject.com
The Prairie Homestead - www.theprairiehomestead.com
Timber Creek Farm - www.timbercreekfarmer.com
Trayer Wilderness - www.trayerwilderness.com
Untrained Housewife - www.untrainedhousewife.com
Upcycle That - www.upcyclethat.com
Urban Farms - www.urbanfarmsonline.com
Urban Overalls - www.urbanoveralls.com
UP Pastured Farms – www.uppastured.com
Vinegar Tips – www.vinegartips.com
Vomiting Chicken – www.vomitingchicken.com
Weed 'Em & Reap – www.weedemandreap.com
Wellness Mama - www.wellnessmama.com
Weston Price - www.westonaprice.org
Whole Family Rhythms – www.wholefamilyrhythms.com
Whole New Mom - www.wholenewmom.com
Woodland Farmhouse Inn – www.woodlandfarmhouseinn.com
You Should Grow – www.youshouldgrow.com

30 Minute Crafts – www.30minutecrafts.com

Afterword

I feel uncomfortable putting only my name on the cover of this book and calling myself the author, simply because so many other voices can be heard throughout the text. In no particular order, and in the surety that I'm inevitably leaving someone out, I'd like to thank the following:

A thank you to the homesteaders who participated in my initial email interviews, and the rambling ones that followed, when I was still trying to get a handle on what needed to be said: Kris Bordessa, Angie Campbell, Bernie Carr, Jenna Dooley, Jessica Lane, Kathie Lapcevic, Donna Miller, Erica Meuller, Lauren Patterson, Brandon Sutter, PJ Schott, Jo Reillme and Teri Page.

For their willingness to tell all about homesteading in an HOA, thanks go to Amber Bradshaw, Sara Mortimer and Julie Sczerbinski.

Anna Esau and Chris MacLaughlin were so helpful when it came to sharing their experience with container gardening. Thank you to Joel Schwartz, Jared Stanley and Amy Stross for their help translating basic permaculture principles into homesteader-speak.

A special thank you to Annie Bernauer, Chris MacLaughlin and Jo Reillme for sharing their wisdom and advice on vermicomposting. Special thanks to Jo for letting me use her comparison graphic. If worms could, they'd sing all ya'll's praises

For their insights into foraging, kudos go to Gregg Carter, Colleen Codekas, Chris Dalziel, LeAnn Edmondson, Jessica Lane, Laurie Neverman, Jane Metzger, Janet Pesaturo, Angi Schneider and Karen Stephenson.

The dairy animal section was benefited by the collective wisdom of LeeAnn Perez, Jenna Dooley and Amy Maus—thank you, my favorite milk maids.

The hunting segment was fleshed out with the help of Tammy Trayer, Jenn Dana and Heather Harris. How ignorant of me that I'd never thought of hunting as a lady's sport. Randall Wilke generously shared his personal insights as well, making sure the men were represented.

Without Rebecca Shirk the Homestead Finances chapter would have consisted of a few muttered phrases about saving your pocket change. Before you were able to benefit from it, I was reading her wisdom and being inspired by it. My dear Rebecca, you are a queen.

For their straight talk about living off-grid, my admiration and thanks go to the Trayer family, Thomas family, Lynn family and Bauer family.

For her assistance with Wonder Oven information, thank you to Megan Smith.

To those poor souls who vetted early material that was haunted by zero editing, vague ideas, hurried spelling, and cobbled together ideas, I give especial thanks for taking the project seriously. You know who you are Tara Bowen, Moroni Leash, Heather Harris, and Allison Hamaker.

For their simple, persisting belief that I could begin and finish a book, my passionate gratitude goes to Kathie Lapcevic, Chris Dalziel, Jessica Lane and Angi Schneider. Ladies, thank you for being true friends in a world of soul-less social media. And for patiently listening to me whine for two years. Only you know how close this book came to never being published and you helped me see it through.

My illustrator, Jennifer Cazzola of Scratchy Pixel, deserves oodles of praise for taking my incoherent inspirations and turning them into pen and ink realities of loveliness. She is responsible for the illustrations both inside and on the cover of the book. The front and back cover design were completed under the masterful hand of Rachel Arsenault.

My editors deserve loads of praise and thanks because editing for me is no small feat. To Moroni Leash, who offered his professional services free of charge simply because he's a sweet soul who's known me since I was a bratty kid, I send a huge thank you. He brought order out of the chaos that was the first draft. Maybe, with his help, I'll eventually learn to use enough commas. My thanks also go to my final editors, Meaghan Lamm and Sarah Dalziel who brought me across the finish line when I was too tired to keep going on my own. Sarah, thank you for your suggestions on text improvement, too—you made things make more sense. Thanks to Kathy Gardiner, who just happens to be my mother, for giving the e-book one last look with her magnificently technical brain. Thank you to Shain Zundel, who just happens to be my husband, for helping me compile and edit the Journal and Unit Study (and for making me emergency omelets). Thank you to Maat Van Uitert for masterfully editing all the bonus material associated with the book.

Thank you to Dean Martin and Nat King Cole who crooned me through the final month of edits.

Thanks to my mom for always letting me garden, cook, and craft (using all those terms very loosely) in the homes of my youth. When I was growing up, she never once told me to clean up before I was done or to hold still when inspiration struck.

Also, thank you to all the homesteaders out there who constantly share their stories in many ways, and who form a community of meaningful work, open sharing and true family.

And finally, thanks be to God. Sanctum est Domino.

Index

Bees
 Beekeeping, 21, 41, 96, 122, 150, 157, 161, 216, 297, 299, 349, 364, 372, 381, 390
Book Club
 100, 163, 336, 360, 362, 363, 364, 365, 366, 367, 405
Broth
 37, 55, 58, 170, 349, 352, 359, 360, 418
Bug Out Bags
 396, 406
Business
 9, 131, 141, 144, 145, 216, 222, 223, 227, 231, 258, 261, 262, 266, 284, 287, 288, 293, 294, 295, 296, 301, 302, 303, 306, 307, 308, 356, 377, 378, 380
Cheese
 25, 41, 44, 47, 51, 52, 55, 63, 83, 84, 85, 86, 88, 170, 237, 242, 271, 320, 333, 362, 372, 374, 418, 424
Chickens
 Poultry, Chicks, 8, 10, 21, 22, 24, 34, 36, 37, 40, 42, 47, 49, 51, 57, 77, 83, 95, 114, 115, 121, 131, 158, 162, 163, 170, 171, 173, 175, 188, 205, 206, 215, 216, 221, 225, 226, 227, 228, 229, 230, 231, 232, 233, 234, 252, 253, 258, 259, 261, 262, 263, 277, 280, 281, 286, 297, 301, 335, 336, 339, 349, 354, 360, 362, 365, 369, 391, 392, 396, 436, 445
Children
 10, 12, 34, 37, 55, 66, 110, 113, 117, 118, 120, 122, 123, 132, 133, 134, 150, 172, 175, 182, 183, 185, 186, 189, 190, 192, 193, 201, 203, 205, 211, 212, 215, 229, 234, 238, 239, 244, 246, 248, 259, 263, 266, 272, 276, 278, 279, 280, 287, 289, 302, 314, 315, 316, 317, 319, 320, 322, 325, 326, 327, 328, 329, 330, 333, 334, 335, 336, 337, 338, 339, 340, 342, 343, 344, 345, 346, 347, 349, 350, 351, 353, 355, 371, 372, 380, 385, 388, 389, 393, 400, 401, 405, 409, 413, 419, 423, 428, 436, 457, 459
Chores
 111, 180, 216, 221, 290, 313, 315, 333, 334, 335, 336, 337, 339, 342, 345, 350, 353, 388, 436
Clothes
 21, 42, 174, 179, 181, 182, 183, 184, 185, 197, 200, 201, 202, 203, 204, 209, 279, 409, 411
Compost, 21, 24, 36, 40, 41, 42, 43, 48, 49, 57, 95, 98, 123, 124, 125, 131, 154, 156, 157, 161, 162, 163, 164, 166, 171, 172, 174, 205, 211, 218, 219, 221, 222, 223, 246, 261, 271, 280, 298, 300, 336, 343, 348
Container Garden
 75, 117, 121, 123, 125, 127, 132, 134, 135
Cows, 115, 236, 237, 238, 239, 240, 241, 242, 248, 253, 305, 445
Crockpot
 Slow Cooker, 47, 200, 276, 283, 429
Dairy, Cultured Dairy, 44, 52, 54, 58, 55, 63, 64, 70, 83, 94, 95, 176, 197, 215, 216, 236, 237, 240, 241, 242, 243, 244, 247, 248, 252, 261, 271, 272, 283, 297, 298, 303, 355, 420, 421, 425, 429, 445, 452
Dehydrating
 36, 40, 74, 75, 76, 82, 83, 86, 88, 257
Dinner, 37, 41, 47, 48, 49, 50, 51, 56, 123, 170, 251, 252, 261, 272, 278, 279, 283, 297, 312, 314, 315, 316, 317, 318, 319, 320, 321, 322, 334, 335, 350, 351, 352, 355, 387, 428
Downsize
 179, 203, 339
Energy
 Power, 8, 11, 21, 23, 29, 76, 124, 154, 155, 157, 159, 160, 164, 168, 173, 177, 179, 180, 182, 183, 197, 198, 199, 200, 201, 202, 203, 205, 206,

207, 208, 209, 210, 211, 212, 229, 260, 276, 277, 279, 284, 291, 298, 302, 312, 315, 316, 317, 327, 352, 362, 366, 377, 427

Fats, 42, 71, 94, 100, 108, 253, 418

Ferments, 34, 58, 63, 69, 70, 97, 352

Forage, 127, 129, 154, 168, 188, 189, 190, 191, 195, 276, 277

Freezing
  38, 51, 54, 82, 170, 199, 201, 244, 245, 260, 352, 411, 418, 419

GAPS, 59, 66, 67, 127, 319, 420, 435

Goats, 11, 25, 36, 56, 63, 121, 129, 150, 176, 187, 195, 236, 237, 238, 239, 240, 241, 242, 243, 247, 248, 249, 253, 255, 261, 271, 272, 297, 299, 303, 304, 305, 309, 336, 349, 354, 438

Grain, 42, 46, 49, 50 55, 56, 58, 63, 64, 66, 71, 92, 93, 94, 97, 101, 108, 109, 110, 113, 114, 115, 160, 205, 215, 230, 236, 252, 254, 255, 277, 280, 305, 324, 403, 415, 417, 418, 419, 420, 422, 424, 426, 430, 448

Herbs
  Wellness Herbs, 25, 29, 41, 50, 54, 76, 88, 99, 107, 114, 125, 126, 134, 135, 138, 139, 140, 141, 145, 146, 147, 148, 149, 150, 155, 158, 159, 187, 188, 189, 190, 191, 193, 194, 276, 277, 297, 298, 320, 346, 348, 349, 352, 387, 388, 402, 405, 419, 457

HOA
  Homeowners Association, CCRs, 20, 21, 22, 206, 296, 303, 410, 452

Homeschool, 179, 182, 185, 192, 205, 247, 286, 305, 315, 333, 336, 371, 391, 393

Hunting
  122, 189, 216, 252, 253, 255, 256, 257, 258, 261, 262, 382, 288, 400, 452

Jam, 34, 36, 40, 44, 73, 81, 87, 89, 188, 359, 365

Journal
  Homestead Journal, 26, 28, 29, 30, 31, 57, 69, 87, 99, 134, 147, 148, 161, 174, 183, 193, 194, 210, 222, 232, 247, 261, 283, 306, 307, 319, 337, 350, 355, 356, 367, 368, 372, 373, 380, 381, 391, 404, 411, 423, 433

Kefir, 42, 52, 53, 63, 64, 69, 70, 71, 97, 352, 418

Hang Drying, 42, 53, 131, 182, 192, 202, 203, 204, 206, 209, 224, 267, 335, 336, 351

Leftovers, 37, 55, 57, 58, 60, 126, 170, 171, 253, 260

Manure
  Poop, 124, 154, 157, 159, 161, 163, 166, 195, 216, 218, 225, 237, 238, 239, 246, 260, 261, 271, 273, 277, 348, 355

Meat, 37, 41, 43, 44, 48, 49, 50, 58, 73, 77, 81, 82, 83, 114, 163, 170, 176, 216, 221, 222, 225, 227, 233, 246, 248, 251, 252, 253, 254, 255, 257, 258, 259, 260, 261, 262, 263, 281, 302, 304, 315, 320, 324, 347, 349, 352, 354, 365, 376, 417, 420, 428

Microgreens, 106, 108, 109, 111, 112, 113, 114, 349

Mulch, 124, 125, 127, 131, 155, 156, 157, 159, 161, 164, 174, 205, 206, 303, 348

Natural Leaven
  Natural Yeast, 64 See Sourdough

Off-Grid
  97, 208, 429, 434

Outdoor
  Outdoor Cooking, 21, 55, 172, 174, 212, 213, 222, 223, 230, 247, 330, 333, 335, 353, 427, 428, 431, 432, 433, 434

Permaculture, 107, 153, 154, 155, 156, 157, 159, 160, 161, 162, 164, 165, 166, 410, 452

Preserve
  12, 40, 46, 56, 73, 75, 76, 81, 83, 86, 87, 89, 131, 132, 260, 268, 276, 288, 294, 316, 343, 346, 352, 365, 397, 410, 419, 423, 426

Raw Milk
  Unpasteurized, 27, 41, 47, 52, 71, 86, 176, 236, 243, 271, 298, 299

Recycling
  Repurposing, 24, 58, 168, 169, 171, 172, 174, 176, 177, 281

Root Cellar, 74, 79, 80, 81, 352, 417, 419

Sauerkraut
  67, 352

Seeds
  Heirloom, 40, 51, 64, 78, 87, 106, 108, 109, 110, 111, 112, 113, 114, 116, 119, 122, 126, 129, 140, 141, 144, 146, 150, 167, 173, 174, 176, 188, 189, 193, 272, 297, 298, 348, 360,

376, 377, 378, 379, 380, 381, 382, 383, 388, 418, 422

Sewing, 174, 184, 201, 233, 262, 301, 429, 433

Soaking
  Pre-soaking, 41, 64, 66, 109, 110, 111, 112, 113, 254, 255, 422

Solar Oven, 40, 42, 82, 200, 260, 278, 405, 421, 424, 428, 429, 433, 434, 435, 436

Sourdough, 41, 50, 51, 59, 64, 65, 66, 67, 69, 70, 71, 93, 97, 101, 170, 219, 315, 319, 347, 352, 421

Special Needs, 353

Sprouting
  49, 64, 108, 109, 110, 111, 112, 113, 114, 115, 255, 305

Storage, 41, 43, 48, 49, 74, 79, 80, 82, 88, 92, 112, 113, 125, 127, 144, 160, 176, 179, 183, 185, 210, 222, 260, 279, 349, 352, 396, 397, 402, 405, 407, 408, 409, 410, 411, 413, 414, 415, 416, 417, 418, 419, 420, 421, 422, 423, 424, 425, 426, 430

Swap, 27, 49, 279, 284, 285, 324, 360, 376, 377, 378, 379, 380, 381, 389

Teaching
  Classes, Mentoring, 150, 224, 263, 290, 327, 334, 339, 345, 354, 370, 371, 372, 373, 390, 438

Traditions, 34, 313, 326, 327, 328, 330, 365

Upcycling, 167, 168, 169, 173, 175, 176, 353, 409

Vermicomposting
  Worms, 175, 216, 218, 220, 221, 222, 223, 300, 346, 452

Water, 21, 37, 39, 40, 41, 42, 43, 44, 45, 48, 49, 52, 53, 64, 65, 68, 70, 74, 78, 80, 81, 82, 85, 86, 91, 95, 98, 102, 106, 109, 110, 112, 113, 114, 124, 126, 127, 130, 131, 134, 140, 149, 155, 156, 157, 159, 160, 161, 162, 163, 171, 173, 177, 191, 194, 199, 200, 202, 203, 204, 205, 206, 209, 211, 219, 220, 221, 222, 228, 230, 231, 236, 244, 245, 253, 260, 263, 305, 346, 349, 352, 353, 370, 378, 387, 396, 404, 405, 407, 408, 409, 410, 411, 412, 413, 414, 418, 419, 422, 424, 428, 431, 432, 435

Wonder Oven, 429, 430, 433, 434, 452

Yogurt, 52, 63, 64, 83, 173, 174, 175, 177, 242, 271, 352, 430

ABOUT THE AUTHOR

Tessa Zundel is the homemaking, homeschooling, homesteading mother of five children and wife to one long-suffering man. Although she has called several states and properties home over the years, she now resides in the great state of Missouri in the middle of acres of fairy forests and hidden ponds. She is the author of several books including *Herbs in the Bathtub* and is the voice behind the blog Homestead Lady. She has also contributed to online and print magazines like Hobby Farms. She is an advanced master gardener and has worked with several community groups in areas of home education, gardening and seed saving.

Most days you'll find her hauling her good natured, adventuresome children around to learn about herbs, small farm livestock, fiber and other lost arts, whole foods and home education. There's always something being tinctured, fermented, built or milked around here - just ask the long-suffering man!

Connect With Me

Feel free to connect with me online—I'd love to hear how your homestead efforts are progressing!

Homestead Lady Blog: www.homesteadlady.com

The Do It Yourself Homestead has its own Facebook Community to provide an immediate network for you as you pursue your homestead education. Join us below:

Facebook Group: https://www.facebook.com/groups/TDIYH/

To make inquiries about continuing education with *The Do It Yourself Homestead* – things like future courses, a children's version of the book and bonus materials – be sure to join our newsletter family. This is a free, emailed newsletter that will arrive in your inbox whenever something is happening with Homestead Lady's books or blog. This newsletter includes access to our members-only free library of homesteading titles, downloads and learning aides. I'd like to tell you the newsletter is bi-monthly, but it ends up a little more scatterbrained than that - especially during spring planting and fall harvest! You probably aren't online much at those times anyway, right?

You can learn about all this and more by becoming a member of our newsletter family. I promise to only send you homestead stuff—no nonsense and no spam. Pinky Swear. To sign up, visit us any time at www.homesteadlady.com.